CAMBRIDGE LIBRARY COLLECTION

Books of enduring scholarly value

Life Sciences

Until the nineteenth century, the various subjects now known as the life sciences were regarded either as arcane studies which had little impact on ordinary daily life, or as a genteel hobby for the leisured classes. The increasing academic rigour and systematisation brought to the study of botany, zoology and other disciplines, and their adoption in university curricula, are reflected in the books reissued in this series.

The World Before the Deluge

Louis Figuier (1819–94) was destined to remain in academia until disagreements with fellow scholars led him to abandon this path and instead pursue 'the idea that scientific knowledge, which until then had been almost exclusively the property of the learned, should be put within the reach of the reading public'. Published in 1863, *La Terre avant le déluge* became a classic of popular science and introduced palaeontology to a wider readership; that this English translation appeared only two years later is an indication of its impact. Figuier wrote that his aim was 'to trace the progressive steps by which the earth has reached its present state … and to describe the various convulsions and transformations through which it has successively passed'. The book was also celebrated for its inclusion of more than 200 illustrations by a pupil of Doré, Édouard Riou (1833–1900), who became famous as Jules Verne's illustrator a few years later.

Cambridge University Press has long been a pioneer in the reissuing of out-of-print titles from its own backlist, producing digital reprints of books that are still sought after by scholars and students but could not be reprinted economically using traditional technology. The Cambridge Library Collection extends this activity to a wider range of books which are still of importance to researchers and professionals, either for the source material they contain, or as landmarks in the history of their academic discipline.

Drawing from the world-renowned collections in the Cambridge University Library and other partner libraries, and guided by the advice of experts in each subject area, Cambridge University Press is using state-of-the-art scanning machines in its own Printing House to capture the content of each book selected for inclusion. The files are processed to give a consistently clear, crisp image, and the books finished to the high quality standard for which the Press is recognised around the world. The latest print-on-demand technology ensures that the books will remain available indefinitely, and that orders for single or multiple copies can quickly be supplied.

The Cambridge Library Collection brings back to life books of enduring scholarly value (including out-of-copyright works originally issued by other publishers) across a wide range of disciplines in the humanities and social sciences and in science and technology.

The World
Before the Deluge

LOUIS FIGUIER
TRANSLATED BY
HENRY W. BRISTOW

CAMBRIDGE
UNIVERSITY PRESS

University Printing House, Cambridge, CB2 8BS, United Kingdom

Published in the United States of America by Cambridge University Press, New York

Cambridge University Press is part of the University of Cambridge.
It furthers the University's mission by disseminating knowledge in the pursuit of
education, learning and research at the highest international levels of excellence.

www.cambridge.org
Information on this title: www.cambridge.org/9781108062473

© in this compilation Cambridge University Press 2013

This edition first published 1865
This digitally printed version 2013

ISBN 978-1-108-06247-3 Paperback

XXXII.—First appearance of Man.

THE WORLD

BEFORE

THE DELUGE.

BY LOUIS FIGUIER.

CONTAINING

TWENTY-FIVE IDEAL LANDSCAPES OF THE ANCIENT WORLD,

DESIGNED BY RIOU.

AND TWO HUNDRED AND EIGHT FIGURES OF ANIMALS, PLANTS, AND
OTHER FOSSIL REMAINS AND RESTORATIONS.

———————

Translated from the Fourth French Edition.

———————

LONDON:
CHAPMAN AND HALL, 193, PICCADILLY.
1865.

LONDON: PRINTED BY WILLIAM CLOWES AND SONS STAMFORD STREET AND CHARING CROSS

TO

SIR RODERICK IMPEY MURCHISON,

K.C.B., D.C.L., LL.D., F.R.S. V.P.G.S., ETC., ETC.,

DIRECTOR-GENERAL OF THE GEOLOGICAL SURVEY OF THE UNITED KINGDOM,
AND PRESIDENT OF THE ROYAL GEOGRAPHICAL SOCIETY,

WHOSE ORIGINAL WORK 'SILURIA' FIRST ATTRACTED THE TRANSLATOR
TO THIS CLASS OF STUDIES,

THIS WORK IS RESPECTFULLY INSCRIBED.

L'ENVOI.

The 'World before the Flood' is a reproduction of one of those *Œuvres de Luxe* in which our neighbours of France delight to exhibit their refined tastes in literature and art. Its object is to give a History of the progressive steps by which the earth has reached its present state, from that condition of chaos "when it was without form and void, and darkness covered the face of the waters," tracing the various convulsions and transformations through which it has passed, until, in the words of the poet, it may be said—

> Where rolls the deep, there grew the tree;
> O Earth, what changes hast thou seen!
> There, where the long street roars, hath been
> The silence of the central sea.

It is no small compliment to the Author, and to the intelligence of our neighbours, to be able to state, that four editions have been demanded, and twenty-five thousand purchasers found for the work in the space of less than two years: a demand, so far as I know, altogether unprecedented for works of a scientific character.

With regard to this reproduction, little need be said; the simple and elegant language in which the Author has expressed himself, and the profound interest inseparable from the subject itself, rendered the task of translating him a labour of love. The only instances in which it has been found necessary to depart from, or rather to enlarge upon, the text have occurred where examples were necessary to elucidate particular theories or series of stratification—these the Author had naturally enough drawn from France: in the translation these are preserved, but, in addition, others drawn from British Geology have been added, either from the writer's own knowledge, or from the works of well-known British writers. It has also been considered desirable, for similar reasons, to enlarge upon the views of British geologists, to whom the French work scarcely does justice, considering the extent to which the science is indebted to them for its elucidation—I may almost say for its existence.

One point more, and I have done. In the original work the chapter on Eruptive Rocks comes at the end of the work, but, as the work proceeded, I found so many unexplained allusions to that chapter that it seemed more logical, and more in accordance with chronological order, if I may use an expression not quite applicable, to place that chapter at the beginning.

In most cases I have explained the technical terms on the first occasion of their being used, so as to render them intelligible to the general reader; but to have done so whenever they occurred would have made the work tedious. I have therefore made the Index an explanatory one, by giving definitions of most of the terms in their alphabetical order.

In dedicating the English version of M. Figuier's work to a well-known geologist, I must not lead the reader to believe that Sir Roderick Murchison entertains views similar to those of the Author on several portions of his great subject; but I presume that he fully admits the value of a volume which, through its popular and attractive character, has roused the minds of many persons for the first time to a due consideration of the noble science of Geology.

W. S. O.

Kensington, May, 1865.

CONTENTS.

CONTENTS.

FULL-PAGE ILLUSTRATIONS.

FRONTISPIECE—FIRST APPEARANCE OF MAN.

A THESIS.

———◇———

I AM about to maintain what may be thought a strange thesis.

I assert that the first books placed in the hands of the young, when they have mastered the first steps to knowledge and can read, should be on Natural History; that in place of awakening the faculties of the youthful mind to admiration, by the fables of Æsop or Fontaine, by the fairy tales of 'Puss in Boots,' 'Jack the Giant Killer,' 'Cinderella,' 'Beauty and the Beast'—or even 'Aladdin and the Wonderful Lamp,' and such purely imaginative productions, it would be better to direct their admiring attention to the simple spectacles of nature—to the structure of a tree, the composition of a flower, the organs of animals, the perfection of the crystalline form in minerals, above all, to the history of the world—our habitation; the arrangement of its stratification, and the story of its birth, as related by the remains of its many revolutions to be gathered from the rocks beneath our feet.

Many readers will protest against this proposition. Is it not the fact, they will say, that fairy tales, fables, and the legends of mythology, have always been the first intellectual food offered to the young? Is not that the natural means of amusing them, as some relaxation from more severe study?

And, they will add, society has been none the worse!

B

It is here I would claim the reader's attention. I think, on the contrary, that many of the evils of society may be traced to this very cause. It is because we cherish this dangerous aliment, that the living generation includes so much that is false, so many weak and irresolute minds, given to credulity, inclined to mysticism—proselytes, in advance, to chimerical conceptions and to every extravagant system.

Intelligence is scarcely awakened, when we do our best to destroy it by our training. Our very first step in this path of folly is to teach the impossible and absurd. We crush, so to speak, good sense in the eggshell, when we concentrate the ideas of the young on conceptions at once dreamy and opposed to fact and reason ; introducing them into a fantastic world, in which are jumbled together gods, demigods, and pagan heroes, mingling with fairies, goblins, and sylphs ; spirits — good and bad—enchanters, magicians, devils, devilkins, and demons ; and all this while no doubts seem to be entertained of the danger likely to arise from this constant presence of ideas so subversive of common sense. At a period of life when intelligence is like soft wax, which takes and preserves the feeblest impressions—when, innocent as yet of all knowledge, the mind is eager to acquire it—when we may bend or break it at pleasure. And yet we are surprised that this intelligence, this wax so soft and ductile, should preserve, at a later period, indelible marks of the absurd course of instruction which has been adopted. Suppose that we were to find a people wise enough to seek, in a reasonable contemplation of nature, the means of amusing and inte- resting the young—a generation which had thus been trained from an early period to examine and study the creation, which should have formed its judgment from the naked truth, its reason from the infal- lible logic of nature, which had learned to comprehend and bless the Creator in his works ; would not such a generation assure to the state honest citizens of a right spirit, firm and enlightened minds, imbued with devotion to God, affectionate to their relations, and love to their country ?

Were I to live a hundred years I could never forget the frightful confusion in which my young head was left by my first lesson in mythology. There was Deucalion, who created the human race by throwing stones over his shoulder, these stones giving birth to man ; Jupiter, who was made to open his head to give birth to Minerva, with all her accessories ; Venus, who was born one fine morning of the wild sea-foam; there was old Saturn, who had the bad habit of devouring his own children, and whose paternal voracity was deceived one day by the substitution of a stone for the last-born. And that Olympus ! where gods and goddesses mingled and were daily guilty of the vilest actions. How is a brain of tender years to resist the entire overthrow of the more simple impulses of good sense ? And is it not deplorable to be compelled thus to enter into the paradise of Knowledge through the gates of Folly ?

To the fantastic vagabonds of the pretended religious legends of Paganism, we have now added those of the fairy tales. The child has scarcely learned to rattle its coral when it is told of the good and the bad fairy; of the magician Rothomagus and the enchanter Merlin ; of the seven-leagued boots; of men changed into mice; of mice changed into princes; of old beggar-women changed by the waving of a wand into young princesses, rustling in silks and covered with precious stones. These are the fine thoughts upon which we exercise the dawning imagination, without reckoning the Chinese shadows, the juggler's cup and balls—which, in the hands of Houdin and Anderson, render still darker the atmospheric clouds which surround the young brain to its destruction. In the midst of this inundation of follies, how are the young to preserve the reason with which Providence has provided them ? Alas, it is never entirely preserved; much of the primitive good sense with which the child was gifted is superseded by the love of the marvellous to which humanity is naturally too prone; excited as it is from the days of earliest infancy, it is never entirely overcome—seldom reduced to reasonable bounds.

Already roused in its cradle, as it were, by the songs of its nurse, which tell of bogies with their escort of devils and demons, its instruction is made to consist of fairies and fairy tales, or of the still more objectionable stories of mythology; by which the love of the

marvellous—that is, of all things opposed by and contrary to reason
—is encouraged. In this manner the love of the marvellous pervades
the mind in the hour of its awakening, leaving it a prey to the worst
superstitions. How, then, should we be astonished at the alternate
appearance of fanatical ignorance and threatening socialism, or, worse
still, with the appearance of those epidemics which, under the name
of animal magnetism, table-turning, and spirit-rapping, come to lead
us periodically back to the superstitious practices of the middle ages.

The proposition that we would defend is thus less paradoxical than
it appears at first. The tales and legends that we give as the teachers
of infancy are dangerous, because they cultivate and excite that
inclination for the marvellous which is already excessive in the human
mind, whereas the first book placed in their hands ought rather to
have a tendency to fortify—to strengthen its young reason.

But, we may be asked, would you then mutilate the human mind
by reducing it to the single faculty of reason—rejecting, as out of its
sphere, all that is imaginative—all the ideal? Would you suppress
thus all poetry, and even all imaginative literature, for both have
their foundation in the love of the marvellous; or, to speak even
more correctly, they are the marvellous itself? A generation educated
in such principles would, no doubt, reason justly; its mind would be
well furnished, but destitute of all ideality—of all imagination—of all
sentiment; the masses so trained would be a mere collection of
calculating machines; whereas the man of cultivated mind should
possess sentiment as well as reason. It is well that he should comprehend the material phenomena by which he is surrounded, but he
ought to learn also to love and to feel: if it is his duty to cultivate
his mind, ought he not also to form his heart?

These are objections which naturally present themselves, and here is
our reply:—

The imaginative faculty, which permits of ideality and of the
abstract,—which forms poets, inventors, and artists,—is inherent in
the mind, and cannot be suppressed; it can only perish with it. It is
the integral part of intelligence. All which concurs to fortify, to

enrich intelligence, and enlarge the sphere of its activity, turns, or ought to turn, at some time, to the profit of the imaginative faculty itself, which is only a part of the whole. It is for this, among other reasons, that it is necessary to supply our intelligence with exact and rigorous notions—to nourish incontestable truths—to keep aloof all sterile, and especially all injurious fiction; in short, to constitute the mind strongly and healthily, and then to exercise, in all its freedom, this fine faculty of imagination, the mother of poetry and the arts, freed from all baleful fetters. Let us begin by laying a solid foundation in the mind from the days of infancy, and we shall never want either poets or artists.

But these rigorous notions, these incontestable truths, with which it is desirable to furnish the minds of the young, are they difficult to find ? Do they impose any great labour on the youthful mind ?

It is necessary to take the youth by the hand, lead him into the fields, and tell him to open his eyes. The birds of the woods, the flowers of the fields, the grass of the meadows, the nightingale which sings upon the last lilac-tree, the butterfly which traces in the air its line of rubies and emeralds, the insect which silently weaves its temporary dwelling under a withered leaf, the rosy dawn, the fruitful rain, the warm refreshing breeze which murmurs through the valley,—these are the ever-varying objects of his simple labour, these his *plan of studies*.

The feeling of insatiable curiosity which possesses the mind in the dawn of life—the eager desire for knowledge which awakens forth reason and desires natural to all ages—is most vivid during youth. Feeling its own want of knowledge, the mind is then eager to acquire it; he throws himself with youthful ardour upon every novelty which presents itself. It would evidently be attended with immense advantage to profit by this disposition, to infuse into the young mind correct notions of the true and the useful; the study of nature responds perfectly to these views. It involves no labour, but is itself, on the contrary, a veritable attraction, appealing, as it does, to all uninfluenced by difference of language or of nationalities.

In habituating themselves to the study—in seeking to comprehend the great and the small phenomena of creation; in reading this admirable book of nature, open to all eyes, and which, when read, ornaments the mind with knowledge at once useful and practical; the youthful student teaches himself, while admiring in its wonders—in its infinitely great as in its infinitely little—the Divine work of the Author of all things. He prepares his mind to receive efficaciously the fruitful seeds of religion—of science—of philosophy. Last, and not least in our eyes, although a negative quality, it extracts the poison sown in his mind by the tales of fairies, by the legends of mythology, by all the dangerous apparatus of infantile marvels to which we have alluded.

We have devoted ourselves to this task—difficult, without doubt, but assuredly fruitful in gentle gratifications—of spreading among the mass of the contemporary public a taste for scientific studies. What we have hitherto done for more matured intellects, we would now attempt for the rising generation, strongly penetrated with the advantages which the study of nature offers to the young.

The study of geology, it has been alleged, is unsuited for the instruction of youth, and in some countries it is even interdicted as anti-religious. These ideas were perhaps legitimate when the science itself was in its infancy—when doctrines now recognized as erroneous—which rendered the earth subject to continual revolutions and cataclysms—prevailed; when, in order to explain the disappearance of organic species, it was thought necessary to invoke a new revolution—to turn the world upside down—to produce a deluge at every epoch. We now know that our globe has, beyond all doubts, been the theatre of frequent catastrophes: immense rents have been torn in its solid crust, and eruptions of diverse kinds have issued from its abyss, which have torn up and displaced the soil, burying continents, excavating deep valleys, and elevating lofty mountains. But all these phenomena, in spite of their power and intensity, have not extended to the entire globe, destroying all living beings. Their action was necessarily local. If, then, the organized beings of one period differ from those which follow, it is not because a revolution of the globe destroyed a whole generation in order to construct on its ruins a new one. This idea is opposed to the facts : it is contradicted by the regularity of the beds enclosing the different species of fossils, and by the fact of the existence of many species through a long series of stratified rocks. Where, then, is the danger of admitting the fact into a work for Popular Reading? The different species have died out quite naturally, and races have died out like individuals. The sovereign Master who created animals and plants has willed that the duration of the existence of species on the surface of the earth should be limited as is the life of individuals : it was not necessary, in order that they should disappear, that the elements should be overthrown, or call to His aid the united fires of earth and the heavens. It is according to a plan emanating from the All-powerful and from His wisdom, that the races which have lived a certain time upon the earth have made way for others, and frequently more perfect races, speaking in the sense which implies a more complex organization. Another important point: geology and biblical revelation are now pretty well agreed. We speak of the existence of the human race at the period of the Great Deluge of Western Asia. It was long alleged that man only appeared upon the earth after the grand geological convulsion, which produced the inundation of the countries situated at the foot of the long Caucasian chain. The recent discoveries of Boucher, Perthes, Lartet, and Lyell, have placed beyond a doubt the existance of man at this epoch; proving that the earth was inhabited by the human race before this Deluge, and justifying in this respect the recital of the sacred historian.

Geology is, then, far from opposing itself to the Christian religion, and the antagonism which formerly existed has given place to a happy agreement. Nothing proves with more certainty than the study of geology the evidences of eternity and divine unity; it shows us, so to speak, the creative power of God in action. We see the sublime work of creation perfecting itself unceasingly in the hands of its divine Author, who has said, "Before the world was, I was." To chaos succeeds a sphere still incandescent, which is modelled into form, and becomes cool enough to permit of organic life. Its burning surface, at first rough and naked, is covered by degrees, and decorated with shrubs and trees. Continents and seas in due course take their definite limits. Rivers · and rivulets flow tranquilly between their banks, and the earth becomes clothed with its present aspect of magnificence and tranquillity. In the case of organized beings, the first animals which appeared were simple in their organization; but as the soil became better adapted for organic life, others succeeded, until God made man—His latest work—to issue from His creative hand, ornamented with the supreme attribute of intelligence, by which he is authorized to rule all nature and subdue it to his laws.

Thus nothing is better calculated to fix in the mind of youth the thought of the All-powerful than the study of the successive revolutions of our globe, and of the generations of beings which have preceded and prepared the way for man. The work leaves us to adore the workman.

THE WORLD

BEFORE THE DELUGE.

GENERAL CONSIDERATIONS.

THE observer who glances over a rich ·and fertile plain, watered by
rivers and watercourses which have, during a long course of ages,
pursued the same uniform and tranquil course; the traveller who con-
templates the walls and monuments of a great city, whose foundations
are lost in the night of ages, witnessing, apparently, to the unchange-
ableness of things and places; the naturalist who examines a mountain
or other locality, and finds the hills and valleys and other accidents of
the soil in the very spot and condition in which they are described by
history and tradition;—neither of these inquirers would at first
suspect that any serious subversion had ever occurred to disturb the
surface. Nevertheless, the spot has not always presented the calm
aspect of stability which it now exhibits; in common with every spot
of earth, it has had its convulsions, its physical revolutions, whose
story we are about to trace. Buried in the depths of the soil, for
example, in one of those vast excavations which the intrepidity of the
miner has dug, in search of coal and other minerals and metals, there
are numerous phenomena which strike the mind of the inquirer, and

carry their own conclusions with them. A striking increase of temperature occurring in these subterranean places is one of the most remarkable of these. It is found that the temperature of the earth rises one degree for every sixty or seventy feet of descent from its surface. Again: if the mine be examined vertically, it is found to consist of a series of layers or beds, sometimes horizontal, but more frequently oblique, upright, or even corrugated and undulating—even folding back upon itself. Then, instances are numerous where horizontal and parallel beds have been penetrated, and traversed at right or oblique angles by veins of metals or minerals totally different in their nature and appearance from the surrounding rocks. All these undulations and changing inclinations of strata are indications that some powerful cause, some violent mechanical action, has intervened to produce them. Finally, if the interior of the beds be examined more minutely, if, armed with the miner's pick and shovel, the surrounding earth is dug up, it is not impossible that the very first efforts at mining may be rewarded by the discovery of some fossil form no longer found in the living state. The remains of plants, and animals belonging to the first ages of the world, are, in fact, very common; entire mountains are formed of them, and, in some localities, the soil can scarcely be touched at a certain depth without yielding fragments of bones and shells, or the impression of fossilized animals and vegetables, the buried remains of extinct creations.

These bones—these remains of animals or vegetables which the pick of the young geologist has torn from the soil—belong probably to some organic species which no longer exists anywhere: it cannot be compared to any animal or plant living in our times; but it is evident that these beings, whose remains are now so deeply buried, have not always been so covered; they lived on the surface of the earth as plants and animals do in our days, for their organization is essentially the same. The beds in which they now repose, then, must in other times have formed the surface; and the presence of these bones and fossils prove that the earth has suffered great mutations in distant times.

Geology explains to us the transformations which the earth has passed through in order to arrive at its present condition. We can determine, with its help, the epoch to which the various beds belong, as well as the order in which others have been superposed upon them; but of all sciences geology is the most recent. It was only at the commencement of our century that it was constituted in a positive manner. It is therefore, of all modern sciences, that which has been modified most profoundly and most rapidly: in short, resting as it does on observation, it has been modified and transformed according to every series of facts recorded. Its applications are numerous and varied, projecting upon many other sciences new and useful lights. We ask of it here only the teachings which serve to explain the origin of the globe. The evidence it furnishes of the progressive formation of the different beds and mineral veins of which it is composed, and the description and restoration of the several species of animals and vegetables which have become extinct, and which form, in the language of naturalists, the *Flora* and *Fauna* of the ancient world.

In order to explain the origin of the earth, and the cause of its various revolutions, modern geologists invoke three orders of facts, or fundamental considerations :—

I. The consideration of fossils.

II. The hypothesis of the incandescence of the globe.

III. The successive deposition of the sedimentary rocks.

As a corollary of these, the hypothesis of the upheaval of the earth's crust follows—upheavings having produced most of the local revolutions. Superposing new materials upon the older rocks, introducing abnormal rocks, called *Eruptive*, upon preceding deposits in such a manner as to change their nature in divers ways. Whence is derived a third class of rocks called *Metamorphic rocks*, of which our knowledge is of very recent date.

FOSSILS.

The name of *Fossil* is given to all vestiges of organized bodies, animal or vegetable, buried naturally in the terrestrial strata, and

belonging to no species known to be living in our days. These fossil
bodies have neither the beauty nor the elegance of the greater part of
living beings; mutilated, discoloured, and often deformed, they seem
to hide themselves from the eyes of the observer who would interrogate
them, and who seeks to reconstruct, with their assistance, the Fauna
and Flora of past ages. Fossil shells, in the more recent deposits,
are found scarcely altered. In other cases only an impression of the
external form is left; sometimes an entire cast of the shell, exterior
and interior. In other cases the shell has left a perfect impression of
its form in the imbedding mud, and has then been dissolved and
washed away, leaving its mould. This mould, again, has sometimes
been filled up by calcareous spar, silica, or pyrites, and an exact cast
of the original shell obtained—a petrified shell, in short. Petrified
wood is equally common.

These remains of the primitive creation had long been examined
and classed scientifically as *freaks of nature,* for so we find them
described in the works of the ancient philosophers who wrote on
Natural History, and in the few treatises on Natural History which
the middle ages have bequeathed to us. Fossil bones, especially those
of elephants, were known to the ancients, giving birth to all sorts of
legends and fabulous histories: the tradition which attributed to
Achilles, to Ajax, and the other heroes of the Trojan war, a height of
twenty feet, was traceable no doubt to the discovery of the bones of
elephants near their tombs. In the time of Pericles we are assured
that in the tomb of Ajax a *patella,* or knee-bone, of that hero was
found, which was as large as a dinner-plate. This was probably only
the fossilized patella of an elephant.

The great French artist, Bernard Palissy, had the glory of being
the first modern author to recognize and proclaim the true character
of the fossilized fragments which are met with in such numbers in
certain formations, particularly in those of Touraine, which had come
more especially under his notice. In his work on 'Waters and
Fountains,' published in 1580, he maintains that the *figured stones,*
as fossils were then called, were the remains of organized beings pre-

served at the bottom of the sea. But the existence of marine shells upon the summits of mountains had already struck the mind of the ancient authors. Witness Ovid, who in Book XV. of the 'Metamorphoses,' tells us he had seen land formed at the expense of the sea, and marine shells lying dead far from the ocean; and more than that, an ancient anchor had been found on the very summit of a mountain.

Vidi factus ex æquore terras,
Et procul a pelago conchæ jacuere marinæ,
Et vetus inventa est in montibus anchora summis.

The Danish geologist Steno, who published his principal works in Italy about the middle of the seventeenth century, had deeply studied the fossil shells discovered in that country. The Italian painter Scilla produced a Latin treatise on the fossils of Calabria, in 1760, in which he established the organic character of fossil shells.

In the eighteenth century, which gave birth to two very opposite theories as to the origin of our globe, the *Plutonian*, or igneous, and the *Neptunian*, or aqueous origin, the Italian geologists gave a serious impulse to the study of fossils: the name of Vallisneri may be cited as the author to whom science is indebted for the earliest account of the marine deposits of Italy, and of the most characteristic organic débris which they contained. Lazzaro Moro continued the studies of Vallisneri, which the monk Gemerelli reduced to a complete system, explaining the ideas of these two geologists as Vallisneri had wished, "without violence, without fiction, without miracles;" Marselli and Donati both studied in a very scientific manner the fossil shells of Italy, and in particular those of the Adriatic, recognizing the fact that they affected in their beds a regular and constant order of superposition.*

In France the celebrated Buffon gave, by his eloquent writings, great popularity to the notions of the Italian naturalists concerning the origin of fossil remains. In his admirable 'Epoques de la Nature' he sought to establish that the shells found in great quantities buried

* Consult Lyell's 'Principles of Geology' and a recent edition of the 'Elements,' with much new matter, for further information relative to fossils during the two last centuries.

in the soil, and even on the summit of mountains, belonged, in reality,
to species not living in our days. But this idea was yet too new not
to find objectors : it counted among its adversaries the hardy philosopher who might have been expected to adopt it with most ardour.
Voltaire attacked, with his jesting and biting criticism, the doctrines of
the illustrious innovator. Buffon insisted, reasonably enough, on the
existence of shells on the summit of the Alps, as a proof that the sea
had at one time occupied that position. But Voltaire asserted that
the shells found on the Alps and Apennines had been thrown there by
pilgrims returning from Rome. Buffon might have replied to his
opponent by pointing out whole mountains formed by the accumulation
of shells. He might have sent him to the Pyrenees, where shells of
marine origin form immense mountains rising six thousand six hundred
feet above the present sea level. But his genius was averse to controversy ; and the philosopher of Ferney himself put an end to a dis-
cussion in which perhaps he would not have had the best of the
argument. "I have no wish," he wrote, "to embroil myself with
Monsieur Buffon for a few shells."

It was reserved for the genius of George Cuvier to draw from the
study of fossils the most wonderful results : it is the study of these
remains, in short, which constitutes in these days positive geology, aided
by mineralogy, a science which is sometimes made too abstract. "It is
to fossils," says the great Cuvier, "that we owe the discovery of the
true theory of the earth ; without them we should not have dreamed,
perhaps, that the globe was formed at successive epochs, and by a
successive series of operations. They alone, in short, tell us with
certainty that the globe has not always had the same envelope ; we
cannot resist the conviction that they must have lived on the surface of
the earth before being buried in its depths. It is only by analogy that
we extend to the primitive soil the direct conclusions which fossils
furnish us with in respect to the secondary formations ; but if we had
only unfossiliferous rocks to examine, no one could maintain that the
earth was not formed all at once."

The method adopted by Cuvier for the reconstruction and restora-

tion of the fossils found in the tertiary rocks of Montmartre, at the
gates of Paris, have served as a model for all other naturalists; let us
listen, then, to his exposition of the vast problem whose solution he
proposed. "In my work on the fossil bones," he says, "I propose to
ascertain to what animals the osseous débris belong; it is seeking to
traverse a road on which we have as yet only ventured a few steps.
As an antiquary of a new kind, it seemed to me necessary to learn at
once to restore these monuments of past revolutions, and to decipher
the direction we were to pursue. I had to gather and bring
together in their primitive order the fragments of which they are
composed; to reconstruct the antique beings to which the fragments
belong; to reproduce them in their proportions and with their characteristics; to compare them, finally, with others now living on the surface
of the globe; an art at present little known, which supposes a science
scarcely touched upon as yet, namely, that of the laws which preside
over the coexistence of the several parts in organized beings. I must,
then, prepare myself for these researches by others upon existing
animals even more laborious. A review, almost universal, of the
actual creation could alone give a character of demonstration to my
account of these ancient inhabitants of the world; but it ought, at
the same time, to give me a great collection of laws, and of relations
not less demonstrable, thus forming a body of new laws to which the
whole animal kingdom could not fail to find itself subject.

"When the sight of a few bones inspired me, more than twenty years
ago, with the idea of applying the general laws of comparative anatomy
to reconstituting and naming fossil species; when I began to perceive
that these species were not quite perfectly represented by those of our
days, which resembled them most; I had not the slightest doubt that I
walked daily upon soil filled with spoils more extraordinary than any
I had yet seen, or that I was destined to bring to light other creatures
altogether unknown to the world, which had been buried, during incalculable ages, at great depths in the earth.

"I had not yet given any attention to the published notices of these
bones, by naturalists who had no pretensions to the recognition of

species. To M. Vaurin, however, I owe the first intimation of the existence of these bones, with which the gypsum swarms. Some portions which he brought me one day struck me with astonishment; I learned, with all the interest that the discovery could inspire me, that this industrious and zealous collector had already furnished some of them to proprietors of cabinets. Received by these amateurs with much politeness, I found in their collections much to confirm my hopes and excite my curiosity. From that time I searched in all the quarries with great care for other bones, offering rewards to the workmen, such as might awaken their attention. I soon gathered a considerable number, and after a few years I had nothing to desire in the shape of materials. But it was otherwise with their arrangement, and with the reconstruction of the skeleton, which could alone lead to any just idea of the species.

"From the first moment of discovery I perceived that, in these remains, the species were numerous. Soon afterwards I saw that they belonged to many genera, and that the species of the different genera were nearly of the same size, but this was likely rather to hinder than aid me. Mine was the case of a man to whom had been given at random the mutilated and imperfect remains of some hundreds of skeletons belonging to twenty sorts of animals; it was necessary that each bone should find itself alongside that to which it had been connected : it was almost like a small resurrection, and I had not at my disposal the all-powerful trumpet; but I had the immutable laws prescribed to living beings to supply its place ; and at the voice of the anatomist each bone and each part of a bone took its place. I have not expressions in which to paint the pleasure I experienced in seeing, that, as soon as I discovered the character of a bone, all the consequences, more or less foreseen, of the character developed itself successively : the feet were found conformable to what the teeth announced ; the teeth to that announced by the feet ; the bones of the ankles, of the thighs, all those which ought to reunite these two extreme parts were found to agree as I expected ; in a word, each species was reproducèd, so to speak, from its elements."

While the Baron Cuvier was thus zealously prosecuting his inquiries in France, assisted by many eminent collaborateurs, what was the state of geological science in the British islands, where the mineralogical treasures exceed in quantity and value those of all Europe besides? About the same time, Dr. William Smith, better known as " the father of English geology," was preparing, unaided, the first geological map of the country. Dr. Smith was a native of Oxfordshire, and a mining engineer in Somersetshire; his pursuits, therefore, brought him in the midst of these hieroglyphics of nature. It was his practice when travelling professionally during many years to consult masons, miners, waggoners, and agriculturists. He consulted the soil; and in the course of his inquiries he came to the conclusion that the earth was not all of the same age; that there was an age of rocks and an age of fossils; and that the rocks were arranged in layers, or strata, superposed on each other. In 1794 he formed the plan of his geological map, showing this superposition of the various beds: for a quarter of a century did he pursue his self-allotted task, which was completed and in 1801 was published, the first attempt to construct a stratagraphical map.

Taking the men in the order of the objects of their investigation, rather than in chronological order, brings before us the patient and sagacious investigator to whom we are indebted for our knowledge of the Silurian system. For many years a vast assemblage of broken and contorted beds had been observed on the borders of North Wales, and stretching away to the east, as far as Worcestershire, and south into Gloucester, now rising into mountains, now sinking into valleys. The ablest geologist considered them as a mere labyrinth of ruins, whose line of induction was entirely lost. "But a man came," as M. Esquiros eloquently writes, "who threw light upon this sublime confusion of elements." Sir Roderick Impey Murchison, then a young and zealous student, having occasion to consult the late Professor Sedgwick as to his line of study, as he himself informs us, had his attention directed to these broken beds. After years of unremitting labour, he was rewarded by success. He established that these masses of sedimentary rocks, torn and penetrated here and there by eruptive veins of igneous origin,

c

formed an unique system, to which he gave the name of *Silurian*, because the rocks which he considered the most typical of the whole were most fully developed in the country of the ancient Silurian tribes who so bravely opposed the Roman invaders of their country. Many investigators have followed in Sir Roderick's steps, but few men have more nobly earned the honours and fame which surround him.

While Murchison, Sedgwick, Lyell, and De la Beche, and a host of others were prosecuting their inquiries into the Silurian rocks, Hugh Miller and many others had their attention occupied with the Old Red Sandstone, or Devonian, which immediately overlie them. After a youth passed in wandering among the woods and rocks of his native Cromarty, the day came when Miller found himself twenty years of age, and, for the first time, a workman in a quarry. A hard fate he thought it at the time, but to him it was the road to fame and success in life. The quarry in which he laboured was at the bottom of a bay formed by the mouth of a river opening to the south, a clear current of water on one side, as he vividly described it, and a thick wood on the other. In this silent spot in the remote Highlands the first fossil of the old red sandstone was revealed to him ; its appearance struck him with astonishment ; a fellow-workman named a spot where many such monuments of a former world were scattered about : he visited the spot, and became a geologist and the historian of the " Old Red." And what strange, fantastic forms did it fall to his lot to describe ! " The figures on a China vase or Egyptian obelisk," he says, " differ less from the real representation of the objects than the fossil fishes of the ' Old Red ' differ from the living forms which now swim in our seas."

The *carboniferous limestone*, which underlies the coal, the *coal-measures* themselves, the *new red*, the *chalk*, and the *lias*, have in their turn found their historians ; but it would be foreign to our object to dwell further here on these branches of our subject.

Some few of the fossilized beings referred to resemble species still found living, but the greater part have become quite extinct. They constituted natural families, of which none of the genera have survived. Such is the *Pterodactylus* among the reptiles ; the *Ammonites* among the mollusks; that of the *Ichthyosaurus* and the *Plesiosaurus* among Saurians. Of other times only, there are lost genera belonging to families which are still living, as the genus *Palæoniscus* among fishes. Finally, we meet with lost species belonging to the genera of existing fauna. The *Mammoth*, for example, is a lost species of elephant.

Some fossils are terrestrial, like the gigantic stag, *Cervus Megaceros;* the Limaçon, or *Helix:* fluivatal or lacustrine, like the *Planorba;* the *Lymna;* the *Physe;* the *Unio:* marine, inhabiting the sea exclusively, as *Cypræa elegans*, and the oyster, *Ostrea Virgula.*

Sometimes fossils are preserved naturally, or only very slightly changed. Such is the state of some of the bones extracted from the more recent caves; such, also, is the condition of the insects found encased in the fossilized resins in which they are preserved from decomposition, and certain mollusks, found in the recent formations, and even in ancient jurassic or cretaceous strata: in some of these, the beds have preserved their colour, as well as the brilliant mother-of-pearl or nacre, of their shells. At Trouvelle, in the Kimmeridge strata, some magnificent ammonites were found in the earth and marl, quite brilliant with the colour of their nacre. In the cretaceous soil at Machéroménil, some *Ancyloceras* and *Hamites* were found still covered with nacre, having a reflected brilliancy of blue, green, and red, retaining an admirable effect. At Glos, near Liseaux, in the coral rag, not only the *Ammonites*, but the *Trigonias* and *Aviculas* have all preserved their brilliant nacre. Sometimes these remains are much changed, the organic matter having totally disappeared; it sometimes happens also, but rarely, that they become petrified, that is to say, the exterior form is preserved, but the primitive organic elements have disappeared, and have been replaced by foreign mineral substances—by silica or by carbonate of lime.

Geology has also enabled us to draw very important conclusions from certain fossil remains whose nature had long remained unknown, but which, under the name of *coprolites*, had given birth to long controversial discussions. Coprolites are the petrified or stony excrements of great fossil animals. The study of these singular remains

c 2

has thrown unexpected light on the manners and physiological organization of some of the great antediluvian animals. Their examination has revealed the scales of fish and teeth, thus permitting us to determine the kind of nourishment in which the animals of the ancient world indulged : for example, the coprolites of the great marine reptile which bore the name of the *Icthyosaurus* contain, with the bones of other animals, remains of the vertebræ, or of the phalanges of other Icthyosauri, showing that this animal habitually fed on the flesh of its own species, as many fishes do in our days, but especially the more voracious ones.

The imprints left upon the earth or sand, which time has hardened and transformed into sandstone, furnish to the geologist another series of precious indications. The reptiles of the ancient world, the turtles in particular, have left upon the sands, which time has transformed into blocks of stone, imprints which evidently represent the exact mould of the feet of these animals. These impressions have sometimes been sufficient for naturalists to determine to what species

Fig. 1.—Labyrinthodon pachygnathus and foot-marks.

the animal belonged which thus left its impress on the wet soil. Some of these exhibit tracks to which we shall have occasion to return, others present traces of the steps of the great reptile known as the *Labyrinthodon* or *Cheirotherium,* whose foot slightly resembles the hand of a man. Another well-known impression, which has been left upon the sandstone of Corncockle Moor in Dumfriesshire, is supposed to have been the impress of the foot of some great turtle.

We may be permitted to offer a short remark on this subject. The historian and antiquary may traverse the battle-fields of the Greeks and Romans, and search in vain for traces of these conquerors, whose armies ravaged the world. Time, which has overthrown the monuments of their victories, has also effaced the imprint of their footsteps; and of millions of men besides, whose invasions have spread desolation over Europe, there is not even a trace of their footsteps. These reptiles, on the contrary, which ranged for thousands of years on the surface of our planet when still in its infancy, have impressed on the soil indelible recollections of their existence. Hannibal and his legions, the barbarians and their savage hordes, have passed over the land without leaving a material mark of their passage, while the poor turtle which drags itself along on the silent shore of the primitive seas has bequeathed to learned posterity the image and imprint of a part of its body. These imprints may be perceived as distinctly marked on the rocks as the traces left in moist sand or in newly-fallen snow by some animal under our own eyes. What grave reflections should be awakened within us at the sight of these blocks of hardened earth, which thus carry back our thoughts to the first ages of the world! and how insignificant the discoveries of the archæologist who throws himself into ecstacies before some piece of Greek or Etruscan pottery, when compared with these veritable antiquities of the earth!

The paleontologist (from παλαιός, ancient, οἶτος, being, λόγος, discourse), who occupies himself with the study of animated beings which have lived on the earth, has also taken careful account of the sort of moulds left by organized bodies in the fine sediment which has enveloped them after death. Many organic beings have left no organic trace of their existence in nature, but we find impressions of them perfectly preserved in the sand and limestone rocks, in the earth or clay, in the coal measures; and these moulds are sufficient to let us know the species to which they belonged. We shall no doubt

astonish our readers when we tell them that there are blocks which
have the distinct impression of drops of rain which had fallen upon
soil of the ancient world.   These imprints, made upon the sands, and

Fig. 2.—Imprints of rain-drops.

preserved there by desiccation, and by being transformed at a later
period into solid and coherent sandstones, have left impressions which
are thus maintained to the present day.   Fig. 2 represents impressions
of this kind upon the sandstone of Connecticut river in America, which
have been reproduced from the block itself by photography.   In a
depression of the granitic rocks of Massachusetts and Connecticut, the
red sandstone occupies an area of a hundred and fifty miles in length
from north to south, and from five to ten miles in breadth.   " On some
shales of the finest texture," says Sir Charles Lyell, "impressions of
rain-drops may be seen, and casts of them in the argillaceous sandstones."   The same impressions occur in the recent red mud of the
Bay of Fundy.   More than that, the undulations left by the passage
of the waters of the sea over the sands of the primitive world are

preserved by the same physical mechanism. Traces of undulations of this kind have been found in the neighbourhood of Boulogne-sur-Mer. Similar phenomena occur in a manner still more striking in some sandstone quarries opened at Chalindrey (Haute-Marne). The banks there present, upon a large surface, traces of the same kind, and along with them impressions of the excrements of marine worms, left on the shore, it is supposed, by the reflux of the tide.

INCANDESCENCE OF THE CENTRAL PART OF THE GLOBE.

The idea of a great central fire is a very ancient hypothesis: admitted by Descartes, and developed by Leibnitz and by Buffon, it has since been confirmed by a crowd of facts, and adopted, at least it is not opposed, by the leading authorities of the age. Herschel, Murchison, Hind, Lyell, Ansted, and the leading English astronomers and geologists, give a cautious admission to the doctrine. Here are the principal facts :—

When we descend into the interior of a mine we feel that the temperature becomes elevated in an appreciable manner as we descend into its depths.

The high temperature of the waters in artesian wells, when these are very deep, testifies to the increase of heat in the interior of the earth.

The thermal waters which issue warm from the earth, whose temperature sometimes rises to 100° Centigrade and upwards, as, for instance, the Geysers of Iceland, are another proof in support of the fact.

The modern volcanoes are a visible demonstration of the existence of a central fire. The heated gases, the liquid lava, the red flame which escapes from their craters, prove sufficiently that the deeper parts of the globe are at a temperature prodigiously elevated.

The disengagement of gas and burning vapour by the accidental fissures in the crust which accompany earthquakes, still tends to establish the existence of a central incandescence in the interior of the globe.

We have already said that the temperature of the globe rises about one degree for every sixty or seventy feet of depth beneath its surface. The exactness of this observation has been verified in a great number of instances—indeed to the greatest depth to which man has penetrated, and taken the thermometer. Now, as we know exactly the length of the radius of the terrestrial sphere, we may reckon on the progression of this temperature; and supposing it to be regular and uniform, then the temperature at the centre of the globe ought to give at the present time a mean temperature of 195,000° Centigrade. No matter could preserve its solid state at this excessive temperature; it follows, then, that the centre of the globe, and all parts near the centre, must be in a state of perfect fluidity.

The works of Werner, Hutton, of Leopold de Buch, of Humboldt, of Cordier, have reduced this hypothesis to a theory, on which has been based, to a considerable extent, the modern science of geology.

Modifications of the Surface of the Globe.

As a consequence of the hypothesis of central fire, it is admitted that our planet has been agitated by a series of local disturbances; that is to say, by ruptures of the crust occurring at intervals more or less distant. These partial revolutions at the surface are produced, as we shall have occasion to explain, by upheavals or depressions of the solid crust, resulting from the fluidity of the central parts, and by the cooling down of the external crust of the globe.

Almost all bodies in passing from a liquid to a solid state are diminished in size in the process. In molten metals which recover the solid state by cooling, this diminution amounts to about a tenth of their volume; but the decrease in size is not equal over the whole mass: the outer envelope, from its greater exposure to radiation, is left much too large for the inner surface of the consolidated mass when entirely cooled. Cracks and hollows consequently occur even in small masses; but the effect of converting such a vast body as the earth from a liquid, or rather molten state, may be imagined. As the interior became solid and concrete, by slow degrees, the crust would be found too large for the interior of the sphere, which has contracted in cooling. The consequence would be furrows, corrugations, and depressions in the crust, producing great inequalities in the surface of the globe; producing, in short, what constitutes *chains of mountains* and valleys, ravines or gorges.

In other places, in lieu of furrows and wrinkles, the solid crust has broken in, producing fissures and ruptures in the exterior envelope, sometimes of immense extent. The liquid substances contained in the interior, with or without the action of the gases they enclose, escape through these openings, and accumulating on the surface become, on cooling and consolidating, *isolated mountains* of various heights.

It would sometimes happen, and always from the same cause, namely, from the interior contraction occasioned by the unequal cooling of the globe, that incandescent liquid matter would be ejected into the smaller interior fissures in the terrestrial crust, filling them up, and forming in the centre of the rocky crust those long contracted lines of foreign substances which we call *veins*.

Finally, it occasionally happens, that in place of molten matter such as granite or metallic compounds, we find escaping through these fractures and crevices of the globe, veritable rivers of boiling water, charged abundantly with mineral salts; that is to say, with silicates, and with calcareous and magnesian compounds. The chemical and mechanical action of the atmosphere, of rain, and the waves, has a tendency to destroy even the hardest rocks, which crumble and are carried off in the waters. These salts and other substances, uniting themselves at first with those already existing in the sea, would at a subsequent period be separated again from the waters, deposited, and thus contribute greatly to extend the dry land. These became the *sedimentary rocks*.

The furrows, corrugations, and fractures in the terrestrial crust, which so changed the aspect of the surface, and for the moment displaced the sea basins, would be followed by periods of calm. During these periods, the débris, torn by the movement of the waters from certain points of the land, would be transported to other parts by the oceanic currents. These heterogeneous materials, when deposited, would accumulate, and form new land, or *transported rocks*.

Such is the brief history of the origin of the earth and some of its phenomena; but it is proper to state that it is not the only hypothesis by which the ingenuity of man has sought to solve the mighty problem; and there are not wanting great names and strong reasoning to support some of them in the present day. The Nebular theory, for instance, which embraces the whole solar system, and, by analogy, the universe, assumes that the sun was the original scene of incandescent matter, the vast body being brought into a state of revolution by the action of laws to which, in His divine wisdom, the Creator has subjected all matter. In consequence· of its immense expansion and rarity, the exterior zone of vapour, expanding beyond the sphere of attraction, is suppcsed to have been thrown off by centrifugal force. This ring of vapour, which may be supposed at one time to have resembled the rings of Saturn, would in time break into several masses, which, coalescing into one great body, would, by the renewed power of attraction which the consolidated body would assume, revolve round the sun, and, from mechanical considerations, have a rotatory motion on its own axis.

This doctrine applies to all the planets, and assumes each to be in a state of incandescent vapour, with a central burning nucleus. As the cooling went on, each of these bodies may be supposed to have thrown off similar masses of vapour, which, by the operation of the same laws, would assume the rotatory state, and, as a satellite, revolve round the parent planet. Such, in brief, was the grand conception of La Place; and surely it detracts nothing from the omnipotence of the Creator that it initiates the creation step by step, and under the laws to which matter is subjected, rather than by the direct fiat of the Almighty. The hypothesis assumes that as the vapoury mass cooled by the radiation of heat into space, the particles of matter would approximate and crystallize.

That the figure of the earth is such as a very large mass of matter in a state of fluidity would assume from a state of rotation, seems to be admitted, thus corroborating the speculations of Leibnitz, that the earth is to be looked on as a heated fluid globe, cooled, and still cooling, at the surface by radiation of its superfluous heat into space.

But Mr. Hopkins has put forth some strong but simple reasons bearing another construction, although he does not attempt to solve the problem otherwise than by inference. We gather from them that :—

If the earth were a fluid mass cooled by radiation, the cooled parts would, by the laws of circulating fluids, descend towards the centre, and be replaced on the surface by matter at a higher temperature.

The tendency of such a mass to consolidate would, therefore, be a struggle for superiority between pressure and temperature, both of which would be at their maximum at the centre of the mass.

At the surface, it would be a question of rapidity of cooling by radiation as compared with the internal condition of heat—relations for comparing which we are without data ; but on the result of which depends whether such a body would most rapidly solidify at the surface by radiation, or at the centre by pressure.

The effect of the first would be solidification at the surface first, followed by condensation at the centre through pressure. There would be two masses, a spherical fluid nucleus, and a spherical shell or envelope, with a large zone of semi-fluid, pasty matter between, continually changing its temperature as its outer surface was converted to the solid state.

If pressure, on the other hand, gained the victory, the centre would solidify before the circulation of the heated matter had ceased, and the solidifying process would proceed through a large portion of the globe, and even approach the surface before that would become solid. In other words, solidification would proceed from the centre until the diminishing power of pressure was balanced by radiation, when the gradual abstraction of heat would allow the particles to approximate and crystallize.

The terrestrial sphere may thus be a solid indurated mass at the centre, with a solid stony crust at the surface, and a shifting viscous, but daily-decreasing, mass, between the two ; a supposition which the diminished and diminishing frequency and magnitude of volcanic and other eruptive convulsions seems to render not improbable.

It is to be observed, however, that although the problem is probably beyond the power of man to solve, either hypothesis would

account for many of the phenomena which surround us. Each alike
serves to explain the origin of *mountains*, of *eruptive rocks*, of *metallic veins*, of *sedimentary rocks*, and *transported rocks*. The various
phenomena—corrugation and upheaval; ejection of igneous matter;
the emergence of thermal waters charged with mineral salts; the
decomposition of superficial rocks by pluvial and sea water, aided by
atmospheric influences; sedimentary deposits produced by gravitation
—have each been the great agents of change during all the geological
periods, and they have continued in action up to our own days.   It is
to this complex series of phenomena that the terrestrial crust owes its
so complicated and variable internal and external structure.   From
these considerations, we may divide the mineral substances of which
the earth is composed into three general groups, under the following
heads :—

1. *Crystallized Rocks.*—That portion of the terrestrial crust primitively liquid, but solidified at its first cooling down.

2. *Sedimentary Rocks.*—Consisting of mineral substances, such as
silica, carbonates of lime and magnesia, deposited by the waters of the
seas.

3. *Eruptive Rocks.*—Crystalline, like the first, but formed in every
geological period by irruption or ejection of the liquid matter occupying the interior of our globe through the primitive rocks.

The mineral masses which constitute the *sedimentary rocks* form
beds, or *strata*, which have among themselves a constant order of superposition which indicate their relative age.   The mineral structure
of these beds, and the remains of organized beings they contain, impress on them that character which permits us to distinguish each
from that which precedes and follows it.

It does not follow, however, that all these beds are met with, regularly superposed, over the whole surface of the globe : under such circumstances geology would be a very simple science, and, so to speak,
under the jurisdiction of the eyes.   In consequence of the frequent
eruptions of granite, of porphyry, serpentines and trachytes, of basalt
and lava, these beds are often broken, interrupted, and replaced by
others.   At certain points, a whole series of sedimentary rocks, and

often many successive ones, have been displaced from this cause; the regular series are, in fact, rarely found complete. It is only by combining the collected observations of the geologists of all countries that we are enabled to arrange, according to their ancient relations, the several beds composing the terrestrial crust as they occur in the following table, which proceeds from the centre towards the superficies:—

Primitive Rocks	Gneiss, &c.
Transition Rocks	Silurian Rocks. Devonian Rocks. Carboniferous Rocks. Permian Rocks.
Secondary Rocks. . . .	Triassic Rocks. Jurassic Rocks. Cretaceous Rocks.
Tertiary Rocks	Eocene Period. Miocene Period. Pliocene Period.
Quaternary	Modern Period.

Under these heads we propose to ourselves to examine the successive transformations to which the earth has been subjected in reaching its present condition : in other words, we propose, both from a historical and descriptive point of view, to take a survey of the several *epochs* which can be distinguished in the gradual formation of the earth—the residence and domain of man, which correspond with the formation of the great groups of rocks in the preceding table, including the living creatures which have peopled the earth, and which have disappeared, from causes which we shall also endeavour to trace : we shall describe the plants and animals belonging to each great phase in the history of the globe. At the same time, we shall not pass entirely in silence the rocks deposited by the waters, or thrown up by irruption during these periods; we shall give, also, a summary of the mineralogical characters and fossil characteristics belonging to each period What we propose, in short, is to give a history of the formation of the globe, and a description of the principal rocks which actually compose it, with a rapid glance at the several generations of animals and plants which have succeeded and replaced each other on the earth, from the very beginning of organized life to the appearance of man.

PRIMITIVE EPOCH.

THE theory which has been developed, and which considers the earth as an extinct sun—as a cooled-down star—as a nebula, or luminous cloud, which has passed from the gaseous to the solid state—this fine conception, which unites so brilliantly the kindred sciences of astronomy and geology, belongs to the French mathematician, La Place, the immortal author of the ‘Mécanique Céleste.’

We have stated, in the preceding chapter, that the centre of the earth is still calculated to be about the temperature of 195,000° Cent., a degree of heat which surpasses all that imagination can conceive. We can have no difficulty in admitting that, in a heat so excessive, all substances which enter into the composition of the globe would be reduced in its origin to the state of gas or vapour.   Our planet, then, must have been an aggregate of aëriform fluids—a mass of matter entirely gaseous; and if we reflect that the substances in their gaseous state would occupy a space eighteen hundred times larger in that state than in the solid, we shall have some conception of the enormous volume of this gaseous mass.   It would be as large as the sun, which is fourteen hundred thousand times larger than the terrestrial sphere. In Fig. 3 we have attempted to give an idea of the vast difference existing between the earth in its solid state and the primitive gaseous

mass.   A represents the former, B the latter.   It is simply a
geometrical figure, putting in relief the comparative size of the globe;

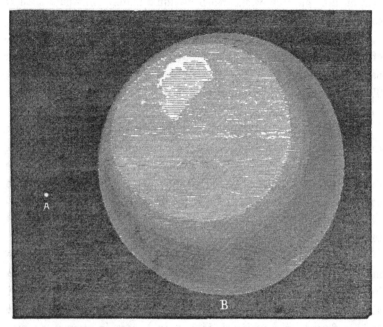

Fig. 3.—Comparative volume of the earth in the gaseous and solid state.

the one being the twentieth part of an inch, the other two inches and
three quarters.

At this excessive temperature, the gaseous mass which we have
described would be borne in space much as the sun may be supposed
to be, and it would shine with the same brilliancy with which, in our
eyes, the fixed stars and planets burn in the serenity of night, as represented on the opposite page (PL. I.).   Circulating round the sun, in-
obedience to the laws of gravitation, this incandescent gaseous mass
necessarily submitted to the laws which regulate other material substances.   As it got cooler, it gradually transferred portions of its
warmth to the glacial regions of the inter-planetary spaces, in the
middle of which it traces the line of its flaming orbit.   Consequent

on its continual cooling, but at the end of a period of time, of which it would be impossible, even approximately, to fix the duration, the star, at first gaseous, would reach its liquid state. It would then be considerably diminished in volume.

The laws of mechanics teach us that liquid bodies, when in a state of rotation, take the spherical form : it is one of the laws of their being, emanating from the Creator, due to the force of attraction. Thus the earth takes the spheroidal form proper to it in common with the other celestial bodies.

The hypothesis of Laplace gives to the sun, and to all bodies which gravitate in what Descartes calls his tourbillon, a common origin. " In the primitive state in which we must suppose the sun to be," he says, " it resembles one of those nebula which the telescope reveals to us, consisting of a *nucleus*, more or less brilliant, surrounded by luminous clouds, which clouds, condensing at the surface, are transformed into a star."

The earth is subject to two distinct movements; namely, a movement of translation round the sun, and a movement of rotation on its own axis—the latter a uniform movement, which produces the regular alternations of days and nights. Mechanics have established the fact, confirmed by experiment, that a fluid mass in movement, produced by the variation of centrifugal force on its different diameters, produces a swelling towards the middle of the sphere, and flattening at the poles or extremities of its axis. In virtue of this phenomenon, the earth, when in its liquid state, became swollen at the equator and flattened at each pole, and has passed from its primitive spherical form to the ellipsoid form—that is, it is flattened at each of its extremities.

This swelling at the equator and flattening towards the poles afford the most direct proofs that can be adduced of the primitive liquid state of our planet. A solid and non-elastic sphere—an ivory ball, for example —might turn during ages upon its axis, but its form would sustain no change; but a liquid ball, of a pasty consistence, would swell out towards the middle, and, in the same proportion, be flattened at the

I.—The Earth circulating in space in a gaseous state.

extremities of the axis. It was upon this principle, namely, by admitting the primitive fluidity of the globe, that Newton announced à priori the expansion of the globe at the equator and flattening at the poles, and he even fixed in advance the proportion of the flattening. The direct measurement, both of the expansion and flattening, by Maupertuis, Clairaut, Camus, and Lemonnier, in 1736, proved how exact the calculations of the great geometrician were. These gentlemen, accompanied by the Abbé Outhier, were sent into Laponia by the Aca-démie des Sciences; the Swedish astronomer, Celsius, accompanied them, and furnished them with the best instruments for measuring and surveying. At the same time, the Academy sent Bouguer and Condamine to the equatorial regions of South America. The measurements taken in both these places established the existence of the equatorial expansion and the polar depression. It was found that the depression was sensibly greater than Newton had estimated in his calculations.

If it did not follow, as a consequence of the partial cooling down of the terrestrial mass, that all the gaseous substances composing it should pass into a liquid state, some of these would remain in the state of gas or vapour, and would form round the terrestrial spheroid, an envelope or *atmosphere*, from the Greek word ἀτμός, vapoury, σφαῖρα, sphere. We shall not have formed a very inexact idea of the atmosphere which surrounded the globe at this remote period, if we compare it with that which surrounds us now; but the extent of atmosphere which surrounded the primitive earth must have been immense; it doubtless reached the moon. It included, in short, in the state of vapour, the enormous mass of waters which now, in their condensed form, constitute the mighty ocean, added to the other substances which preserved their gaseous state at the temperature then exhibited by the incandescent earth, and it is certainly no exaggeration to place this at 2000° Centigrade. The atmosphere would participate in this temperature; and, acted on by this excessive heat, the pressure that it would exercise on the earth would be infinitely more considerable than that which it exercises at the present time. To the gases which form the compo-

D

nent parts of the air—namely, azote, oxygen, and carbonic acid gas
—to the enormous masses of watery vapour, we should have to add
vast quantities of mineral substances, metallic or earthy, reduced to
the gaseous state, and maintained in that state by the temperature of
this gigantic furnace. The metals, the chlorides—metallic, alkaline,
and earthy, the sulphurets, and even the earthy bases of silica, of
aluminum, and of lime; all at this temperature would exist in a
vapoury form in the atmosphere surrounding the primitive globe.

It is thought that, under these circumstances, the different substances
composing the atmosphere would be ranged round the globe in the order
of their density; the first layer—that nearest to the surface—being
formed of the heavier vapours, as the metals, iron, platinum, copper,
mixed doubtless with clouds of fine metallic grains produced by the
partial condensation of their vapours. This first and heaviest zone,
and the thickest also, would be opaque, although the surface of the
earth was still at a red heat. Next in order would come the more
vapourizable substances, such as the metallic and alkaline chlorides,
particularly the chloride of sodium or marine salt, of sulphur and
phosphorus, with all the volatile combinations of these substances.
The upper zone would contain matter still more easily vapourized,
such as watery vapour or steam, united with bodies naturally gaseous,
as oxygen, azote, and carbonic acid. This order of superposition,
however, would not maintain itself constantly. In spite of their unequal density the three atmospherical beds would often be mixed,
producing formidable storms, violent ebullitions; throwing down,
tearing, upheaving, and confounding these incandescent zones.

As to the globe itself, without being so much agitated as its fiery and
mobile atmosphere, it would be no less the prey of perpetual storms,
occasioned by the thousand chemical processes which were in action in
its molten mass. On the other hand, the electricity resulting from
these powerful chemical operations, conducted on a scale so unlimited,
would provoke frightful detonations, the echoes of thunder adding to
the horror of this primitive picture, which no imagination, no human
pencil could trace, and which constitutes the gloomy and disastrous

chaos of which the legendary history of every ancient race has transmitted the tradition. In this manner would our globe circulate in space, carrying in its train the burning streaks of its multiplied atmosphere, unfitted, as yet, for living beings, and impenetrable to the rays of the sun, around which, nevertheless, it describes its gigantic curve.

The temperature of the planetary regions is infinitely low: according to Laplace, it cannot be estimated at less than a hundred degrees below zero. These glacial regions, which would be traversed in its course by the glowing incandescent globe, would necessarily cool it step by step; at first, superficially, when it would take a pasty consistence. Nor must it be forgotten that the earth in its liquid state would be obedient in all its mass to the action of flux and reflux which proceeds from the attraction of the sun and moon, but to which the sea alone is now subject. This action, to which all liquid and moveable molecules are subject, would singularly accelerate the preludes to solidification in the terrestrial mass : it would thus gradually assume that sort of consistence which iron attains when it is first withdrawn from the furnace for puddling.

In its cooling process beds or strata of a concrete substance would be produced, which, floating at first in isolated masses on the surface of the semi-fluid matter, would float together, consolidate, and form continuous banks such as we now see icebergs form on the shore of the Polar Seas; and, finally, when washed by the agitation of the waves, the masses would coalesce and form banks more or less moveable. By extending this phenomenon to the whole surface of the globe, the total solidification of its surface would be produced. A solid, but still thin and fragile crust would thus surround the whole earth, enclosing entirely its liquid interior, whose entire consolidation would necessarily be a much slower process—one which is very far from being completed at the present time; for it is estimated that the actual thickness of the earth's crust does not exceed thirty miles, while the mean radius or distance from the centre of the terrestrial sphere approaches four thousand, the mean diameter being in fact 7,912·409 miles; so that the portion of our planet supposed to be solidified represents only a very small fraction of its total mass. We may indeed express, after a very simple but just fashion, the relation between the liquid and concrete portions of our globe, if we imagine an orange to represent the earth, when the thickness of a sheet of paper placed round it will represent the solid crust which now forms the superficies. Fig. 4 represents the proportionate relations of the solid and liquid masses between the terrestrial crust and the interior still liquid. The terrestrial sphere having a radius of 3,956 miles, and a diameter of 7,912, and the solid crust about thirty miles, which represents $\frac{1}{260}$ of the diameter, or $\frac{1}{130}$ of the radius, the engraving represents these proportions with tolerable exactness.

To determine, even

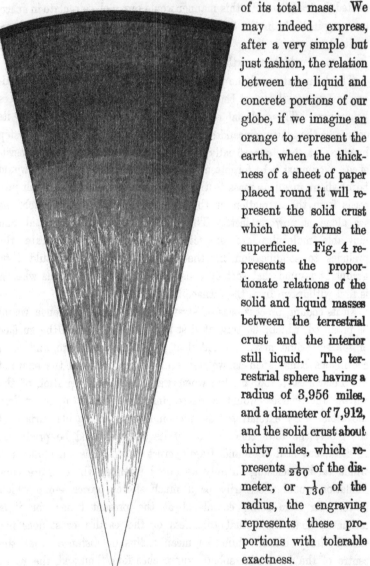

Fig. 4.— Relative volume of the solid and liquid mass of the globe. approximately, the time

such a vast body would take in cooling so as to permit of the formation of a solid crust, or fix the duration of the transformations which we are describing, would be an impossible task.

The first terrestrial crust formed as indicated would be unable to resist the waves of the ocean of internal fire, which would depress and raise it up at its daily flux and reflux in obedience to the attraction of the sun and moon. And who can trace, even in imagination, the fearful rendings, the gigantic inundations which would result from them? Who would dare to paint the horrors of these first and mysterious convulsions of the globe? Amid torrents of molten matter, mixed with gases, upheaving and piercing the scarcely solid crust, large crevasses would be ripped open, and through these gaping wounds waves of liquid granite would be ejected, and left again to cool and consolidate on the surface. Fig. 5 represents the formation of

Fig. 5.—Formation of primitive granitic mountains.

one of these primitive granitic mountains by the eruption of matter from the interior, ejected to the surface through an opening in the crust. In some of these mountains, Ben Nevis for example, three different eruptions can be traced. In this manner would the first mountains be

formed. In this manner also would metallic veins be ejected through
the smaller openings, true injections of eruptive matter produced
from the interior of the globe, traversing the primitive rocks and constituting the precious depository of metals, such as copper, zinc, antimony, and lead. The accompanying engraving (Fig. 6) represents the

Fig. 6.—Metallic veins.

formation of one of these metallic veins. In this case the fracture is
only a cleft in the rock, which was soon filled with the injected matter,
often of different substances, which in crystallizing completely fills the
cleft, or *fault*, as such interruptions in the line of stratification are
called.

But some eruptions of granitic and other substances ejected from
the interior never reach the surface at all. In such cases the clefts
and crevasses—longitudinal or oblique—are filled; but the overlying
crust is too thick to be penetrated by the slender column of molten

matter. Fig. 7 represents an eruption of granite through a mass of
sedimentary rock: the granite ejected from the centre fills all the

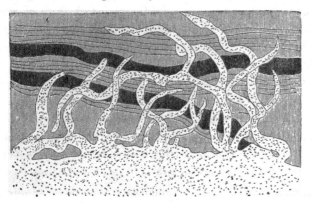

Fig. 7.—Eruption of granite.

clefts and fractures, but it has not been sufficiently powerful to penetrate to the surface.

On the surface of the earth, then, which would be at first smooth
and connected, there were formed, from the very beginning, swelling
eminences, hollows, corrugations, and crevasses, which would materially
change its primitive aspect; its arid and burning surface bristled with
rugged eminences or was furrowed with clefts and cracks. Nevertheless, as the globe continued to cool, a time arrived when its temperature
became insufficient to maintain in a state of vapour the vast masses of
water which floated, suspended and vapourized, in the atmosphere.
These vapours would pass into the liquid state, and now the first rain-
drops fell upon the earth. Let us here remark that these were veritable
rain-drops of boiling water; for, in consequence of the very considerable
pressure of the atmosphere, water would be condensed and become
liquid at a temperature very superior to 100° Centigrade.

The first drop of water which fell upon the still-burning terrestrial
sphere marked a new period in its evolution—a period the mechanical
and chemical effects of which it is important to analyse. The contact
of the condensed water with the consolidated surface of the globe opens

up a series of modifications which science enables us to comprehend and
explain with some confidence—at least, with more positive elements of
appreciation than any we possess for the period of chaos of which we
have attempted to represent some of the features, leaving much
to the imagination, which the reader must interpret after his own
fashion.

The first water which fell in the liquid state upon the gradually
cooling surface of the earth would be rapidly reduced to steam by the
elevation of its temperature. Thus rendered much lighter than the
surrounding atmosphere, these vapours would rise to the utmost limits
of the upper atmospheric zone : thus circumstanced, they would radiate
towards the glacial regions of space, and, again condensing, they would
again descend to the earth in a liquid state, to reascend as vapour and
fall again in a state of condensation. But these alternate changes in
the physical condition of water could only be maintained by a very
considerable temperature on the surface of the globe, which these
alternations of heat and cold were rapidly diminishing : the excess of
heat was being dissipated in the regions of celestial space.

This phenomenon extending itself by degrees to the whole mass of
watery vapour existing in the atmosphere, the waters in increasing
quantities covered the earth ; and as the conversion of all liquids into
vapour is provocative of a notable disengagement of electricity, a vast
quantity of electric fluid necessarily resulted from the conversion of
such masses of water into vapour. Bursts of thunder, and bright
gleams of lightning, were the necessary accompaniments of this extraordinary struggle of the elements—a state of things which M. Maurando
has attempted to represent on the opposite page (PL. II.).

How long did this struggle for supremacy between fire and water,
with the incessant noise of thunder, continue ? All that can be said in
reply is, that a day came when water was triumphant. After having
covered, to a vast extent, the basins and hollows of the earth, it finally
occupied and covered the whole surface ; for there is good reason to
believe that at a certain epoch, at the commencement, so to speak, of
its evolutions, the earth was covered by the waters in its whole extent.

II.—Condensation and rain-fall on the primitive globe.

Ocean was universal. From this moment our globe entered on a regular series of revolutions, interrupted only by the outbreaks of internal fire still concealed under its imperfectly consolidated crust.

In order to comprehend the complex actions, some of them mechanical, others chemical, which the waters, still boiling, exercised on this solid crust, let us consider what were the components of this crust. The rocks which formed its first *stratum*, the framework of the earth, the foundation upon which all others repose, is a compound which in different proportions forms gneiss and granite.

What is this gneiss, this granite, speaking of it in its mineralogical character? It is a combination of silicates, with a base of aluminum, of potash, and soda: *quartz, felspar*, and *mica* form, by their simple aggregation without cement, *granite*: it is thus a combination of three minerals.

Quartz is a silica more or less pure, and often crystallized. *Felspar* is a white crystalline substance, composed of *silicate* of aluminum, of potash, or of soda; potassic felspar is called *orthose*, felspar of soda, *albite*. *Mica*, a silicate of aluminum and of potash, containing magnesia and the oxide of iron: it takes its name from the Latin *micare*, to shine or glitter.

Granite, from the Italian *grano*, being granulated in its structure, is, then, a compound rock, formed of felspar, quartz, and mica, and the three elements are crystalline. *Gneiss*, a word drawn from the Saxon, is a variety of granite, composed of *felspar* and *mica*, but in which the latter predominates: its foliated structure leads sometimes to its being called *stratified granite*.

The felspar which enters into the composition of granite is a mineral which is easily decomposed in water, either cold or boiling, or in carbonic acid contained in the atmosphere. The chemical action of air and water, and the action at once chemical and mechanical of the hot water which was universal in the primitive seas, powerfully modified the granitic rocks which lay beneath these seas. The warm rain which fell upon the mountain peaks and granitic pinnacles, the torrents of rain which fell upon the slopes or in the valleys, dissolved the several silicates which constitute the felspar and mica, and their débris were swept away to form elsewhere banks of argillaceous and quartzose sand : thus were the first modifications in the primitive rocks produced by the united action of air and water, and thus were the first sedimentary rocks deposited by the oceanic waters.

The argillaceous soil produced by this decomposition of the felspathic and micaceous rocks would participate in the still burning temperature of the globe—would be again subjected to partial fusion—and when they became cool again, they would take, by a kind of semi-crystallization, that foliated form which is called schistose structure, from the Greek word σχιστός, easy to divide, or σχιστειν, to divide, a structure of which the roofing slates which separate naturally into thin leaves give an exact idea.

In this manner would the first argillaceous and schistose rocks be formed ; and thus it is that a thick bed of *schist*, the earliest sedimentary rocks known, rests immediately upon gneissic rocks of supposed igneous origin.

At the end of this first phase of its existence, the terrestrial sphere was, then, covered in nearly its whole extent with hot and muddy water, forming extensive but shallow seas. A few islets, raising here and there their granitic peaks, would form a sort of archipelago, surrounded by seas filled with earthy débris held in suspension. During a long series of ages, the solid crust of the globe would go on increasing in thickness as the process of solidification of the liquid matter nearest to the surface proceeded ; but this process was not altogether one of tranquil progress : the solid surface had not yet attained the consistency necessary to resist the pressure of the heated, expansive gases and boiling liquids covered and compressed by its elastic crust. The waves of this internal sea triumphed more than once over the feeble resistance opposed to it ; enormous dislocations and breaches were made in its thickness ; immense upheavals of the solid crust raised the bottoms of seas far above their former level ; mountains rose out of the ocean, and these were not now exclusively granitic, but were composed, besides, of those schistose rocks which had been deposited under the ocean after being long held in suspension in the muddy seas.

On the other hand, the earth, as it continued to cool and settle, would also be diminished in size; and this process of contraction, as we have already explained, was another cause of dislocation at the surface, producing either considerable ruptures or simple fissures in the continuity of the crust. These fissures would be filled at a subsequent period by jets of the molten matter occupying the interior of the globe—by eruptive granite, that is to say, or by some of the many metallic compounds; or they opened a passage to those torrents of boiling water charged with mineral salts, with silica, bicarbonate of lime, and magnesia, which, mingling with the waters of the vast primitive ocean, were deposited on the basins of the seas, thus helping to increase the mass of the mineral substances of the globe.

These eruptions of mineral or metallic matter—these vast outpourings of mineral waters through the fractured surface—would be of frequent occurrence during the primitive epoch we are contemplating : it should not, therefore, be matter of surprise to find the more ancient rocks almost always dislocated, reduced in dimensions, and often interrupted by veins containing metals, metallic oxides, such as oxide of copper and tin, or the sulphurates, such as those of lead, of antimony, or of iron, the objects of the miner's art.

Terebratula alata.

THE PRIMITIVE ROCKS.

In sketching the *first epoch* of the earth's history, we have endeavoured to prepare the reader for comprehending the very summary description we shall give of the rocks which constitute the base of all the mineral beds of the globe, which were formerly designated *primitive rocks*. These rocks, which are of very limited extent in Central Europe, occupy in the north of Europe and America, and wherever there are lofty mountain ranges, spaces of great extent. They comprehend the granites, porphyries, and other rocks called *Plutonic*, as one class, and in the other they consist of *crystalline strata, slates,* and *schists*, which are found to underlie the older sedimentary rocks in every quarter of the globe; and along with this rock may be grouped micaceous schist, hornblende schist, and some other slaty rocks which have no very well-defined place in the system. "This great division of rocks," says Sir Charles Lyell, "contains no pebbles, or sand, or scoriæ, or angular pieces of embedded stone, and no traces of organic bodies, and they are often as crystalline as granite, yet are divided into beds corresponding in form and arrangement to those of sedimentary formations, and are therefore said to be stratified. The beds sometimes consist of an alternation of substances varying in colour, composition, and thickness, precisely as we see in stratified fossiliferous deposits. According to the Huttonian theory, which I adopt as the most probable, the materials of these strata were originally deposited from water in the usual form of sediment, but they were subsequently so altered by subterranean heat as to assume a new texture. It is demonstrable, in some cases at least, that such a complete conversion has actually taken place, fossiliferous strata having exchanged an earthy for a highly-crystalline texture for the distance of a quarter of a mile from their contact with the granite." Sir Charles adds : " How far hot water or steam, permeating stratified masses under great pressure, has operated to produce the crystalline texture, may be matter of speculation ; but it is clear that the Plutonic influence has sometimes pervaded entire mountain masses of strata." Arguing from these premises, he divides rocks into four great classes—the aqueous, the volcanic, the Plutonic, and the transformed rocks in question, to which he applies the term " metamorphic." " It is not true," he adds, " that all granites, together with the crystallinear metamorphic rocks, were first formed, and therefore entitled to be called ' primitive.' " They are of all ages, and he suggests the substitution of *Hypogine*, a word implying the theory that gneiss, granite, and the other crystalline formations, are alike formed underground, and have not assumed their present form and structure at the surface. Such, however, are the rocks which lie at the base of our system—the rocky framework of the earth : it supports all the sedimentary rocks, and in its composition three distinct mineral compounds are recognised, namely, 1. *Micaschists ;* 2. *Gneiss ;* 3. *Chloritic Schists.*

Beds of the first present, as an essential element, the brightly shining, foliated, elastic, and transparent mineral which bears the name of *mica*, which, when found in great masses, receive the name of *micaschists.*

In contact with the granite, a formation quite distinct from the preceding appears ; in consequence of the felspar, which it contains in variable quantities, it is known as *gneiss.* On the much-controverted question of the origin of this substance, Professor Fournet says that it sometimes appears to be only granite laminated in consequence of the action to which it has been subjected, either in flowing or in being thrust between other rocks in its molten state. Sometimes it is the result of capillary infiltration of the granite, in a pasty state, between the foliations of the micaschist, effected under the influence of a high temperature, in which the foliated tissue has been given; but the tissue is subject to modifications; the leaves being regular and parallel, or much contorted and rumpled. Its stratiform structure, generally very characteristic, giving to *gneiss* a certain resemblance to rocks of sedimentary formation, although they originate in causes so very different. "Gneiss," says Sir Charles Lyell, "may be called stratified, or by those who object to that term, foliated granite, being formed of the same materials, namely, felspar, quartz, and mica:" but in the example he gives these occur in layers — a white consisting almost exclusively of granular felspar, the dark layers of grey quartz and black mica, with occasional grains of felspar.

The beds of micaschists and gneiss would appear to form a fourth or fifth of the solid crust of the earth. In Great Britain, gneiss, micaschists, and hornblende-schists, are confined to the country north of the Forth and Clyde. In France they are found in the Lyonnais, the Limousin, the Lozere, the Cevennes, Auvergne, Brittany, La Vendée, the Vosges. Poor districts for agriculture, they are fruitful for the miner : the stratum being rich in metals, because it is penetrated in all directions and in all ages by veins. Gold and silver are found in it in Saxony; copper, in Sweden, at Fahlun, with oxide of tin, of iron, and even precious stones, as garnets, rubies, &c. After the *micaschists* and *gneiss*, the primitive rocks include the *chloritic schists*.

This stratum has *chlorite* for its chief characteristic : it is a scaly substance, of a more or less sombre green, coloured by a silicate of iron. In Languedoc, the chloritic schists appear in holes and corners, but they acquire a fine development in the Lyonnais, where they are designated by the inhabitants *green horns* or *red horns*, according to their colour, which depends on the felspar absorbed. The *chloritic schists* are traversed by numerous veins. In the *Alps*,

E

in several localities, such as Allevard, Allemont, Cogne, Saint-Marcel, Chessy, and Sain-Bel, Alagna, and Pastarena, they include veins of copper, of oxide of iron, of spathic carbonated iron, of platinum, some silver, gold, and manganese.

ERUPTIVE ROCKS.

Nothing is more difficult than to write a chronological history of the revolutions and changes to which the earth has been subjected during the ages which preceded the historic times. The phenomena which have concurred to fashion its enormous masses, and give them their present form and structure, are so numerous, so varied, and sometimes so nearly simultaneous in their action, that the records defy the powers of observation to separate them. The deposit of the sedimentary rocks has been subject to interruption during all ages of the world. Violent igneous eruptions have penetrated the sedimentary beds, elevating them in some places, depressing them in others, and in all cases disturbing their order of superposition, and ejecting masses of crystallized rock from the incandescent centre to the surface. Amidst these perturbations, sometimes stretching over a vast extent of country, anything like a rigorous chronological record becomes impossible; the phenomena are continual and complex: it is no longer possible to distinguish the fundamental from the accidental, or either from the secondary, either in the causes or the results.

In order to render the subject somewhat clearer, the great facts relative to the progressive formation of the terrestrial globe are divided into epochs, during which the sedimentary rocks are deposited in due order by the seas of the ancient world, and along with the mud and sand, which was deposited day by day, the newly-made soil became the cemetery of the myriads of beings which lived and died in the bosom of the ocean. The rocks thus deposited were called *Neptunian* by the older geologists.

But while the seas of each epoch were thus building up, grain by grain, and rock by rock, the new formation out of the ruins of the old, other influences were at work, sometimes, to all appearance, impeding, sometimes advancing the great work. The *Plutonian rocks*—to use an expression now somewhat superannuated—the *igneous rocks* of modern geology, as we have seen above, were the great disturbing agents, and these disturbances occur in every age of the earth's history. We shall have occasion to speak of these eruptive formations while describing the phenomena of the several epochs. But it is thought that the narrative will be rendered more homogeneous and clearer by grouping this class of phenomena into one chapter, and we place it at the commencement, inasmuch as the constant reference to the eruptive rocks will thus be rendered more intelligible.

The rocks which issued from the centre of the earth in a state of fusion would mix themselves with the stratified masses of every epoch, more especially with the more ancient. The formations to which these rocks have given birth possess great interest; first, because they are necessary constituents of the terrestrial crust; afterwards, because they have impressed on the soil in the course of their eruption some of the characteristics of its configuration and structure; finally, because by their means the metals which are the objects of human industry have been brought near to the surface. According to the order of their appearance, as nearly as can be ascertained, we shall class the eruptive formations in the following chapter in two groups :—

I. The *Plutonic eruptions*, which have produced the extremely varied kinds of granite, the syenites, the protogines and porphyries. These differ from the volcanic rocks in their more compact crystalline structure, absence of tufa, pores and cavities, from which it is inferred that they have been formed at considerable depths in the earth, and have cooled and crystallized under great pressure.

II. The *volcanic eruptions*, of more recent origin, which have given birth to a succession of trachytes, basalts, and lavas. These, being of looser texture, are presumed to have cooled more rapidly, and at or near the surface.

PLUTONIC ERUPTIONS.

The great eruptions of *ancient granite* are supposed to have
occurred during the primitive epoch, and chiefly in the carboniferous
period. They present themselves sometimes in considerable masses,
for the crust being still thin, probably only partially consolidated,
and permeable, it was prepared as it were for absorbing the granitic
masses. In consequence of its weak cohesion, the primitive crust would
be torn and penetrated in all directions, as represented in the following
section of Cape Wrath, in Sutherlandshire, in which the veins of

Fig. 8.—Injected veins in the gneiss of Cape Wrath.

granite ramify in a very irregular manner across the gneiss and
hornblende-schist, here associated with it.

Granite when it is sound furnishes a fine building stone; and we
need not add that it deserves the character of extreme hardness with
which the poets have gratuitously gifted it. Its granulated texture
renders it unfit for road-stone, where it gets reduced too quickly to

dust. With his hammer the geologist easily fashions its fragments; and in the Russian War, in the bombardment of Bomarsund, the bullets from our ships demonstrated that ramparts of granite were demolished as easily as those constructed of limestone.

Granite varies very much in its composition, and occasionally it passes insensibly from fine to coarse grained granite, and the finer is even found imbedded in the coarser grained; but it forms no conglomerate : its component minerals are felspar, quartz, and mica in varying proportions; felspar being generally the most abundant in quantity, and quartz exceeding the quantity of mica. The whole being united in a confused and irregular kind of crystallization.

Granite is supposed to have been "formed at considerable depths in the earth, where it has cooled and crystallized slowly under a pressure so great that the contained gases could not expand." "The influence," says Lyell, "of subterranean heat may extend downwards from the crater of every active volcano to a great depth below, perhaps several miles or leagues; and the effects which are produced deep in the bowels of the earth may, or rather must, be distinct: so that volcanic and Plutonic rocks each differ in texture, and sometimes even in composition, and may originate simultaneously, the one at the surface, the other far beneath it." Other views, however, of its origin are not unknown to science : Professor Ramsay and some other geologists consider granite to be metamorphic. "For my part," says the Professor, "I believe that in one sense it is an igneous rock; that is to say, that it has been completely fused; but in another sense that it is a metamorphic rock, partly because it is impossible in many cases to draw any definite line between the gneiss and granite : they pass into each other by insensible gradations, and granite frequently occupies the space that ought to be filled with gneiss, were it not that the gneiss has been fused. I believe therefore that granite and its allies are simply the effect of the extreme of metamorphose. In other words, when the metamorphose has been so great that all traces of the same crystalline laminated structure has disappeared, a more perfect crystallization has taken place." It is obvious that the very

result on which the Professor founds his theory, namely, the difficulty
" in many cases," of drawing a line between the granite and gneiss,
would be produced by the sudden injection of the fluid minerals into
gneiss, composed of the same materials. Moreover, it is only in some
cases that the difficulty exists ; in many others the line of separation is
definable enough.

Syenite, in which a part of the mica is replaced by amphibole or
hornblende rock, has to all appearance surged to the surface after the
granite, and very often alongside that rock. Thus the two extremities of the Vosges towards Belfort and Strasburg are eminently
syenitic, while the intermediate part towards Colmar is as markedly
granitic. In the Lyonnais the southern region is granitic; the
northern region, from Arbresle, is in great part syenitic. The syenite
also shows itself in the Limousin.

The mineralogical constitution of syenite, into which felspar, often
rose-coloured, enters, forms a rock much finer than the amphibole or
hornblende, its green colour now nearly black by contrast. This rock
is a valuable adjunct for architectural ornament ; it is the rock out of
which the ancient Egyptians shaped their celebrated monumental
columns, their sphinxes, and many of the sarcophagi ; and the most
perfect specimens are quarried not far from the city of Syene in that
country, from which it derives its name. The obelisk of Luxor, now in
Paris, and several of the Egyptian obelisks in Rome, and the celebrated
sphinxes of which copies may be seen in front of the Egyptian Court
of the Crystal Palace, the pedestal of the statue of Peter the Great
at St. Petersburg, and the facing of the sub-basement of the column
in the Place Vendôme in Paris, are of this stone, of which there
are quarries in the neighbourhood of Plaucher-les-Mines in the
Vosges.

Syenite disintegrates more easily than granite, and it contains
some nodules very highly condensed, which often remain in the form
of large round balls, in the middle of the débris produced by disinte-
gration. It remains to be added that the syenitic masses are often
highly heterogeneous ; the amphibole sometimes fails, and we can only

recognise an ancient granite. In other instances the amphibole pre-dominates so much that a *diorite*, or greenstone, results. The geologist should be prepared to observe these transformations, which are apt to lead him into error if passed over.

Protogine, or talcose granite, is another kind of this rock, composed of felspar, quartz, and talc, *chlorite* taking the place of mica. Ex-cessively variable in its texture, the protogine passes from the most perfect granitoid aspect to the porphyritic, in such a manner as to present continual subjects of uncertainty, rendering it very difficult to determine its geological age. Nevertheless, it is believed that it came to the surface before the coal period : in short, at Creusat, protogine overlies the coal fields so completely that it is necessary to sink the pits through the protogine, in order to penetrate to the coal beds, and the rock has so manifestly acted on the coal fields, as to have contorted and metamorphosed them. Something analogous to this manifests itself near Mont Blanc : the colossal mass which dominates over this chain, and the pinnacles which belong to it, are composed of protogine. But as no such action can be traced in the rocks in jux-taposition with the triassic period, it may be assumed that in the epoch of the new red sandstone the protoginous emissions had ceased.

It is necessary to add, however, that if the protogine raises its bold and hardy pinnacles round Mont Blanc, the circumstance only applies to the vertical parts of the mountain, and is influenced by the excessive rigour of the weather, which demolishes and reduces the parts of the rock which have been decomposed by the atmospheric agents. Where protogine occurs in a milder climate,—around Creusat, and at Pierre-sur-Autre, in the Forez chain, for instance,—the mountains show none of the scarped and bristling peaks exhibited in the chain of Mont Blanc. Single projecting rocks occasionally form rocking-stones, so called because they repose a convex base upon a pedestal equally convex, but in a contrary way. It is easy to move these badly-placed blocks, but from their vast size it would require very consi-derable force to displace them. This tendency to fashion themselves into rounded or ellipsoid forms belongs to other granitic rocks, and

even to some of the sandstones. The rocking-stones have often been
the subjects of legends and popular myths.

The great eruptions of granite, of protogine and porphyry, took
place, according to M. Fournet, during the carboniferous period, for
porphyritic stones are found in the conglomerates of that period. "The
granite of Dartmoor, in Devonshire," says Lyell, "was formerly supposed to be one of the most ancient of the Plutonic rocks, but it is now
ascertained to be posterior in date to the culm measures of that county,
which from their position, and as containing true coal plants, are
regarded by Professor Sedgwick and Sir R. Murchison as members
of the carboniferous series. This granite, like the syenite of Christiania, has broken through the stratified formations without much
changing their strike. Hence on the north-west side of Dartmoor the
successive members of the culm measures abut on the granite, and
become metamorphic as they approach. The granite of Cornwall is
probably of the same date, and therefore, like it, as modern as the
carboniferous strata, if not newer."

The *ancient granites* show themselves in France, in the Vosges,
in Auvergne, at Espinouse in Languedoc, at Plan-de-la-Tour in
Provence, in the chain of the Cevenole, at Mont Pilat, near Lyon, and
in the southern part of the Lyonnaise plain. They rarely impart boldness or grandeur to the landscape, as might be expected from its
crystallized form and hardness. Probably when first exposed to the
atmosphere at its high temperature, while the globe was yet consolidating, the rock assumed a uniform character, forming hills of a
peculiar rounded form; clothed in a scanty vegetation, the surface
of the rock in a crumbling state, the corners rounded by the action
of air and water, it is only where recent dislocations have broken
them up that they assume a picturesque character.

The Christiania granite alluded to above was at one time thought
to have belonged to the Silurian period. But in 1813 Von Buch
announced that the strata in question consisted of limestones containing
orthoceretis and trilobites; the shales and limestone being only penetrated by granite veins, and altered for a considerable distance from the

point of contact. The same granite is found to penetrate the ancient gneiss of the country on which the fossiliferous beds rest—unconformably, as the geologists say; that is, it rests on the edges of the gneiss, from which other stratified deposits had been washed away or "denuded" before the sedimentary beds were deposited. "Between the origin, therefore, of gneiss and the granite," says Lyell, "there intervened, first, the period in which the strata of gneiss were denuded; secondly, the period of the Silurian deposits. Yet the granite produced after this long interval is often so intimately blended with the gneiss at the point of junction that it is impossible to draw any other than an arbitrary line of separation between them: when this is not the case, tortuous veins of granite pass freely through the gneiss, ending sometimes in mere threads, as if the older rock had offered no resistance to their passage." From this example, Sir Charles concludes that it is impossible to conjecture whether certain granites, which send veins into gneiss and other metamorphic rocks, have been so injected while the gneiss was scarcely solidified, or at some secondary or tertiary period. As it is, no single mass of granite can be pointed out more ancient than the fossiliferous deposits; no Lower Cambrian stratum is known to rest immediately on granite; no pebbles of granite are found in the conglomerates of the Lower Cambrian. On the contrary, granite is usually found, as in the case of Dartmoor, in immediate contact with secondary formations with every sign of elevation subsequent to their formation. Porphyritic pebbles are found in the coal measures: they continue during the triassic period; since, in some parts of Germany, veins of porphyry are found traversing the new red sandstones, or *grés bigarées* of French geologists. Syenites have especially reacted upon the Silurian deposits, and other older sedimentary rocks, up to those of the lower carboniferous period.

Porphyry is a variety of granite, the elements of which—quartz, felspar, and mica—are the nucleus, set in a non-crystallized paste, which unites the mass in a manner which will be familiar to many of our readers who remember the granite of the Land's End, Cornwall, where the crystals of quartz, sometimes of considerable size, are sprinkled over the surface of the block of felspar, mixing with silvery specks of mica, as if the quartz and mica already crystallized had been set in the felspar while yet in a viscous condition. In granite, on the contrary, large quantities of the several minerals would seem to have been incorporated together, and crystallized under the same conditions. The paste of porphyry is essentially composed of felspar, and a quantity, more or less considerable, of silica, in which the felspathic crystals, more or less voluminous, assume their natural form. The variety of their mineralogical characters, the admirable polish which can be given to them, which renders them eminently useful for ornamentation, give to the porphyries an artistic and industrial importance, which would be greatly enhanced if the difficulty of working the hard material did not render the price so high.

The porphyries, taken apart, present, as we have said, a paste essentially composed of compact felspar, in which the orthoclose crystals—that is, felspar with a potash base—arrange themselves, sometimes with great regularity, and in well-developed designs. Alongside these orthoclose crystals, quartz is frequently implanted, sometimes in globules, sometimes in crystals, terminating at both ends in twelve equal-sided pentagons. To these elements add micaschist of a chloritous appearance. Finally, the name of *quartziferous porphyry* is reserved. for those which present siliceous crystals; the other varieties are simply named *porphyry*.

The porphyries possess various degrees of hardness and compactness. When a fine dark-red colour, which contrasts well with the white of the felspar, is united with hardness, a magnificent stone is the result, susceptible of taking a polish, and fit for any kind of ornamental work; for the decoration of buildings, for the construction of vases, or for columns. The red porphyry of Egypt, called *antique*, was particularly sought after by the ancients, who made sepulchres, baths, and obelisks of it. The grandest mass of this porphyry which exists is the obelisk of Sextus V. at Rome. In the Musée du Louvre at Paris some magnificent basins and statues made of the same stone may be seen.

In spite of its compact form, porphyry disintegrates like other rocks when exposed to air and water. One of the sphinxes transported from Egypt to Paris, being accidentally placed under the waterspouts of the Louvre, was not slow to exhibit marks of exfoliation, while it had remained sound for ages under the climate of Egypt. In our country and even in France, where the. climate is so much drier, the porphyries are frequently decomposed so as not to be recognizable. The rocks crop out in various parts of France, but they are only abundant in the north-east part of the central plateau, and in some parts of the south. They form mountains of a conical form, presenting nearly always considerable depressions on their flanks. In the Vosges they attain the height of from 3,000 to 4,500 feet.

The *Serpentine* rocks are a compound diallage, garnet, chlorite, oxydulous iron and chromate of iron, forming compact *talc*, which owe their fatty and unctuous structure to the silicate of magnesia. Their soft character permits of their being turned and fashioned into vases of various forms. Even stoves are constructed of this substance, which bear the fire. The serpentine quarried on the banks of Lake Como, which bears the name of olaves, or pot-stone, is excellently adapted for this purpose. Serpentine shows itself in the Vosges, in the Limousin, in the Lyonnais, and in the Var : it occupies an immense surface in the Alps, as well as in the Apennines. Mona marble is an example of serpentine, and the Lizard Point, Cornwall, is a mass of it. A portion of the stratified rocks of Tuscany, and also those of the Island of Elba, have been upheaved and overturned by its eruptions.

In the British Islands the Plutonic rocks are extensively developed in Scotland, where the Lower Cambrian and Silurian rocks of gneissic character, with quartz rock at the base, associated here and there with great bosses of granite and syenite, in which hornblende prevails, form by far the greater part of the region known as the Highlands. In the Isle of Arran a vast mass of coarse-grained granite protrudes through the schists of the north part of the island, while in the southern part a finer-grained granite and veins of porphyry and coarse-grained granite have penetrated the schists. In Devonshire and Cornwall there are four great bosses of granite: in the southern parts of Cornwall the mineral axis is defined by a line drawn through the centre of the several bosses from north-east to east; but in the north of Cornwall, and extending into Devonshire, it strikes nearly east and west. The great granite mass in Cornwall lies on the moors north of St. Austell, and indicates the existence of more than one disturbing force. " There was an elevating force," says Professor Sedgwick, " protruding the St. Austell granite; and if I interpret the phenomenon correctly, there was a contemporaneous elevating force acting from the south, and between these two forces, the beds spread over the surface between St. Austell and the Dodman and Nazehead, now broken, contorted, and thrown into their present disturbed position, affecting, I believe, the whole transverse section from the headlands near Fowey to those south of Padstow." This great granite axis was upheaved in a line commencing at the west end of Cornwall, rising through the slate rocks of the older Devonian group; continuing in association with them as far as the boss north of St. Austell, producing much confusion in the stratified masses : the granite mass between St. Clear and Camelford rose between the Petherwin and Plymouth group ; lastly, the Dartmoor granite rose, partially moving the adjacent slates in such a manner that its north end abuts against and tilts up the base of the culm trough, mineralizing the great culm limestone, while on the south it does the same to the base of the Plymouth slates. These facts prove that the granite of Dartmoor, which was formerly thought to be the most ancient of the Plutonic rocks, is of a date subsequent to the Culm measures of Devonshire, which are regarded as of the true carboniferous series.

VOLCANIC ERUPTIONS.

Considered as a whole, the volcanic masses may be grouped into three distinct formations, of which we shall speak in the following order, namely, 1. *Trachytic* ; 2. *Basaltic* ; 3. *Lava formations,* which is that of their relative antiquity.

TRACHYTIC FORMATIONS.

Trachyte, derived from τραχυς, *rough,* belongs to the class of volcanic rocks, having a coarse, cellular appearance, and forming a rough and gritty paste. The eruptions of trachyte seem to have occurred towards the middle of the tertiary period, and to have continued up to the close of that epoch. They present considerable analogy in their composition to the felspathic porphyries, but their mineralo-gical characters are different. Their tissues are porous: they form a white, grey, black, sometimes yellowish paste, forming crystals of amphibole, or hornblende and mica, mixed with felspar. In its external appearance, trachyte is very variable. It forms three of the most elevated central mountain ranges in France ; the Cantal group, Mont D'Ores, and the Velay chain.

Fig. 9.—A peak of the Cantal chain.

The Cantal group is a series of irregular cones, arched and hollowed out at its centre, with a base nearly circular, which occupies a surface of nearly fifteen leagues in diameter. The trachytic portion of the group occupies the centre, and is composed of high mountains, supported by spurs, which gradually decrease in height, terminating in plateaux, more or less inclined to the neighbouring country. These central mountains attain a height varying between 3,300 and 5,500 feet above the level of the sea. A variety of trachyte, called *phonolite*, or clinkstone, remarkable for its tendency to foliation, with an unusual proportion of felspar, or, according to Gmelin, felspar and mesotype, forms the steep trachytic escarpments at the centre, which encloses the principal valleys with its abrupt peaks, giving a remarkably picturesque appearance to the landscape. In the accompanying engraving, the slaty, foliaceous character of the phonolite is well represented in one of the peaks of the Cantal group. The Pié de Lancy, represented on the opposite page, representing the so-called gold mountains of Auvergne, gives an excellent idea of the general appearance of the trachytic mountains. The group, of which Mont D'Ores is the central cone, occupies a space nearly circular, of five leagues in diameter. The massive trachytic rock, of which this mountainous mass is chiefly formed, is of the mean thickness of 1,200 to 2,500 feet; comprehending in that range, beds of fine dust and pumiceous conglomerate, of which, with beds of lignite, or imperfectly fossilized woody fibre, tufa forms the base; the whole superposed on a primitive plateau of about 3,000 yards in height. Torn and parcelled out by deep valleys, the viscous mass was gradually upheaved, until in the peak of Lancy (Pl. III.) it attained the height of 6,120 feet.

Upon the same plateau with Mont D'Ores, and about seven miles north of its last slopes, the trachytic formation is repeated in four rounded domes, one of them well known: those of Le Puy, of the Sarcoucy, of the Clierzou, and of the Petit Suchet. The rock takes a peculiar physiognomy, from the argillaceous *domite* which is found here. Le Puy presents another fine and very striking example of the eruptive trachytic rock.

The chain of the Velay forms a zone, composed of independent

III.—De Lancy Peak in the Mont l'Ores group, Auvergne.

plateaux and peaks, which forms upon the horizon a long and strangely intersected ridge. The nakedness of the mountains, their forms—pointed or rounded, sometimes terminating in scarped plateaux —give to the whole landscape an outline at once picturesque and characteristic, of which the peak of Mesenc, which rises 4,750 feet above the sea, forms the culminating point of the chain. The phonolites of which it consists have issued from fissures which present themselves at a great number of points, ranging from north-north-west to south-south-east.

On the banks of the Rhine, and in Hungary, the trachytic formation presents itself in features identical with those which indicate it in France. In America it is principally represented by some immense cones, superposed on the chain of the Andes; the colossal Chimborazo being one of those trachytic cones.

BASALTIC FORMATION.

Basaltic eruptions seem to have occurred during the secondary and tertiary periods. Basalt, according to Dr. Daubeny, " is composed of an intimate mixture of augite with a zeolotic mineral, which appears to have been formed out of Labradorite (felspar of Labrador), with an addition of water—the presence of water being in all *zeolites* the cause of that bubbling-up under the blow-pipe to which they owe their appellation." M. Delesse, and other mineralogists, are of opinion that the idea of augite being the prevailing mineral in basalt must be abandoned, although its presence gives the rock its distinctive character, as compared with trachytic, and most other trap rocks, but still their principal element is felspar. The basaltic lava consists essentially of *pyroxene*, a combination of augite, diallage, and hypersthène, a black and compact rock which predominates in this formation. It contains a smaller quantity of silica than the trachyte, and a larger proportion of lime and magnesia. Hence, independent of the iron in its composition, it is heavier in proportion as it contains more or less silica. Both basalts and trachyte contain more soda and less silica in their composition than granites : some of the basalts are highly fusible ; the alkaline matter and lime in their composition acting as a flux to the silica. Basalt exists in well-determined currents, which attach themselves to craters still apparent, as to the igneous character of which there can be no doubt. One of the most striking of the basaltic craters is furnished by that of La Coupe, in the valleys of the Vivarais, in the south of France, represented on the opposite page (PL. IV.) Upon the flank of this mountain, the traces left by the current of liquefied basalt are still seen at the foot of the mountain, occupying the bottom of a narrow valley, except at those places where the river Volant has cut away portions of the lava; the whole resting on a region of gneiss. Basaltic eruptions sometimes form a plateau, as represented in Fig. 10, where the process of formation is shown theoretically, and in a manner which renders further explanation unnecessary. Many of these basaltic table-lands form plateaux of very considerable extent and thickness; others form fragments of the same, more or less dislocated; others, again, present themselves in isolated knolls, far removed

Fig. 10.—Theoretic view of a plateau.

from congenerous formations. In short, basaltic rocks present themselves in veins or dykes, more or less powerful in most countries, of which Central France and the banks of the Rhine offer many striking examples. These veins present very evident proofs that the matter has been introduced from below, and in a manner which could only result from injection from the interior of the earth. Such are the proofs presented by the basaltic veins of Villeneuve-de-Berg, which terminate in slender filaments, sometimes bifurcated, which gradually lose themselves in the rock which they traverse. In several parts of the north of Ireland, chalk formations, with flints, are traversed by basaltic

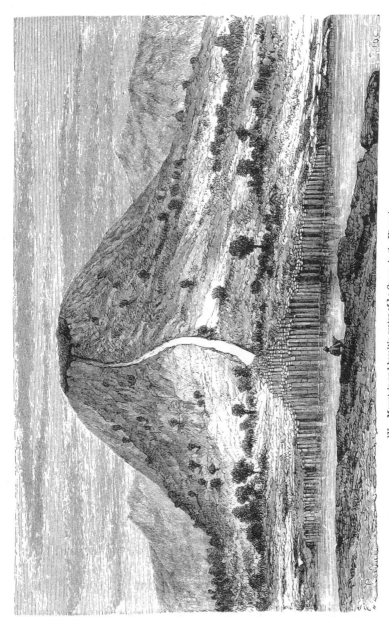

IV.—Mountain and basaltic crater of La Coupe, in the Vivarais.

dykes, the chalk being converted into granular marble near the basalt; the change sometimes extending eight or ten feet from the wall of the dyke, and being greatest near the point of contact. In the Island of Rathlin, the walls of basalt traverse the chalk in three veins or dykes; the central one a foot thick, that on the right twenty feet, and on the left thirty-three feet thick, and all, according to Buckland and Conybeare, within the breadth of ninety feet.

One of the most striking characteristics of basalt is its prismatic and columnar structure: the lava being homogeneous and of very fine grain, the laws which determine the direction of prisms, consolidated from a fluid to a solid state, become here very manifest—they are supposed to be at right angles with the cooling body. The basaltic rocks have been at all times remarkable for the picturesque arrangement of their parts. They usually represent columns of regular prisms, having five, six, or seven, and sometimes as many as twelve sides, whose disposition is always perpendicular to the cooling surface, and divided transversely, as in Fig. 11, at nearly equal distances like the joints of a wall, composed of regularly arranged equal-sided pieces braced together, and frequently extending over a space more or less considerable, placed generally in the form of an amphitheatre. The name of Giant's Causeway has been given from time immemorial to these curious dispositions of the basalts. In France, in

Fig. 11.—Basalt in prismatic columns.

the Vivarais, and in the Velay, are many such basaltic causeways. That of which Fig. 12 is a sketch lies on the banks of the small river Volant, in the department of the Ardèche. Ireland has always been celebrated for its Giant's Causeway, which extends over the whole of the northern part of Antrim, covering all the pre-existing strata of chalk, greensand and Permian formation; the prismatic columns ex-

F

tend for miles along the cliffs, projecting into the sea at the point particularly designated the Giant's Causeway.

Fig. 12.—Giant's Causeway, on the banks of the Volant, in the Ardèche.

These columnar formations vary considerably in length and diameter. McCulloch mentions some in Skye, which "are about four hundred feet high, while others in Morven do not exceed an inch (vol. ii. p. 137). In diameter those of Ailsa Crag measure nine feet, and those of Morven an inch or less." The Grotto of Fingal, in the Isle of Staffa, is renowned among basaltic rocks, although it was scarcely known on the mainland a century ago, when Sir Joseph Banks heard of it accidentally, and was the first to visit and describe it. Fingal's Cave has been hollowed out by the waves through a gallery of immense prismatic columns of trap, which are continually beat by the waves. The columns are usually upright, but sometimes curved and slightly inclined. Fig. 13 represents another aspect of the Staffa formation.

Grottos are sometimes formed by basaltic eruptions on land, followed by their separation into regular columns. The Grotto of Cheeses, between Treves and Coblentz, is a remarkable example of this kind, being so called because its columns are formed of round pieces super-

Fig. 13.—Trappian grotto of Staffa—exterior.

posed on each other in such a manner as to resemble a pile of that product of the dairy.

If we consider that in the basaltic table lands the lower part is compact, often divided into prismatic columns, and the upper part porous, cellular, scoriated, and irregularly divided—that the points of separation on which they rest are small beds presenting fragments of the porous stony concretions known under the name of *Lapilli*—that the lower portion of these masses present a multitude of points which penetrate into the rocks on which they repose, which denotes that some fluid matter had moulded itself into its crevices—that the neighbouring rocks are often calcined to a considerable thickness, and the vegetable remains they contain carbonized—no doubt can exist as to the igneous origin of basaltic rocks. On examining certain crevices, the fluid basalt is observed to have extended itself. flooding, as it were, the horizontal surface of the soil, otherwise it could not have taken the uniform surface and constant thickness in which it generally exhibits itself.

Lavic Formations.

The *lavic* formation comprehends at once the extinct and active volcanoes. "The term," says Lyell, "has a somewhat vague significa-tion, having been applied to all melted matter observed to flow in streams from volcanic vents. When this matter consolidates in the open air, the upper part is usually scoriaceous, and the mass becomes more and more stony as we descend, or in proportion as it has con-solidated more slowly and under greater pressure."

The earliest volcanic regions in France are those situated in the ancient province of Auvergne, of the Cantal, the Velay, and the Viva-rais, but principally those of about fifty volcanic cones of the height of from seven hundred to a thousand feet, composed of scoria and pozzuolana, arranged upon a granite plateau which overlooks the city of Clermont-Ferraud, and which seem to have been produced in conse-quence of a longitudinal fracture in the terrestrial crust running from north to south. It is the chain of *Le Puys*, of nearly twenty miles in length. By its cellular and porous structure, which is also granulated and crystalline, the lava from these feldspathic or pyroxenic volcanoes is readily distinguished from the analogous lava which proceeds from the basaltic or trachytic formations. Their surface is irregular, and bristles with asperities, formed by heaped-up angular blocks.

The volcanoes of the chain of *Le Puys*, represented on the opposite page (Pl. V.), are so perfectly preserved, their lava is so frequently superposed on basaltic castings, and presents a composition and texture so distinct, that there is no difficulty in establishing the fact that they are posterior to the basaltic formation, and of a recent period. Never-theless, they do not appear to belong to the historic ages, for no tradition attests their eruption. Lyell places these eruptions in the Lower Miocene period, and their greatest activity in the Upper Miocene and Pliocene eras. "Extinct quadrupeds of those eras," he says, "belonging to the genera mastodon, rhinoceros and others, were buried in ashes and beds of alluvial sand and gravel, which owe their preserv-ation to overspreading sheets of lava."

V —Extinct volcanoes forming the Le Puy Chain, in Auvergne

Everything which concerns volcanoes is explainable on the theory we have already indicated, of fractures in the solid crust resulting from its cooling. The several phenomena which existing volcanoes present to us are, as Humboldt has said, " the result of the reaction of the internal fluid nucleus of our planet upon the exterior crust." We designate as volcanic all conduits which establish a permanent communication between the interior of the earth and its surface—a conduit which gives passage at intervals to eruptions of lavic matter or *lava*, and in Fig. 14 we have represented in an ideal section the theory of volcanic eruptions. The volcanoes on the surface of the globe known

Fig. 14.—Section of a volcano in action.

to be in an occasional state of ignition number about three hundred, and these may be divided into two groups : the *isolated* or *central*, and the volcanoes which belong to a *series*.

The first are active volcanoes, around which there may be established secondary eruptive mouths, but always in connection with some principal crater. The second may be disposed like the chimneys of a forge along clefts which extend themselves over a considerable space. Twenty, thirty, and even a greater number of volcanic cones may rise above one such rent in the terrestrial crust, the direction of which will be indicated by their linear course. Sometimes the rent occurs on the crest of a dislocated chain of elevated mountains, as, for example, in the chain of the Andes, in South America, and of the remarkable range of volcanoes near Quito. Darwin relates that on the 19th of March, 1835, his attention was called to something like a large star, which gradually increased in size till about 3 o'clock, when it presented a very magnificent spectacle. By the aid of a glass, dark objects, thrown up and falling down in constant succession, were seen in the midst of a glare of red light, sufficient to cast a very bright reflection on the water: it was the volcano of Osorno in action. Mr. Darwin was afterwards assured that Aconcagua, in Chili, 480 miles to the north, was in action on the same night, and that the great eruption of Coseguina, 2700 miles to the north, accompanied by an earthquake felt over a thousand miles, occurred within six hours; and yet Coseguina had been dormant for six-and-twenty years, and Aconcagua rarely shows signs of action.

On the sea, the *series* of volcanoes show themselves in groups of islands disposed in longitudinal series.

Among these may be ranged the volcanic series of *Sonde*, which, according to the report of the matter ejected, and the violence of the eruptions, seem to be among the most remarkable on the globe; the series of the Moluccas and of the Philippines; those of Japan; of the Marianne Islands; of Chili, of the double series of volcanic summits near Quito, those of the Antilles, of Guatemala, and of Mexico.

Among the central, or isolated volcanoes, we may rank the Lipari Islands, which have Stromboli, in permanent activity, for their centre; *Mount Etna, Vesuvius,* the volcanoes of the *Azores,* of the Canaries, of the Cape de Verde Islands, of the *Gallapagos,* the *Sandwich* Islands, the *Marquesas*, the *Society* Islands, the *Friendly* Islands, *Bourbon*, and, finally, *Ararat*.

The mouths of the volcanic chimneys are almost always placed near the summit of a conical mountain, more or less isolated : they usually consist of an opening in the form of a funnel, which is called the *crater*, which descends into the interior of the chimney. But in the course of ages the crater has been extended and enlarged, until, in some of the older volcanoes, it has attained incredible dimensions. In 1822, the crater of Vesuvius was two thousand feet deep, and of a very considerable circumference. The crater of Kelanea, in the Sandwich group, is an immense chasm, a thousand feet deep, and its outer circle no less than from two to three miles in diameter, in which lava

Fig. 15.—Existing crater of Vesuvius.

is usually seen, Mr. Dana tells us, to boil up at the bottom of a lake, the level of which varies continually according to the active or quiescent state of the volcano. The cone which supports these craters is composed for the most part of lava, the products of ejection, which is designated the *cone of ejection* or *scoriæ*. Some volcanoes consist only of a *cone of scoriæ*. Such is the Barren Isle, in the Bay of Bengal. Others, on the contrary, present a cone of small dimensions compared to the height of the volcanic chain. The new crater of Vesuvius, represented in Fig. 15, was produced in 1829, within the old crater. But the frequency and intensity of the eruptions bear no proportions to the dimensions of the volcanic mountain. The eruption is usually announced by a subterranean noise, accompanied by shocks, quivering of the soil, and sometimes by real shocks of an earthquake. These noises which usually proceed from a great depth, make themselves heard sometimes over a great extent of country, resembling a well-sustained fire of artillery, accompanied by the rattle of musketry. Sometimes it resembles the heavy rolling of underground thunder, and crevices are

Fig. 16.—Fissures near Locarno.

frequently produced during the eruptions, extending over a considerable radius, as represented in the accompanying engraving of the fissures of Locarno (Fig. 16), where they present a singular appearance; the clefts extending from a centre in all directions, not unlike the star in a

VI.—Air volcano at Turbaco, South America.

broken square of glass. The eruption begins by a shock which shakes the whole interior of the mountain : masses of vapour and fluids begin to ascend, revealing themselves in some cases by the melting of the snow upon the sides of the cone of ejection ; at the same moment when the final shock triumphs over the last resistance offered by the solid crust, a considerable body of gas and watery vapour rises from the mouth of the crater.

The steam, it is important to remark, is essentially the cause of the terrible mechanical effects which generally accompany volcanic eruptions. Granite, porphyries, trachytes, and sometimes even basaltic matter, have reached the surface without producing any of those violent explosions, or ejections of rocks and stones which usually accompany modern volcanic eruptions : the older substances were discharged without violence, because steam did not accompany them, which explains sufficiently the comparative calm which attended the ancient as compared with the modern eruptions. Well-established scientific observations enable us to explain the cause of the tremendous mechanical effects which attend modern volcanic eruptions.

In the first moments of a volcanic crisis, the accumulated masses of stones and ashes which cover the crater are projected into the air by the suddenly and powerfully developed elasticity of the steam. This steam, which has been disengaged by the heat of the fluid lava, assumes the form of great round bubbles, which are ejected into the air to a great height above the crater, where they expand in all directions, in clouds of dazzling whiteness ; assuming that appearance which Pliny the Younger compares to a stone-pine rising over Vesuvius. The masses of clouds finally condense and follow the direction of the wind.

These volcanic clouds are grey or black, according to the quantity of *ashes*, that is, of the pulverulent matter, mixed with watery vapour, which they convey. In some eruptions, it has been remarked that these clouds, descending to the surface of the soil, spread all round an odour of hydrochloric acid or sulphur, and traces of both these acids are found in the rain which proceeds from the condensation of these clouds.

The condensed clouds of vapour which issue from the volcanoes are
streaked with lightning, and followed by continuous peals of thunder : in
condensing, they throw off disastrous showers, which sweep the sides of
the mountain. Many eruptions, known as *mud volcanoes* and *watery
volcanoes*, are nothing more·than these heavy rains, bringing down
with them showers of ashes, stones, and scoriæ, more or less mixed
with water.

Passing on to the phenomena of which the crater is the scene, it is
stated that, at first, there is an alternate movement of ascension and
depression of the lava which fills the interior of the crater. This incessant movement is often interrupted by violent explosions of gas.
The crater of Kelauea, in the Island of Hawaii, contains a lake of
molten matter sixteen hundred feet broad. This lake is subject to this
double movement of elevation and depression. Each of the vapoury
bubbles which issue from the crater presses the molten lava upwards,
till it rises and bursts with great force at the surface. A portion of the
lava, half-cooled and reduced to scoriæ, is thus projected upwards, and
the several fragments are darted violently in all directions, like those
of a shell when it bursts.

The greater number of the fragments being thrown vertically into
the air, they fall back into the crater. Many accumulating on the
edge of the opening add more and more to the height of the cone of
eruption. The lighter and smaller fragments, as well as the finer
ashes, are drawn upwards by the spiral vapours and carried by the
winds over an extent of country sometimes quite incredible.

In 1794 the ashes from Vesuvius were carried as far as the bottom
of Calabria. In 1812 the volcanic ashes of Saint Vincent, in the
Antilles, were carried to the east of Barbadoes, spreading such
obscurity over the island that in open day passengers could scarcely
see their way. Finally, vast masses of lava, mixed with scoriæ,
ashes, and vapour, are projected into the air during an eruption, where
centrifugal force unites them in the rounded shape in which they
are called *volcanic bombs*.

We have said that the lava, which in a fluid state fills the crater and the interior of the volcanic chimney, is forced upwards by the gases and the steam which have been generated from the water, mixed with the lava. In some cases the mechanical force of this compound vapour is immense, throwing clouds of vapour, stones, and lava to an immense height above the crater; while the great body of burning lava flows over the edge, or, forcing its way through some vent in the wall of the crater, it soon covers the mountain side with the boiling fluid. In very lofty volcanoes it is not unusual for the lava thus to find itself a vent near the base of the mountain, through which the fiery torrent discharges itself upon the surrounding country. In such circumstances, the lava cools very rapidly, presents a scaly crust on the surface, while the vapour escapes, in jets of steam, through the interstices. But under this superficial crust the lava retains for some time its fluid state, cooling slowly in the interior, while the thickening mass moves sluggishly along, clogged in its progress by the wall of débris which the burning river drives before.

The rate at which a current of lava moves along depends upon its mass, and, of course, upon its degree of fluidity, and that on its temperature. It has been stated that currents of lava have traversed more than a thousand yards in an hour; but the rate at which it travels on a level plain is usually much smaller, varying, however, with its dimensions. The most considerable current of lava from Etna had, upon some points, a thickness of thirty-eight yards, and the breadth of a geographical mile and a half. The largest lavic stream which has been recorded issued from the Skaptor Jokul, in Iceland, in 1783. It formed two currents, whose extremities were distant twenty leagues the one from the other, and which from time to time presented a breadth of three leagues, and a thickness of 650 feet.

An effect quite peculiar, and which only simulates volcanic activity, is observable in localities where the *mud volcanoes* exist. Volcanoes of this class are, for the most part, conical hills of low elevation, with a hollow or depression in the centre, and they flow outwardly from the centre; the mud being pressed upwards by the gas and steam. The temperature of the ejected matter is generally low. The mud, grey-coloured, with the odour of petroleum, is subject to the same alternate movement already ascribed to the fluid lava in the volcanoes, properly so-called. The gases which act upon this argillaceous fluid mixture of salts, gypsum, naphtha, sulphur, sometimes even ammonia, is, habitually, carbonated hydrogen and carbonic acid. Everything leads to the belief that these compounds proceed, at least in great part, from the reaction produced among the various elements of the subsoil, under the influence of the water, infiltrated through the bituminous clay, complex carbonates, and probably carbonic acid, from acidulated fountains. M. Fournet saw in Languedoc, near Roujan, traces of some of these formations; and not far from that neighbourhood is the bituminous fountain of Gabian.

The mud volcanoes, or salsis, exist in numerous localities. Many are found in the neighbourhood of Modena. They are seen in Sicily, between Arragona and Gergenti. Pallas observed them in the Crimea, in the peninsula of Kertch; in the Isle of Tamàn. Von Humboldt has described and figured a group of them in the province of Carthagena, in South America. Finally, they have been observed in the Isle of the Trinity and in Hindostan; and on the opposite page is represented the mud volcano of Turbaco, in the province of Carthagena (PL. VI.), which is described and figured by Von Humboldt in his 'Voyage to the Equatorial Regions of America.'

In certain countries we find small hillocks of argillaceous formation, resulting from the ancient dejections of mud volcanoes, from which the gas and water have been quite disengaged, and volcanic action has long ceased. Sometimes, however, the phenomenon returns and resumes its interrupted course with great violence. Slight shocks of earthquakes are felt, blocks of dried earth are projected from the ancient crater, and new waves of mud flow over its edge, and spread itself over the neighbouring hill.

Returning to the ordinary volcanoes. At the conclusion of a lavic eruption, when the violence of the volcanic action begins to subside, the matter discharged from the crater is confined to the disengagement

of the gases, mixed with steam, more or less abundant, which make
their escape by multitudes of fissures in the soil.

The greater number of volcanoes which have thus become extinct form
what are sometimes called *solfataras*. The sulphuretted hydrogen which
disengages itself from the fissures of volcanic soil is decomposed on contact with the air, water being formed by the action of the oxygen of the
atmosphere, and sulphur deposited in the neighbourhood of the crater,
and in the fissures of the soil. Such is the origin of the sulphur
which is gathered at Pozzuoli, near Naples, and in many other similar
regions ; a substance which plays a most important part in the industrial occupations of the world. It is from the products of the *solfatara*
that sulphuric acid is prepared ; and sulphuric acid becomes the fundamental agent—one of the most powerful elements—of the manufactur-
ing productions of both worlds.

The last phase of volcanic activity is the disengagement of carbonic
acid without any increase of temperature. In places where these
emanations of carbonic acid gas manifest themselves, the phenomena
may be recognized as the termination of some ancient volcano, and
nowhere does the phenomenon exhibit itself in a more remarkable
fashion than in Auvergne, where a multitude of acidulated fountains,
that is to say, springs charged with carbonic acid, exist. During the
time when he was opening the mines of Pontgibaud, M. Fournet had
to struggle against emanations which sometimes exhibited themselves
with great explosive power. Jets of water were thrown to great
distances in the galleries, roaring with the noise of steam escaping
from a locômotive steam-engine. The liquid which filled an abandoned
mine was, on two separate occasions, upheaved with great violence, half
emptying the pit, while vast volumes of the gas overspread the whole
valley, suffocating a horse and a flock of geese, while the miners were
compelled to fly in all haste at the moment when the gas belched out,
carrying themselves as upright as possible, to avoid plunging their
heads into the carbonic acid gas, which was now occupying the lower
stratum of the air in the galleries. It represented on a small scale
the effect of the *Grotto del Cana*, which excites such surprise among

the ignorant near Naples—passing also for one of the marvels of
nature all over the world. M. Fournet states that all the minute
fissures of the metalliferous gneiss, near Clermont, are quite saturated
with free carbonic acid gas, which rises plentifully from the soil there,
and in many parts of the surrounding country. The elements of the
gneiss, with the exception of the quartz, are softened by it; and new
combinations of the acid with lime, iron, and manganese are continually
taking place. In short, long after volcanoes have become extinct, hot
springs continue to flow in the same area, charged with mineral
ingredients.

The same facts manifest themselves with even greater intensity in
Java, in the so-called valley of poison, which is an object of veritable
terror to the natives. In this renowned valley, the soil is said to be
covered with skeletons and carcases of tigers, of goats, of stags, of
birds, and even with human bones, for asphyxia or suffocation, it seems,
strikes all living things which venture into this desolate place. In
the same island a stream of sulphurous water, as white as milk, issues
from the volcanic mountain of Indienne, on the east coast; and on
one occasion, as cited by Nozet in the 'Journal de Geology,' a great
body of hot water, charged with sulphuric acid, was discharged from
the same volcano, inundating and destroying all the vegetation of a
large tract of country by its noxious properties.

It is known that the alkaline waters of Plombières, in the Vosges,
have a temperature of 160° Fahr. For 2,000 years, according to
Daubrée, through beds of concrete, of lime, brick, and sandstone,
these hot waters have percolated until they have originated calcareous
spar, arragonite, and fluor spar, with such siliceous minerals as opal,
which are found among the interstices of the bricks and mortar. From
these and other similar statements, "We are led," says Sir Charles
Lyell, "to infer that when in the bowels of the earth there are large
volumes of molten matter, containing heated water, and various acids,
under enormous pressure, these subterraneous fluid masses will
gradually part with their heat by the escape of steam and various
gases, through fissures producing hot springs, or by the passage of

VII.—Great Geyser of Iceland.

the same through the pores of the overlying and injected rocks. Although," he adds, " we can only study the phenomena as exhibited at the surface, it is clear that the gaseous fluids must have made their way through the whole thickness of the porous or fissured rocks, which intervene between the subterranean reservoirs of gas and the open air. The extent, therefore, of the earth's crust, which the vapours have permeated, and are now permeating, may be thousands of fathoms in thickness, and their heating and modifying influence may be spread throughout the whole solid mass."

The fountains of boiling water, known under the name of *Geysers*, are another emanation attaching to ancient craters. They are either continuous or intermittent. In Iceland we find great numbers of these gushing sources—in fact, the island is one entire eruptive rock. Nearly all the volcanoes are situated upon a broad band of trachyte, which traverses the island from south-west to north-east. It is furrowed with immense fissures, and covered with masses of lava, such as no other country presents. The volcanic action, in short, goes on with such energy, that certain paroxysms of Mount Hecla have lasted six years without interruption. But the Great Geyser, represented on the opposite page (PL. VII.), is, perhaps, even more an object of curiosity. This water volcano projects a column of boiling water, eight yards in diameter, charged with silica, to the height, it has been said, of about 150 feet, depositing vast quantities of silica as it cools after it reaches the earth.

The volcanoes in actual activity are, as we have said, very numerous ; the best known are those of Vesuvius, near Naples ; of Etna, in Sicily ; and Stromboli, in the Lipari Islands. The island of Java alone contains about fifty, which have been mapped and described by Dr. Junghahn : a rapid sketch of a few of these may interest the reader.

Vesuvius is of all volcanoes that which has been most closely studied : it is, so to speak, the classical volcano. Few persons are ignorant that it opened, after a period of quiescence extending beyond the memory of living man :—in the year 79 of our era. This eruption cost the life of

Pliny the Elder, who fell a sacrifice to his desire to witness one of these most imposing natural phenomena. After many mutations the actual crater of Vesuvius consists of a cone, surrounded on the side opposite the sea by a semicircular crest, composed of pumiceous matter, foreign to the Vesuvius properly speaking, and we believe that Mount Vesuvius was primitively the mountain to which the name of *Somma* is now given. The cone which now bears the name was probably formed during the celebrated eruption of 79, which buried under its avalanches of pulverulent pumice-stone the cities of Pompeii and Herculaneum. This cone is terminated by a crater, the form of which has changed much with time. It has, since its origin, vomited dejections of a most varied character along with torrents of lava. In our days the eruptions of Vesuvius have only been separated by intervals of a few years.

The Lipari Isles, which include the volcano of Stromboli, are continually in a state of ignition : it forms the natural pharo of the Tyrrhenian Sea : such it was when Homer mentioned it, such it was seen before old Homer, and such it still appears in our days. Its eruptions are continuous. The crater whence they issue is not situated on the summit of the cone, but upon one of its sides, at nearly two-thirds of its height. It is in part filled with fluid lava, which is continually subjected to elevation and depression—a movement provoked by the ebullition and ascension of bubbles which mount to the surface, projecting upwards tall columns of ashes. During the night, these clouds of vapour shine with a magnificent red reflection which lightens up the whole isle and surrounding sea with a lurid flame.

Situated on the western coast of Sicily, Etna appears at first glance to have a much more simple structural form than Vesuvius. Its slopes are less steep—more uniform on all sides : its vast base nearly represents the form of a buckler. The lower portion of Etna, or the cultivated region of the mountain, is inclined about three degrees. The middle region, or that of the forests, is steeper, and has an inclination of about eight degrees. The mountain terminates in a cone of an elliptic form, of thirty-two degrees of inclination, which bears in the middle, above a terrace nearly horizontal, the cone of eruption, with its circular crater. The crater is ten thousand seven hundred feet high. It gives no issue to lava, but only vomits out gas and vapour ; while streams of lava issue from sixteen smaller cones which have been formed on the slopes of the mountain. The observer may, by looking at the summit, convince himself that these cones are disposed in rays, and are placed upon clefts or fissures which converge towards the crater as towards a centre.

But the most extraordinary volcanic exhibitions occur in the Pacific Ocean, in the Sandwich Islands, and in Java. Mounts Loa and Kea, in Owyhee, are mere flattened cones, fourteen thousand feet high. According to Mr. Dana, these huge, featureless hills sometimes throw out successive streams of lava, not very far below their summits, often two miles in breadth and six-and-twenty in length ; and that not from one vent, but in every direction, from the apex of the cone down slopes varying from four to eight degrees of inclination. The lateral crater of Kelauea, on the flank of Mount Loa, is from three to four thousand feet above the level of the sea—an immense chasm a thousand feet deep, and its outer circuit two to three miles in diameter. At the bottom, lava is seen to boil up in a molten lake, the level of which rises or falls according to the active or quiescent state of the volcano ; but in place of overflowing, the column of melted rock, when the pressure becomes excessive, forces a passage through subterranean galleries leading to the sea. One of these outbursts was observed by Mr. Coan, a missionary, which took place at the ancient wooded crater, six miles east of Kelauea. Another indication of subterranean progress took place a mile or two beyond this, in which the fiery flood spread itself over fifty acres of land, and then found its way underground for several miles further, to reappear at the bottom of a second ancient wooded crater, which it partly filled up.

The volcanic mountains of Java constitute the highest peaks of a mountain range running through the island from east to west, on which Dr. Junghahn described and mapped forty-six conical eminences, ranging from four to eleven thousand feet high. At the top

G

of many of the loftiest of these Dr. Junghahn found the active cones
of small size, and surrounded by a plain of ashes and sand, which he
calls the " crater wall," sometimes exceeding a thousand feet in vertical
height, and many of the semicircular walls enclosing spaces of four
geographical miles in diameter. From the highest parts of these
depressions rivers flow, which, in the course of ages, have cut out deep
valleys in the mountain's side.

To this rapid sketch of the actually existing volcanic phenomena we
may add a brief view of submarine volcanoes. If these are known to
us only in small numbers, the circumstance is explained by the fact
that their appearance, surging from the bosom of the sea, is almost invariably followed by their immediate disappearance, more or less completely: at the same time, such very striking and visible phenomena
demonstrate the power and continued persistence of volcanic action
beneath the sea basin. At various times islands have suddenly appeared amid the ocean upon points where the navigator had not before
observed them. In this manner, we have witnessed the island called
Graham's Ferdinanda, or Julia, which suddenly appeared on the south-
west coast of Sicily in 1831, and was swept away by the waves two
months afterwards. At several epochs also, and notably in 1811, new
islands were formed at the Azores, raising themselves above the waves
by repeated efforts all round the islands, and on many other points.

The island which appeared in 1796, at ten leagues from the northern
point of Unalaska, one of the Aleutian group, is specially remarkable.
We first see a column of smoke issuing from the bosom of the ocean,
afterwards a black point appears, from which bundles of fiery sparks
seem to rise over the surface of the sea. During the many months
that this phenomena continued, the island increased in breadth and in
height. In a short time the smoke is no longer seen ; at the end of
four years, this last trace of volcanic convulsion has altogether ceased.
The island continues, nevertheless, to enlarge and to increase in height,
and in 1806 it formed a cone, surmounted by four smaller ones.

In the space comprised between the isles of Santorin, Tharasia, and
Aspronisi, in the Mediterranean, there arose, a hundred and sixty-six

years before our era, the island of *Hyera*, which was enlarged by upheaval of islets on its banks during the years 19, 726, and 1427. Again, in 1773, Micra-Kameni, and in 1707, Nea-Kameni, appear. These islands are increased successively in 1709, in 1711, in 1712. According to the ancient writers, Santorin, Tharasia, and Aspronisi made their appearance many ages before Christ, following an earthquake of great violence.

Occulina Axillaris.

TRANSITION EPOCH.

AFTER the terrible tempests—the grand disturbances of the mineral kingdom during the primitive period—Nature would seem to have collected herself in sublime silence preparatory to the grand mystery of creation—that of living beings.

During the primitive epoch the temperature of the earth was much too high to permit of the appearance of living beings on its surface. The darkness of thickest night concealed the birth of the world ; the atmosphere, in short, was so charged with vapours of various kinds that the sun's rays were powerless to pierce its opacity. Upon this burning soil, and in this continuous night, organic life could not manifest itself. No plant, no animal, then, could exist upon the silent earth, and in the seas of the period are deposited only unfossiliferous strata.

Nevertheless, our planet continued to be subjected to a gradual abstraction of heat on the one hand, and, on the other, the continuous rains were purifying its atmosphere. From this time, then, the sun's rays, less veiled, could reach its surface with less interruption, and under their beneficent influence life soon disclosed itself. " Without light," said the illustrious Lavoisier, " nature was without life ; it was dead and inanimate. A beneficent God, in apportionating light, has spread on the surface of the earth organization, sentiment, and thought." We begin accordingly to see upon the earth—its temperature being nearly that of the equatorial zone—a few plants and a few animals. These first generations of life will be succeeded by others of a higher organization, finally culminating in the creation of man, gifted with the supreme attribute we call intelligence. " The word *progress*, which we think, belongs to humanity, and even to modern times," said Albert Gaudry, in a recent public lecture, " was pronounced by the Creator the day he created the first living organism."

Did plants appear before animals? We cannot tell; but such would appear to have been the order of creation. It is certain that in the sediment of the first seas, and in the vestiges which remain to us of the first period of organic life, that is to say, in the argillaceous schists, and in the *grauwackes* which cover them, we find at once plants and animals, and even animals of advanced organization. But, on the other hand, during the greater part of the transition period— in particular during the carboniferous age—the plants are very considerable in number, and the terrestrial animals scarcely show themselves: this would lead us to think that the plants preceded animals. It may be remarked, besides, that from their cellular nature, their looser tissues and composite elements, so readily affected by the air, the first plants could be easily destroyed without leaving any material vestiges: from which it may be concluded that in these primitive times an immense number of plants existed of which no traces now remain.

We have stated that, in the first ages of the world, the waters covered great part of its surface, and it is in the waters that we find the first appearance of life: in this medium was accomplished the grand mystery of creation. When the waters were sufficiently cool to permit of organized existence, creation was developed, and exercised itself with great energy, for it manifests itself by the appearance of numerous and very different species.

The most ancient organic remains belong to the Brachiopode Mollusks, in particular to the genus Lingula, which still exists in the present seas; to the Trilobites, a family of Crustaceans, belonging exclusively to this period; then come the Orthoceretites, the Productas, and the Terebratules, another genera of mollusks. The Polypids, which appear in due time, seemed to have lived in all ages, and have been preserved to our days.

Side by side with these animals, vegetation of inferior organization have left their imprint upon the schists; they are the Algæ, aquatic plants; but as the continents advance upon the ocean, plants of a loftier sort make their appearance—the Equisetacea and herbaceous ferns, and other plants of the air, which we shall have occasion to note in studying the periods which constitute the Transition Epoch, which consists of three periods: the Silurian, the Devonian, and Carboniferous.

Paradoxis Tessini—Upper Cambrian.

THE SILURIAN PERIOD.

The first period of the transition epoch is the *Silurian*, a designation the origin of which we have already briefly explained. The reader may find the nomenclature odd, and even inconvenient, as applied to the vast range of rocks which it represents in all parts of the Old and New World, but it describes with sufficient exactness the region in our own country in which the system predominates—reasons which led to the designation being adopted at a time when its extent was not suspected.

On this subject, and on the principles which have guided geologists in their classification of rocks, Professor Sedgwick has some excellent remarks in one of his papers in the Journal of the Geological Society. "In every country," he says, "which is not made out by a pre-existing type, the physical groups are to be determined, and their relations established, by its natural sections. Next, by the fossils found in the successive groups of rocks; and, as a matter of fact, the natural groups of fossils are generally found to be nearly co-ordinate with the physical group, each successive group resulting from certain conditions which have modified the distribution of the organic types. In the third place comes the collective arrangement of groups into systems: the establishment of the Silurian system is an admirable example of the whole process. The groups called Caradoc, Wenlock, and Ludlow were physical groups determined by natural sections. The successive groups of fossils were also determined by sections. Then followed a collective view of the whole series, and the establishment of its nomenclature, in which the whole system, as well as each subordinate group, was defined by a geographical name referring to a local type within the limits of Siluria; in this respect adopting the principle of grouping and nomenclature applied by W. Smith to our secondary rocks. At the same time, the older slate rocks of Wales, which are inferior, and lie at the base of the Silurian rocks, were

VIII.—Ideal landscape of the Silurian Period.

called *Cambrian*. Soon afterwards, the next great collective group of rocks superior to it was changed to *Devonian*, although the familiar name of Old Red Sandstone still clings to it. In this way a perfect congruity of names was established, geographical in principle, and representing the actual development of our older rocks, and giving them a new value and meaning." The period, then, for the purposes of scientific description, may be divided into three sub-periods—the Cambrian and Lower and Upper Silurian.

The characteristics of the period, of which we give an ideal view on the opposite page (PL. VIII.), are supposed to have been shallow seas of great extent, with barren submarine reefs, and isolated rocks rising here and there out of the water, covered with algæ, and frequented by mollusks and articulated animals.

The earliest traces of vegetation belong to the humble *Thallogens*, flowerless plants of the Algæ class, without leaves or stems, which are found among the Lower Silurian rocks. To these succeed the *Acrogens* — spore-like bodies, according to Dr. Hooker, belonging to the Lycopods, which are found sparingly in the Upper Silurian beds. Among animals, the *Orthoceretites*, belonging to the former, led the life of sea-rovers in the Silurian seas.

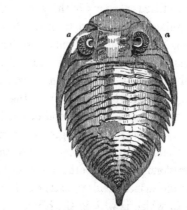

Fig. 17.—Back of Asaphus caudatus (Dudley, Mus. Stokes), with the eyes, *a a*, well preserved. (Buckland.)

Fig. 18.—*a*, Side view of the left eye of the same, magnified. (Buckland.) *b*, Magnified view of a portion of the eye of Calymene macrophthalmus. (Hœninghaus.)

Their organization indicates that they preyed upon other animals, pur-

suing them into the deepest abysses, and strangling them in the embrace
of their long arms. The *Trilobites*, whose wonderful eyes, consisting of
four hundred spherical facets placed on a horny surface, permitted
them to see in every direction (Figs. 17 and 18). Add to this a pale
sun, struggling to penetrate through a dense atmosphere, and yielding
a dim and imperfect light to the first created beings as they left the
hand of the Creator, with organization often rudimentary, but sufficiently advanced to indicate progress towards a more perfect creation.
Such is the picture which the artist has attempted to embody.

<h2 style="text-align:center">CAMBRIAN AND LOWER SILURIAN.</h2>

The Silurian rocks have been traced from Cumberland to the Land's
End, at the southern extremity of the island. They lie at the base of the
southern Highlands of Scotland, from the North Channel to the North
Sea, and they range along the entire western coast of that country.
It is probable that the Silurian seas beat against the hills of Abberley
and the Malvern range as their most easterly limits. In a westerly
direction they extended to the sea, where the mountains of Wales—the
Alps of the great chain—would stand out in bold relief, some of them
facing the sea, others in detached groups ; some clothed with a stunted
vegetation, others naked and desolate; all of them wild and picturesque.
But an interest surpassing all others belongs to these mountains. They
are the most ancient sedimentary rocks which exist on our globe—a page
of the book in which is written the history of the antiquities of Great
Britain—of the world.

On the borders of Wales, near Shrewsbury, in a somewhat barren
region, are the Longmynds, a range of mountains rising some sixteen
hundred feet above the level of the sea. Deep clefts and fissures,
ravines and rugged grassy slopes, and insignificant watercourses, characterise it ; but here lies the typical region of the Cambrian formation. The researches of geologists have failed to discover traces of
animal life in the rocks which form the base of the system in England,
but greater success has attended their efforts elsewhere. In Ireland,
where the Cambrian formation acquires considerable development—in

the picturesque rocky tract of Bray Head on the coast south and east of Dublin—we find in masses unquestionably of the same age as the Longmynd rocks, a peculiar zoophyte has been found there, named *Oldhamia*, after its discoverer, by Dr. Forbes. In the eyes of science, this fossil represents the first inhabitant of the ocean, which then floated over the greater part of the British isles. " In the hard, purplish, and schistoze rocks of Bray Head," says Dr. Kinahan, "as well as other parts of Ireland which are recognized as Cambrian rocks, markings of a very peculiar character are found. They occur in masses, and are recognised as palæozoic or hydrozoic animal assemblages. They have regularity of form, abundant, but not universal, occurrence in beds, and permanent in character even when the beds are at a distance from each other, and dissimilar in chemical and physical character." In the course of his investigations, Dr. Kinahan discovered at least four species of the Oldhamia, which he has described and figured, thus placing the fact of their existence in these ancient rocks beyond a doubt.

Other discoveries of organic remains followed. In Shropshire and Wales three zones of life have been established. In rocks of three different ages *Graptolites* have left the trace of their obscure existence. Another fossil characteristic of these ancient rocks is the *lingula*, a zoophyte with shelled plate. This shell is horny, and slightly calcareous ; and it has been thought that the covering of the bivalve is adapted to the condition of a sea, the bottom of which was composed of mud and sand, but which contained very little lime. The family to which the lingula belongs has left spoils so abundant in the Welsh mountains, that Sir R. Murchison has used them to designate a geological era. The age of the brachiopodes closes the first Silurian horizon.

In the Llandeilo beds, which are next in succession to the brachiopodes, other animal forms present themselves : thirty different forms of animals can be traced in the whole Lower Silurian rocks. They mark, however, only a very ephemeral passage over the globe, and soon disappear altogether. They are zoophytes, articulata, and mollusks.

The vertebrated animals leave no trace till we approach the upper beds of the system, and then the vertebral traces are very unsatisfactory; the most zealous advocates of the presence of fish-bones on the Ludlow beds having doubts of their being such.

The class of *Crustaceans*, of which the lobsters, shrimps, and crabs of our days are the representatives, was that which was dominant in this epoch, so to speak, of rudimentary animal life. Their forms were most singular, and different from all living crustaceans. They

consisted mainly of the *Trilobites*, a family which has entirely disappeared, but in whose nicely-jointed shell the armourer of the middle ages might have found all his contrivances anticipated, with not a few besides which he has failed to discover. They present, in general, the form of an oval buckler, composed of a series of articulations or rings, as represented in Fig. 19, the anterior ring carrying the eyes, which seem to have been reticulated, like those of some insects, having the mouth forward. The feet were probably numerous and fleshy, but no traces of them are preserved. Many of these crustaceans could roll themselves into balls, like the woodlouse, swimming on their backs, and living far from the shore, but in shallow water, and in numerous families.

Fig. 19.—Oxygia Guettardi. Natural Size.

During the middle and later Silurian ages, whole rocks were formed almost exclusively of their remains; during the Devonian period they seem to have gradually died out, disappearing altogether in the. Carboniferous age. The trilobites are unique as a family, marking with certainty the rocks in which they occur; "and yet," says Hugh

Miller, "how admirably do they exhibit the articulated type of being, and illustrate that unity of design which pervades all nature amid its endless diversity!" Among other beings which have left their traces on the Silurian rocks is *Nereites Cambrensis*, a species of annelide, whose articulations are very distinctly marked in the ancient rocks.

Besides the trilobites, the different orders of mollusks were numerously represented in the Silurian seas. Among the Cephalopodes were the *Gyroceras* and *Lituites cornu-arietis* (Fig. 20), which are formed

Fig. 20.—Lituites cornu-arietis.
One-third natural size.

Fig. 21.—Hemicosmites pyreformis.
One-third natural size.

of two distinct parts, having the body and head furnished with arms or tentacula, and whose living representatives are the nautilus and cuttlefish of every sea. The genus *Bellerophon*, with many others, represented the gasteropodes, which, like snails, crept on the ground by means of the fleshy part of its belly. The gasteropodes, with the Lamellibranches, of which the oyster is a living type, are intermediate between the brachiopodes, whose congeners may still be detected in the *terebratula* of our Highland lochs and bays, and the *lingula* of the southern hemisphere, all cephalopodes, of an obsolete type. The lamellibranchiopodes are without the head, and almost entirely destitute of power of motion. Among the zoophytes we may cite the *Hemicosmites*, of which *H. pyreformis* may be considered the type (Fig. 21).

The rocks of the Lower Silurian age in France are found in Languedoc, on the coast of Neffiez and of Bedarrieux. They occupy, also, great part of Brittany. They are found in Bohemia, in Spain, in Russia, and in the New World, where the Cambrian slates are extensively represented by the *Laurentian* formation in Canada and the United States. The calcareous rocks, sandstones, and slaty schists of Angers enter into its mineral composition.

LAURENTIAN GROUP.

Formation.	Prevailing Rocks.	Thickness.	Fossils.
I. Lower Laurentian . . .	Gneiss, quartzite, hornblende, and micaceous schists . .	18000	Foraminifera.
II. Upper Laurentian . . .	Stratified, highly-crystalline rocks, and feldspar . . .	12000	None.

CAMBRIAN GROUP.

III. Lower Cambrian	Llanberis slates, with sandy strata.	3000	Annelides.
	Harlech grits	6000	Oldhamia.
IV. Upper Cambrian	Lingula flags	6000	Trilobites; Olenus, Conocoryphe paradoxus; Brachiopodes; Cystideans.
	Tremadoc slates	2000	Trilobites; Bellerophon orthoceretites; Theca.

LOWER SILURIAN GROUP.

V. Llandeilo formation (Lower)	Quartzose sandstones and grits; argillaceous slates .	1000 to 1500	Trilobites; Graptolites, larger species; Heterapoda; large Cephalopodes
„ (Upper)	Dark-coloured slates, calcareous flags, sandstones .		
VI. Caradoc . .	Shelly sandstones : conglomerate, and shales; Bala limestone	12000	Brachiopodes; Lamellibranchi; Pteropods; Cystidæ; Graptolites; Trilobites.

UPPER SILURIAN PERIOD.

During the Upper Silurian sub-period, the seas probably contained some genera of fishes which were unknown in the previous period. The so-called fish-bones have been the subject of considerable doubt. Between the upper Ludlow rocks opposite the Castle and the next

ascending stratum, there occurs a thin bed of soft earthy shale, immediately over the Ludlow rock of fine, soft, yellowish greenstone: just
below this a remarkable animal deposit occurs, called the Ludlow bone
bed, and sometimes the Ludlow well, because the bones of small
animals were sometimes washed out of the open joints of the incumbent rocks. Old Drayton treats these bones as a great marvel:—

————————With strange and wondrous tales
Of all their wondrous things; and not the least, in Wales,
Of that prodigious spring, him neighbouring, as he feasts,
That little fishes' bones continually doth cast.

The poet thus post-dates a great geological discovery in these very
rocks.

Above the yellow greenstone, or Downton stone, as it is called,
organic remains are extensively diffused in the argillaceous strata.
There is then a lower bone bed, with the exception of a few shells.
This bed yielded what was supposed to be fragments of fishes' bones
and terrestrial plants, being the earliest trace yet found of vertebrated
life. "They formed," says Sir R. Murchison, "a band three or four
inches thick, dwindling to a quarter of an inch, exhibiting, when my
attention was first directed to it, a matted mass of bony fragments,
for the most part of small size and very peculiar character: some of
the fragments were of a mahogany hue, but others of so brilliant a
black, that when first discovered the impression was that the bed was
a heap of broken beetles."

The fragments thus discovered were, after examination on the spot,
supposed to be fishes; but after further investigation, the bulk of them
were found to be crustaceans. The icthyic nature of some of them
was supposed to be established. Professor Owen felt himself justified
in declaring that the remains of fishes had been found. "Although
no trace of a well-ossified vertical column appears," he infers "that a
well-developed fish's brain and reproductive system were combined
with the cartilaginous backbone." It may be added, that although
Professor McCoy has declared the onchus, or fish's defence (the principal fragment), to be the portion of a crustacean, Sir R. Murchison is

not convinced of the correctness of the statement. "Their texture" (speaking of their jaws and teeth) "is solid and bony," he says, "and retains the jet-like lustre."

Among the remains of this period may also be remarked a greater number of trilobites, which then attained their largest development. Among others, *Calymena Blumenbachii*, some *Phragmoceras*, and some Brachiopodes; among them, *Pentameras Knighti, Orthis rustica*, and some polypids, as *Halysites labyrinthica* (Fig. 22).

Two crustaceans of a very odd form, and in no respect resembling the trilobites, have been discovered in the Silurian rocks of England

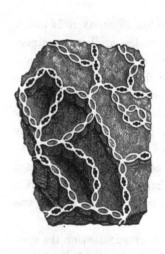

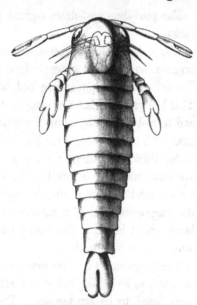

Fig. 22.—Halysites labyrinthica.   Fig. 23. –Pterygotus bilotus.

and America—the *Pterygotus* and the *Eurypterus* (Figs. 23 and 24). They are supposed to have been the inhabitants of fresh water. They were called "Seraphim" by the Scotch quarry-men, from the wing-like form and feather-like ornament of the thoracic appendage, the part most usually met with. Agassiz recognized their crustacean character, and figured them in his 'Fossil Fishes of the Old Red Sandstone.' It has also been found in the Downton sandstone and Upper Ludlow rocks.

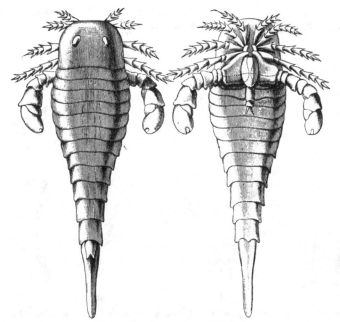

Fig. 24.—Eurypterus remipes. Natural size.

The trilobites, we have already said, were able to roll themselves up into a ball, like the woodlouse, doubtless as a means of defence against the attack of their enemies. In Fig. 25 one of these creatures, *Calymene Bowningii*, is represented in that form, rolled up on itself. The seas were evidently abundantly inhabited at the end of the Upper Silurian period, for naturalists have examined more than 1500 species, animal and

Fig. 25.—Calymene Bowningii, partially rolled up.

vegetable, belonging to the period, and the number of species classified and arranged for public inspection round the galleries of the Museum of Practical Geology in Jermyn Street cannot be much short of that number.

Among the marine plants which have been found in the rocks corre-

H

sponding with this sub-period are some species of algæ, and others belonging to the lycopods — spore-like bodies, which become still more abundant in the old red sandstone. Fig. 26 represents some examples from the impressions they have left.

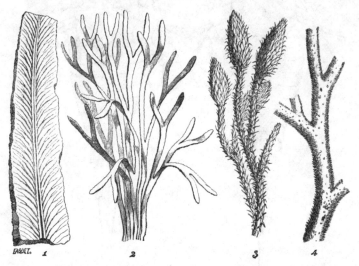

Fig. 26.—Plants of the Silurian Epoch. 1 and 2, Algæ, natural size ; 3 and 4, Lycopods, natural size.

Silurian rocks of this period are found in the department of La Manche, of Calvados, and of the Sarthe. In Languedoc, the Silurian formation has occupied the attention of Messieurs Graff and Fournet, who have traced along the base of the Espinouse, the chloritic schists, green and primordial, surmounted by clay slates, which become more and more pure as the distance from the masses of granite and gneiss increases, and the valley of the Jour is approached. Upon these beds the Silurian system rests, sinking towards the plane under the secondary and tertiary formations. In Great Britain—their typical country— they are found enormously developed in the west and south Highlands of Scotland, on the western slopes of the Pennine chain and the mountains of Wales, and in the adjoining counties of Shrewsbury—its most typical region—and Worcestershire. In Spain, in Germany, on the banks of the Rhine, in Bohemia—where, also they are largely developed, especially in the neighbourhood of Prague—in Russia, in the Oral Mountains especially; and in America, in the neighbourhood of New York,—in all these countries they are more or less abundant.

We may add, as a general characteristic of the Silurian system as a whole, that of all formations it is the most dislocated and disturbed. In the countries where it prevails, it only appears as fragments which have escaped destruction from the numerous eruptions which have traversed it during the first ages of the world. The beds, originally horizontal, are sometimes turned up, contorted, folded over, sometimes even vertical, as in the slates of Angers, of Llanberis, and the Irleth slates. Alc. d'Orbigny found the Silurian beds with their fossils in the American Andes, at the height of sixteen thousand feet above the level of the sea. What a vast upheaval of soil must have been required to carry these rocks to such a height!

In the Silurian period the sea still occupied the earth almost entirely; it covered the greater part of Europe: all the space comprised between Spain and the Oural Mountains was under water. In France only two islands had emerged from the primordial ocean. One of them formed the granitic rocks of Brittany and La Vendée, the other was constituted by the great central plateau, and consisted of the same rocks. The north part of Norway, of Sweden, and of Russian Laponia, formed a vast continental surface. In America, the emerged lands were more extensive. An island extended over eighteen degrees of latitude, now called New Britain. Another island extended for twenty degrees of latitude on the coast of the Pacific, now known as California, Utah, and the Oregon territory. In South America, on the Pacific, Chili formed one elongated island along the coast. Upon the Atlantic, a portion of Brazil, to the extent of twenty degrees of latitude, was raised above the waters. Finally, in the Equatorial regions, Guienne formed the later island in the vast ocean which still covered all other parts of the New World.

There is, perhaps, no scene of greater geological confusion than that presented by the eastern flank of the Pennine chain. A line drawn lon-

gitudinally on about 3° west of Greenwich would include on its western
side Cross Fell, in Cumberland, and the greater part of the Silurian
rocks belonging to the Cambrian system, in which the Cambrian and
Lower Silurian rocks are pretty well established; while the upper
series are so metamorphosed by eruptive granite and the effects of
denudation, as to be scarcely recognizable. "With the rare exception of a sea-weed and a zoophyte," says the author of 'Siluria,' "not a
trace of a fossil has been detected in the thousands of feet of strata,
with interpolated igneous matter, which intervene between the slates of
Skiddaw and the Coniston limestone, with its overlying flags; at that
zone only do we begin to find anything like a fauna: here, judging
from its fossils, we find representations of the Caradoc and Bala
rocks." This much-disturbed district Professor Sedgwick, after several
years devoted to its study, has attempted to reconstruct, the following
being a brief summary of his arguments. The region consists of:—

I. Beds of mud and sandstone, deposited in an ancient sea, apparently without the calcareous matter necessary to the existence of
shells and corals, and without any traces of organic forms—these were
the elements of the Skiddaw slates.

II. Plutonic rocks were, for many ages, poured out among the
aqueous sedimentary deposits; the beds were broken up and re-
cemented—plutonic silt and other matter finely comminuted were deposited along with the igneous rocks: the process was again and again
repeated, till a deep sea was filled up with a formation many thousands
of feet thick by the materials forming the middle Cambrian rocks.

III. A period of comparative repose followed. Beds of shell and
bands of coral were formed upon the more ancient rocks, interrupted
with beds of sand and mud: processes many times repeated; and
thus, in a long succession of ages, were the deposits of the upper series
completed.

IV. Towards the end of the period, mountain masses and eruptive
rocks were pushed up through the older deposits. After many revolutions, all the divisions of the slate series were upheaved and contorted
by movements which did not affect the newer formations.

V. The conglomerates of the old red sandstone were now spread out by the beating of an ancient surf, continued through many ages, against the upheaved and broken slates.

VI. Another period of comparative repose: the coral reefs of the mountain limestone, and the whole carboniferous series, were formed, but not without many oscillations between the land and sea levels.

VII. An age of disruption and violence succeeds, marked by the discordant position of the rocks, and by the conglomerate of the new red. At the beginning of this period the great north and south "Craven fault," which rent off the eastern calcareous mountains from the older slates, was formed. Soon afterwards the disruption of the great "Pennine fault," which ranges from the foot of Stanmore to the coast of North Cumberland, occurred, lifting up the terrace of Cross Fell above the plain of the Eden. About the same time some of the north and south fissures, which now form the valleys leading into Morecombe Bay, may have been formed.

VIII. The more tranquil period of the new red sandstone now dawns, but here our facts fail us on the skirts of the Lake Mountains.

IX. Thousands of ages rolled away during the secondary and tertiary periods, in which we can trace no movement. But the powers of nature are never still: during this age of apparent repose many a fissure may have started into an open chasm, many a valley been scooped out upon the lines of "fault."

X. Close to the historic times we have evidence of new disruptions and violence, and of vast changes of level between land and sea. Ancient valleys probably opened out anew or extended, and fresh ones formed in the changes of the oceanic level. Cracks among the strata may now have become open fissures, vertical escarpments formed by unequal elevations on the lines of fault; and subsidence may have given rise to many of the tarns and lakes of the district.

Such is the picture which one of our most eminent geologists gives as the probable process by which this region has attained its present appearance, after he had devoted years of study and observation to its peculiarities; and his description of one spot applies in its general scope to the whole country. At the close of the Silurian period our island home was probably an archipelago, ranging over ten degrees of latitude, like many of the island groups now found in the great Pacific Ocean; the old gneissic hills of the western coast of Scotland, culminating in the granite range of Ben Nevis, and stretching to the southern Grampians, forming the nucleus of one island group; the south Highlands of Scotland, ranging from the Lammermoor hills, another; the Pennine chain and the Malvern hills, the third, and most easterly group; the Shropshire and Welsh mountains, a fourth; and Devon and Cornwall stretching far to the south and west. The basis of the calculation being that every spot of this island lying now at a lower elevation than eight hundred feet above the sea was under water at the close of the Silurian period, except in those instances where depression by subsidence has since occurred.

There is, however, another element to be considered, which cannot be better stated than in the picturesque language of M. Esquiros, an eminent French writer, who has given much attention to British geology. "The Silurian mountains," he says, "ruins themselves, contain other ruins. In the bosom of the Longmynd rocks geologists discover conglomerates of rounded stones which bear no resemblance to any rocks now near them. These stones consequently prove the existence of rocks more ancient still; they are fragments of other mountains, of other shores, perhaps even of continents, broken up, destroyed, and crumbled by earlier seas. There is, then, little hope of our discovering the origin of life on the globe, since this page of the Genesis of the facts has been torn. For some years geologists loved to rest their eyes, in this long night of ages, upon an ideal limit beyond which plants and animals would begin to appear. Now, this line of demarcation between the rocks which are without vestiges of organized beings and those which contain fossils is nearly effaced among the surrounding ruins. On the horizon of the primitive world we see vaguely indicated a series of other worlds which have altogether disappeared: perhaps it is necessary to resign ourselves to the fact that the dawn of life is lost in this silent epoch, where age succeeds age till they are clothed in the garb of eternity. The river of creation is like the Nile, which, as Bossuet says, hides its head. The endless speculations opened up by these and similar considerations led Lyell to say, Here I am almost prepared to believe in the ancient existence of the Atlantis of Plato."

UPPER SILURIAN GROUP.

	Lithological Characters.	Thickness.	Fossils.
Llandovery Rocks	Hard sandstone slates and conglomerate beds. . . .	600 to 1000	Crinoides and corals
	May Hill sandstone, limestone nodules	800	Cystidæ.
	Calcareous sandstone and coarse grits.	1000	Brachiopodes.
	Purple shales.		Pentamares.
Wenlock Rocks .	Woolhope limestone and shales, felspathic sandstone and grit. Argillaceous shales . . . . Concretionary and thick-bedded limestone	3000	Marine Mollusks. Radiata; Trilobites Eurypterides. Graptolites.
Ludlow Rocks .	Shale with calcareous limestone.	1000	Marine Mollusks of many orders.
	Argillaceous limestone . . .	50	Brachiopodes.
	Micaceous grey sandstone and mudstone	700	Crinoides. Ganoid and Placoid fishes.
	Downton sandstone, at the base of the bone bed	80	Sea-weeds.

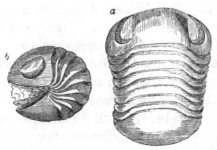

Nileus armadillo.—*a*, from above; *b*, profile; animal rolled up.

DEVONIAN PERIOD.

ANOTHER great period in the Earth's history opens on us—the Devonian; so called because the "Old Red Sandstone" is very distinctly traced and of great extent in that county.

The continents deposited during the Devonian period exhibit some species of animals and vegetables of a much more complex organization than those of the primitive period. We have seen, during the Silurian epoch, life opening, and the first plants appearing, of a very simple organization, namely, zoophytes, articulated animals, and mollusks, among animals, and algæ and lycopodes, among plants. We shall see, as the globe grows older, that organization becomes more complicated. Vertebrated animals, represented by numerous fishes, succeed the zoophites, the articulated trilobites, and the mollusks. Soon afterwards, the reptiles appear, then the mammifera and birds; until the time comes when man, His supreme and last work, issues from the hands of the Creator, to be king of all the earth;—man, who has for the sign of his superiority, intelligence—that celestial gift, the emanation of God.

Vast seas, covered with a few islets, form the ideal of the Devonian period. Upon the rocks of these islets the mollusks and articulata of the period exhibit themselves, as represented on the opposite page (IX.).

IX.—Ideal landscape of the Devonian Period.

Stranded on the shore we see a cuirassed fish, of strange form. A restored group of shrubs, *Asterophyllites coronata*, covers one of the islets, mixed with plants nearly herbaceous, resembling mosses, though the true mosses did not appear till much later. *Encrinites* and *lituites* occupy the rocks in the foreground on the left hand. The vegetation is still humble in its development, for forest-trees are altogether missing. The asterophyllites rise singly to a considerable height, with tall and slender stem. The light, still pale, seen through the semi-opaque atmosphere, only permits of a vegetation essentially cellular, sluggish, and vascular. Cryptogames, of which the mushrooms convey some idea, would be the chief vegetation; but in consequence of the softness of their tissues, their want of consistence and of woody fibre, no vestiges of them have come down to us.

The vegetable forms belonging to the Devonian period differ much. We may observe this in the cellular plants of the present day. They resembled the mosses and lycopodes, which are flowerless cryptogamii, of an inferior organization. The lycopodes are herbaceous plants, playing a very secondary part in the actual vegetation of the globe; but in the first ages of organic creation, they were the pre-eminent occupants of the vegetable kingdom, both as to the dimensions of the individual and the number and variety of the species.

The shrub which bears the name of *Asterophyllites coronata*, and which is represented on Plate IX., belongs to a family now extinct, and to the botanical division of *Dicotyledons*, which now comprehend the *conifers* and *cycades*. The acute pointed leaves of the asterophyllites spread themselves out in whorls upon the top of the stem.

In the engraving (Fig. 27) we have represented three species of aquatic plants belonging to the Devonian period; they are, 1, *Fucoids*, or Algæ; 2, a *Zostera*; 3, *Psilophyton*. The fucoid has a close resemblance to its modern congener; but with the first indications of terrestrial vegetation we pass from the *Thallogens*, to which the *Algæ* belong—plants of simple organization, without flower or stem—to the *Acrogens*, which throw out their branches at the extremity, and bear in the axils of their leaves minute circular cases, which form the

receptacles of their spore-like seeds. "If we stand," says Hugh
Miller, "on the outer edge of one of those iron-bound shores of the

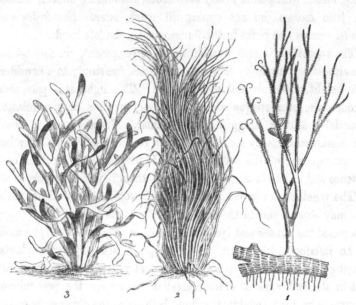

Fig. 27.—Plants of the Devonian Epoch. 1, Algæ; 2, Zostera; 3, Psilophyton, natural size.

Western Highlands, where rock and skerries are crowned with sea-
weeds; the long cylindrical lines of *chordafilium*, many feet in
length, lying aslant in the tideway; long shaggy bunches of *Fucus
serratus* and *F. nodosus* droop from 'the sides of the rock; the flat
ledges bristling with the stiff, cartilaginous, many-cleft fronds of at
least two species of *Chondrus;* now, in the thickly-spread fucoids of
this highland scene we have a not very improbable representation
of the Thallogen vegetation. If we add to this rocky tract, so
rich in fucoids, a submarine meadow of pale shelly sand, covered by a
deep-green swathe of *Zosterii*, with jointed root and slim flowers,
unfurnished with petals, it would be more representative still."

Let us now take a glance at the animals belonging to this period.

The class of fishes seem to have held the first rank and importance in the Devonian *fauna*; but their structure was very different from those of recent times : they were provided with a sort of cuirass, and thence were called *Ganoïds*, or cuirassed fishes. Numerous fragments of *Pterichthys cornutus* are now found in geological collections : it is a fish of strange form, completely covered with a cuirass of many pieces, with a small head, and furnished with wing-like fins.

Let any one picture to himself the surprise he would feel should he, on taking his first lesson in geology, and on breaking his first stone—a pebble, for instance, exhibiting every external sign of a water-worn surface—find, to appropriate Archdeacon Paley's illustration, a watch, or any other delicate piece of mechanism, in its centre. Now this, thirty years ago, would not be more likely to excite surprise than what happened to Hugh Miller, on the first days of his appearance in the sandstone quarry of Cromarty. He was at work on the side of a hill on that stormy coast, a hundred feet below the surface. He had turned out a large flag, which a blow from his hammer laid in two, and behold! an exquisitely-modelled ammonite was displayed before him. It is not surprising that henceforth the half mason, half sailor, altogether poet, became a geologist. He sought for information, and found it; he found that the rocks among which he laboured swarmed with the relics of a former age. He pursued his investigations, and found while working in this strata, all around the coast, that a certain class of fossils abounded; but if he worked in a higher zone, these familiar forms disappeared, and others made their appearance.

He read and learned that in other lands—lands of more recent formation—strange forms of animal life had been discovered;—forms which in their turn had disappeared, to be succeeded by others, more in accordance with beings now living. He came to know that he was surrounded, in his native mountains, by the sedimentary deposit of other mountains; he became alive to the fact that these grand mountain ranges had been built up grain by grain, until the ocean was filled up, and the mountains raised to their present level, by upheaval of one part or subsidence of another. The young geologist now ceased to wonder that each bed, or series of beds, should contain in its bosom records of

its own epoch; it seemed to him as if it had been the object of the
Creator to furnish the inquirer with records of His wisdom and power,
which could not be misinterpreted.

Among the fishes of the Devonian seas, the *Coccasteus*, 1, was only
partially cuirassed; the upper part of the body down to the fins was
defended by scales. *Pterichthys*, 2, a strange form, with jointed arms,
like paddles, and a mouth placed far behind the nose, on the lower
part of the body, was entirely covered with scales. The *Cephalaspis*, 3,
which has a considerable resemblance to fishes of the present age, was
nevertheless protected only on the anterior part of the body.

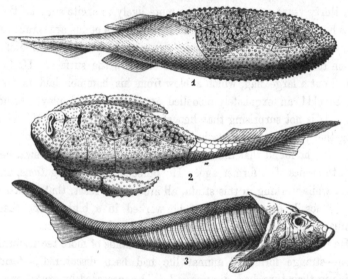

Fig. 28.—Fishes of the Devonian epoch. 1, Coccasteus, one-third natural size; 2, Pterichthys,
one-fourth natural size; 3, Cephalaspis, one-fourth natural size.

Other fishes present themselves without any such cuirass, properly so
called, but still with strong resisting scales, which envelop the whole
body. Such were the *Acanthoides*, 1, the *Climatius*, 2, and the *Diplocanthus*, 3, represented in Fig. 29.

Among the organic beings of the period we find vermiform animals,
such as *Annelides tubicola*, protected by an external shell, which now

appears for the first time, and which are probably represented by the
Serpula. Among the crustaceans the *Trilobites* are still very
numerous, especially in the lower rocks of the period. We also find

Fig. 29.—Fishes of the Devonian Epoch. 1, Acanthoides; 2, Climatius; 3, Diplocanthus.

there many different species of mollusks, among which the *Brachiopodes* form more than one half. We may say of this period that it is
the reign of brachiopodes: they revel in it with their extraordinary
forms, and in numbers their species are immense. Among the most
curious we may instance the enormous *Strigocephalus Burtini*, *David-
sonia Verneuilli*, *Uncites gryphus*, and *Calceola Sandalina*, mollusks of
most fantastic shape, differing entirely from all known forms. Among
the most characteristic forms of the first strata *Atrypa reticularis*
(Fig. 30) may be mentioned, with *Spirifera concentrica*, *Calymenia
Sedgwickii* (Fig. 31), and *Leptœna Murchisoni*. In the ascending
strata also we begin to find *Productus subaculatus*. Among the
Cephalopodes, the fine family of the *Goniatites*, with their congeners,
the ammonites, which characterise the second epoch, lived also during
the Devonian period.

Among the Radiata of this epoch, the order of the Crinoides appear
in the earlier formations. The Encrinites, under which name the

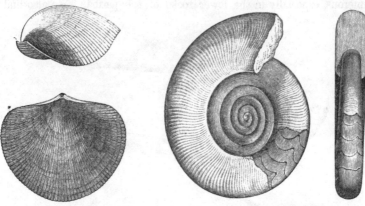

Fig. 30.—Atrypa reticularis.   F.g. 31.—Calymenia Sedgwickii.

whole of these animals are sometimes included, live attached to rocky
places and on deep banks, with mouth extended, watching for their prey,

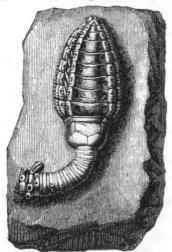

and resembling a small tree of stone.
In some of these primitive crinoïdes
the forms are of great beauty, and
varied with richly graved and ribbed
capitals, on slender columns, the
ancient corals being stars of four
rays or the multiples of four. The
stem which supported the head was
composed of numerous articulations, and is supposed to have been
attached to the rock by a sort of
root ; but specimens have generally
been torn up with violence, and in
consequence the mode of attachment is not known. It is, how-

Fig. 32.—Cupressocrinus crassus.

ever, owing to this attachment to the rock, and to the faculty the
animals have of working with their arms and closing them, like the

calyx of a flower, that the name of *stone lilies* is sometimes given to them.

The Encrinites, as we have seen, existed among Silurian rocks in a very simple form, in the *Hemicosmites*, but they had greatly increased in numbers in the Devonian period. They diminish also as we retire from that geological age, until they are now only represented by two genera of these animals, whose forms were so numerous and varied in the primitive seas.

DEVONIAN ROCKS.

The Devonian Rocks are composed of schists, of sandstone, and of some varieties of limestone. The line of demarcation between the Silurian rocks and those which succeed them may be followed in many places by the eye; but on a closer examination the exact limits of the two systems become more difficult to fix. The beds of the one system pass into the other in graduated shades, for nature admits of no violent contrasts, and few sudden transitions. By-and-by, however, the change becomes very decided, and the contrast between the dark-grey masses lying at the base and the superincumbent masses of yellow and red, become sufficiently striking. In fact, the uppermost beds of the Silurian rocks are the transition beds of the overlying system, consisting of embedded flagstones, occasionally reddish, and called in some districts "tile-stones." Over these lie the Old Red Sandstone conglomerate, the Caithness flags, and the great superincumbent mass which forms the upper formation of the system. Though less abrupt than the eruptive and Silurian mountains, the Devonian is, nevertheless, distinguished by its imposing outline, assuming bold and lofty escarpments in the Fans of Brecon, in Grongar Hill, near Carmarthen, and in the Black Mount, Monmouthshire, in the centre of a landscape which wood, rock, and river combine to render perfect. But it is in the North of Scotland where the rock assumes its grandest aspect, wrapping its mantle round the loftiest mountains, and rising out of the sea in rugged and fantastic masses, as far north as the

Orkneys.   In Devon and Cornwall, where the rocks are of a slaty, calcareous, and sometimes schistose character, from their frequent proximity
to igneous intrusive rocks, by which they have been partially metamorphosed, they are sufficiently extensive to have given a name to the
series, which is recognised all over the world.   They are largely
developed in Herefordshire, Gloucestershire, and in Shropshire; and
at one time, there is evidence, that they formed a belt round the
Grampians, and filled up the valley lying between these mountains
and the South Highlands.   In short, wherever the Silurian seas beat,
there is evidence that the Devonian rocks were deposited by its
surging waters, until it was filled up by its own action, and became
dry land.   It exists in the north, in the west, and in the south of
France; in Belgium; in Russia; in Spain; and in America.
The Devonian rocks contain the most ancient combustibles known,
for such probably are the coal measures of the department of the
Loire-Inferieure and of Marne-et-Loire, in France, and of the Asturias,
in Spain.   It constitutes the Old Red Sandstone of the earlier
geologists.

Sir R. Murchison thinks, however, that the name Devonian, which
has been applied to it from its extensive development in that county, is
erroneous.   No geologist, he thinks, who has examined the Old Red
Sandstone of the east coast of Scotland would be disposed to consider
that otherwise than as a local formation, not having the full equivalents
of the formation in other countries; and he recommends that the
classical name of Old Red Sandstone should be retained.

We have said that the Devonian rocks consist of sandstone, schists
and limestone: the presence of the two first, sandstone and argillaceous schists, in this formation will surprise no one, but the presence
of lime may, and requires some explanation.   Among the mineral
substances of which we have spoken as composing the crust of the
globe, granite has been mentioned.   Now granite is a mixture of the
silicates of potash, of soda, and of magnesia: it is, then, by the decomposition of granite, that certain rocks of the Devonian age are formed, a

remark which applies also in some degree to the Silurian period; although the carbonate of lime which composes the calcareous element scarcely shows itself till now in the formations which lie at the base of the minerals of our globe. Apart, however, from the Silurian and Devonian rocks, carbonate of lime forms an essential part of its upper strata. We may ask what is the origin of this carbonate of lime? Whence proceeds this substance which exists in such quantities in the Devonian rocks?

The fractures and dislocations of the crust of the globe were extremely frequent in the first ages of the world. It was not alone granite in a fluid state which was ejected through these fissures; water in a state of ebullition also escaped, holding in solution bicarbonate of lime, mixed sometimes with bicarbonate of magnesia. In short, whole rivers of lime-water gushed from the centre of the globe, the grand and inexhaustible reservoir which has supplied all which is now visible on the surface. As the sea then covered nearly the whole extent of the terrestrial sphere, these rivers of boiling limewater necessarily discharged themselves into its waves. In this manner, the seas were charged with the salts of lime for a considerable part of the Silurian and Devonian period; and from the same cause the sedimentary rocks deposited by Silurian seas are found to contain large quantities of the carbonate of lime. The same phenomena continued with increased power after the Devonian period: we shall find the calcareous rocks increasing in number and importance in the geological ages which follow. During the jurassic and cretaceous period these deposits will cover immense spaces over the whole earth; they will form rocks of many hundreds of yards in thickness— rivalling, in short, the Devonian period, which is estimated to be seven or eight thousand feet thick. In England the Devonian formation was long thought to be barren of organic remains.

In Herefordshire, Worcestershire, Shropshire, and South Wales, it is widely developed, and sometimes attains the thickness of from 8,000 to 10,000 feet, divided into: 1. Conglomerate; 2. Brown-stone, with eurypteris; 3. Marl and cornstones, with irregular courses of

I

concrete limestone, in which are spines of fish and remains of
cephalaspis and *pteraspis;* 4. Thin olive-coloured shales and sandstone, intercalated with beds of red marl, containing *cephalaspis* and
auchenaspis. In Scotland, south of the Grampians, a yellow sandstone occupies the base of the system; conglomerate, red shales, sandstone and cornstones, containing *holoptychius* and *cephalaspis,* and the
Arbroath paving stone, containing what Agassiz recognized as a
huge crustacean. The whole of the North of Scotland, from Cape
Wrath to the southern flank of the Grampians, H. Miller describes as
consisting of a nucleus of granite, gneiss and other igneous rocks, set
in a frame of Devonian.

Some of the phenomena connected with the older rocks of Devonshire are difficult to unravel. The older Devonian, it is now under-
stood, are the equivalents in another area of the Upper Silurian beds
themselves, and in Cornwall and Devonshire they lie directly on the
Silurian strata, while, elsewhere, the fossils of the Upper Silurian are
almost identical with those in the Devonian beds. Now, although
the Devonian and old red, if there be a distinction between them, are
both of dates that come between the Upper Silurian and carboniferous
era, it does not follow that they are precisely of the same geological
age; for, while in South Wales and Shropshire, the old red succeeds
and seems to pass into Ludlow beds, in the south-west of England
the Devonian rocks rest on Lower Silurian strata. This difficulty of
drawing a satisfactory line of demarcation between different systems is
sufficient to dispel the idea which has sometimes been entertained that
special *fauna* were created and annihilated in the mass at the close of
each epoch. There was no close : each epoch silently disappears in
that which succeeds it, and with it the animals belonging to it, much
as we have seen them disappear from our own fauna almost in our own
times.

Fig. 33.—Tremulens Lloydii.

CARBONIFEROUS PERIOD.

To the Devonian succeeds, in the history of our globe, the carboniferous period, in whose bosom was concealed during many long ages the coal measures, which have done so much to enrich and civilize the world in our own age. This period divides itself into two great sub-periods: 1. The *carboniferous limestone*; and 2. The *coal measures*. The first, a period which gave birth to most important marine deposits; the second, to the great coal deposits. The former underlies the coal-fields in all directions; in England, in Belgium, in France, and America. The latter constitutes only a small portion of the vast thickness which the whole formation attains.

In certain parts of Wales the traveller distinguishes all around him mountains of the old red sandstone; but if he were to examine the southern slopes of these mountains, he would see in the distance rocks surmounted by masses of rock of another shade. These are the transition rocks, as we shall call them, which lie between those of the Devonian and carboniferous age; the boundary line between the two geological provinces. The carboniferous rocks develop themselves with great boldness in the basin of Glamorgan, of Carmarthen, and Monmouth. On some coasts they are elevated into cliffs of an aspect quite cyclopean, presenting a wild and picturesque barrier to the sea. In Derbyshire, and especially in Ireland, the wayfarer is struck by the grand features of this new order of ruins. There is, indeed, no country in the world in which the system shows itself so rich in pic- turesque rocks and so poor in coal as in Ireland.

The limestone mountains which form the base of the coal system, attain, according to Mr. Phillips, the thickness of two thousand five hundred feet. They are of marine origin, as is apparent by the multitude of fossils they contain of zoophites, radiata, cephalopodes and fishes. But its chief characteristic is that here, for the first time, we find traces of a terrestrial flora. Remains of vegetation now become as common as they were rare in all previous formations, announcing great increase of dry land. There was a time when our island was a sea of unlimited extent: we now come to a time when it was a forest, or, rather, endless group of forests, which cover the sur- face of groups of islands thickly studding the sea.

The monuments of this grand forest epoch discover themselves in the rich coal measures of England and Scotland. These give us some idea of the rich vestments which covered the nakedness of the earth newly risen from the bosom of the waves. It was the paradise of ter- restrial vegetation. The grand *Sigillaria*, the *Stigmaria*, and the plant especially typical of this age—the arborescent fern—formed the woods, which were left to grow undisturbed, for as yet no living mam- mifera violated the silent solitude : everything announces a uniformly warm, humid temperature, the only climate in which the gigantic ferns of the coal-measures could have attained their magnitude. In Fig. 34 the reader has a restoration of the arborescent and herbaceous ferns of the period. Conifers have been found of this period with concentric rings, whence it has been concluded that seasons existed ; but these rings are more slightly marked than in existing trees of the same family, from which it is reasonable to think that the seasonal changes were less marked than they are with us.

Everything announces that the duration of the carboniferous lime- stone was of prodigious length. Professor Phillips calculates that, at the ordinary rate of progress, it would require 122,400 years to accu- mulate only sixty feet of coal. English geologists believe, moreover, that the upper coal-fields, where beds have been heaped upon beds, for ages upon ages, were formed in a condition of comparative tranquillity,

Fig. 34.—Ferns restored. 1 and 2, Arborescent ferns; 3 and 4, Herbaceous ferns.

but that the end of this period was marked by violent convulsions—by ruptures of the terrestrial crust, when the masses of coal were broken up, dislocated, and thrown in masses into separate basins, and that upon this theatre of ruins, a fourth age of nature—the Permian era—was entered on.

Coal, then, is the substance of the vegetation of the most remote ages of the world. Buried under an enormous thickness of rocks, they have been preserved to our days, after being modified in their inward nature and in their exterior. Having lost certain of their constituents, they have been transformed into a species of carbon, and impregnated with bituminous or tarry substances, the ordinary product of the slow decomposition of organic matter.

Thus coal, which feeds our manufactures and our furnaces, which is the fundamental agent of our productive industry and economy—the coal which warms our hearths and furnishes the gas which lights our streets and dwellings—is the substance of the plants which formed the forests, the grass, and the marshes of the ancient world at a period too distant for human chronology to assign a date to it. We shall not say with some persons that all in nature was made for man, and who thus form for themselves a very imperfect idea of the immensity of creation. Nor shall we even say that the vegetables of the ancient world have lived and multiplied only to prepare for man the agents of his economic and industrial occupations; which would be presumptuous. Let us rather direct the attention of our readers to the powers of science, which can thus, after an interval of time so prodigious, trace the precise origin, and state with exactness, the genera and species of the plants to which these substances belonged, of which no identical representatives now remain.

Let us pause for a moment, and consider the general character which belonged to our planet during the carboniferous period. Excessive heat and extreme humidity were then the attributes of its atmosphere. The congeners of the species which formed its vegetation are now only found under the burning latitudes of the tropics; and the enormous dimensions in which we find them in the fossil state prove, on the other hand, that the atmosphere was saturated with humidity. The intrepid traveller, Dr. Livingstone, who has in our days made such important observations on the interior of Africa, tells us that continual rains, added to intense heat, are the climatric characteristic of Equatorial Africa, where the vigorous tufted vegetation so pleases and delights the eye.

The remarkable circumstance is, that this elevated temperature and constant humidity does not seem to have been limited to any one part of the globe: the heat seems to have been the same in all latitudes. From the equatorial regions up to Melville Island, in the Arctic Ocean, where in our days the frosts are eternal—from Spitzbergen to the centre of Africa, the carboniferous flora presents an identity. When we find almost the same fossils at Greenland and in Guinea, when the same species, now extinct, are met with under the same degree of development at the equator and the pole, we cannot but admit that at this epoch the temperature of the globe was alike everywhere. What we now call *climate* was, then, unknown in geological times. There seems to have been only one climate over the whole globe. It was only at a later period, that is, in the tertiary epoch, by the progressive cooling of the globe, that the cold began to make itself felt at the polar extremities. Whence, then, proceeds this uniformity of temperature which we now look upon with so much surprise? It proceeded from the excessive heat of the terrestrial sphere. The earth was still so hot in itself that its innate temperature rendered superfluous and inappreciable the heat which reached it from the sun.

Another circumstance, which is established with much less certainty than the preceding, relates to the chemical composition of the air during the carboniferous period. Seeing the enormous mass of vegetation which then covered the globe, extending from one pole to the other; considering, also, the great proportion of carbon and hydrogen which exists in the bituminous matter of coal, it has been thought, and not without reason, that the atmosphere of the period would be richer in carbonic acid than the atmosphere of our days. It has even been thought that the great proportion of carbonic acid gas in the atmosphere was an explanation of the small number of animals, especially aerians, which then lived. This, however, is pure deduction, totally deficient in proof. Nothing proves that the atmosphere of the period was richer in carbonic acid than the atmosphere of our days. We can then only utter here vague conjectures, and cannot profess with any confidence, the opinion, that the atmospheric air of the period contained more carbonic acid gas than that which we breathe. What we can remark on with certainty as a striking characteristic in the vegetation of the globe during this phase of its history was the prodigious development its vegetation assumed. The ferns, which in our days and in our climate are only small herbaceous plants, in the carboniferous age presented themselves sometimes under a lofty form and port.

The marshy herbaceous plants with cylindrical stem, hollow, channeled, and articulated; whose articulations are furnished with a membranous, dentated sheath, which bears the name of "mares-tail," their fructification forming a sort of catkin composed of many rings of scales, which bear on their lower surface sacs full of *spores* or seeds—these humble *Equisetii*, with fluted stem and whorled branches, represented during the coal period by herbaceous trees—immense asparagus—as it were, of forty to five-and-forty feet high, and four to six inches in diameter, their trunks hollowed and channeled longitudinally, and divided transversely by the lines of articulation, have been preserved: they bear the name of *Calamites*. The engraving (Fig. 35) represents one of these gigantic mares-tails, or calamites, of the coal period, restored under the directions of M. Eugene Deslongchamps. It is represented with its fronds or leaves, and its organs of fructification. They seem to have grown somewhat after the manner of our asparagus, by an underground stem, whence issued at intervals new jets, as represented in the engraving, while new buds issued from the ground, as represented. The *Lycopods* of our age are humble plants, rarely exceeding a yard in height, and most commonly creepers; but the Lycopodiaceæ of the ancient world were trees of eighty to ninety feet in height. It was the *Lycopodendrons* which filled the forests. We present a restoration of one of these giants of the primeval forests, which has been traced by the pencil of M. Eugene Deslongchamps, in which this tree is represented with its stem, branches, its fronds, and organs of fructification. Their leaves were sometimes twenty inches long, and their trunks a yard in diameter.

Such are the dimensions of some specimens of *Lycopodendron cannatum* which have been found. Another lycopod of this period, the *Lomatophloyos crassicaule*, attained still more colossal dimensions. The *Sigillarias* sometimes attain the height of a hundred feet and upwards. The herbaceous ferns also abounded. They grew in the shade of the gigantic trees. It was the combination of these trees of immense size and such shrubs, if we may so call them, which formed the forests of the carboniferous age. The gigantic herbaceous tree which bears the name of *Calamites* abounded in the fo-

Fig. 35.—Calamite restored. 30 to 40 feet high.

rests of the carboniferous era, and especially in the later portion

of it : the trunks of these gigantic trees are represented below, reduced
respectively to one-fifth and one-tenth the natural size.

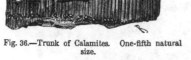

Fig. 36.—Trunk of Calamites. One-fifth natural   Fig. 37.—Trunk of Sigillaria. One tenth natural
size.   size.

What could be more surprising than the whole of this exuberance of
exuberant vegetation!   These immense Sigillarias, which reigned over
the forest! these Lycopodendrons, with flexible and slender stem!
these Lomatophylos, which present themselves as *herbaceous* trees of
gigantic size, furnished with verdant leaflets! these Calamites, rising
forty feet high! these elegant arborescent ferns, with aërial foliage, as
finely shaped and cut as the most delicate lace!   Nothing can now
convey to us an idea of the vast extent of verdure which clothed the
islands of the sea, from one pole to another, under a burning temperature of which the whole terrestrial globe partook.   In the
thickets of these inextricable forests parasite plants were suspended
from the trunks of the great trees, in tufts or garlands, like the wild
vines of our tropical forests.   They would be nearly all pretty
fern-like plants—*Sphenopteris* or *Hymenophyllites;* they would

attach themselves to the stems of the great trees like the orchids and *Bromeliaceas* of our times.

The edge of the waters would also be covered with vegetation, the leaves light and whorled, belonging perhaps to the dicotyledons, and divided into pairs like *Annularias fertilis*, the *Sphenophyllites* and *Asterophyllites*.

How this vegetation, so imposing at once from the dimensions of the individual trees and the space which they occupied, so fantastic in its form, and yet so simple in its organization, must have differed from that which now embellishes the earth and charms our eyes! It certainly had the privilege of size and rapid growth; but was it rich in species, or uniform in appearance? No: no flowers yet decorated the foliage and varied the tone of the forest. Eternal verdure clothed the branches of the ferns, the lycopods and equisetii, which formed, in great part, the vegetation of the age. The forests presented an innumerable collection of individuals, but very few species, and all belonging to these lower types of vegetation: no fruit appears, fit for nourishment; none would appear to have been on the branches. In short, no terrestrial animals yet appear to have existed: animal life was confined to the sea, the vegetable kingdom exclusively occupied the land. Probably a few winged insects, some coleoptera, orthoptera and neuroptera gave animation to the air, while exhibiting their variegated colours.

But for what eyes, we might ask, for whose thoughts, for whose wants did these solitary forests grow? For whom these majestic and infinite shades? For whom these spectacles? What mysterious being contemplated these marvels? A question not to be solved. In its presence we are overwhelmed, and our powerless reason is silent: its solution rests with Him who said, "Before the world was I am!"

CARBONIFEROUS LIMESTONE. (SUB-PERIOD.)

The vegetation which covered the numerous islands of the carboniferous sea consisted, then, of ferns, of Equisetacea, of Lycopodiaceæ and

dicotyledonous Gymnosperms. The Annularia and Sigillaria belong
to families completely extinct of the last-named class.

The *Annularia* were small herbaceous plants which floated on the
surface of fresh-water lakes and ponds; their leaves were verticillate,
that is, arranged in a great number of whorls, at each articulation of
the stem with the branches. The *Sigillaria* were, on the contrary,
great trees, consisting of a simple trunk, surmounted with a bunch or
panicle of slender leaves, drooping at the extremity, the bark often
channeled, and preserving impressions or markings of the old leaves,
which, from their resemblance to a seal, gave origin to their name,
sigillum. Fig. 38 represents the bark of one of the Sigillaria, which
is often met with in coal-mines.

Fig. 38.—Sigillaria lævigata. One-third natural size.   Fig. 39.—Stigmaria. One-tenth natural size.

The *Stigmaria*, according to many palæontologists, were cryptogamia,
of subterranean fructification: we only know the long roots which
carry the reproductive organs, which in some cases are as much as
sixteen feet long. This was suspected by Brongniart, on botanical
grounds, to be the roots of Sigillaria, and recent discoveries have confirmed this impression. Sir Charles Lyell, in company with Dr.
Dawson, examined several erect *Sigillariæ* in the sea cliffs of the South
Joggins in Nova Scotia, and found that from the lower extremities of
the trunk they sent out *Stigmariæ* as roots, which divided into four

parts, and these again threw out eight continuations, and these again divided into pairs.

Two other gigantic trees filled the forests of this period : these were *Lepidodendron carneatum* and *Lomatophloyos crassicaule*, both belonging to the family of Lycapodiaceæ, which includes in our age only very small species. The .trunk of the Latomophloyos threw out numerous branches, which terminated in thick tufts of linear and fleshy leaves.

The *Lepidodendrons*, of which there are about forty species known, have a cylindrical stem or trunk bifurcated in the branches, that is, the branches were evolved in pairs, or *archotomous*. The extremities of the branches were terminated by a fructification in the form of a

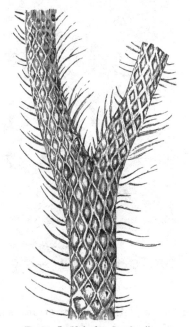

Fig. 40.—Lepidodendron Sternbergii.   Fig. 41.—Lepidostrobus variabilis.

cone, formed of linear scales, to which the name of *Lepidostrobus* has been given. In many of the coal-fields fossil cones have been found,

to which this name has been given by earlier palæontologists. They
sometimes form a nucleus of concrete balls of clay ironstones, and are
well preserved, having a conical axis, surrounded by scales compactly
imbricated.   The opinion of Brougniart is now generally adopted,
that they are the fruit of the Lepidodendron.   At Coalbrookdale and
elsewhere these have been found as terminal tips of a branch of a
well-characterized Lepidodendron.   Both Hooker and Brongniart place

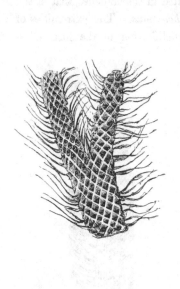

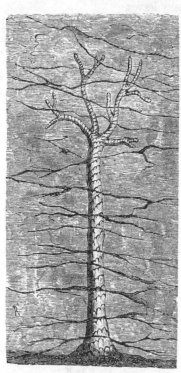

Fig. 42.—Lepidodendron elegans.

Fig. 43.—Lepidodendron Sternbergii.

them with the lycopods, having cones with similar spores and
sporangia, like that family. Nevertheless many of these branches
seem to have been sterile, simply terminating in fronds or elongated
leaves. Most of them were large trees. One tree of *L. Sternbergii*,
nearly fifty feet long, was found in the Jarrow Colliery, near Newcastle,

lying in the shale parallel to the plane of stratification. Fragments of others found in the same shale indicated, by the size of the rhomboidal scars which covered them, a still greater size. Lepidodendron Sternbergii (Fig 43) is represented as it was found under the schists in the colliery of Swina, in Bohemia. *L. elegans* (Fig. 42) is represented in the margin with the portion of a branch, furnished with leaves. M. Eugene Deslongchamps has successfully attempted the restoration of the tree discovered in the Bohemian colliery, which is here represented with its stem, its branches, fronds, and organs of fructification (Fig. 44). The ferns composed a great part of the vegetation of the carboniferous pe-

Fig. 44.—Lepidodendron Sternbergii restored.   40 feet high.

riod, both in the herbaceous and arboraceous form.

The ferns differ chiefly in some of the details of the leaf. Pecopteris, for instance, have the leaves, once, twice, or thrice pinnated, with the leaflets adhering either by their whole base or by the centre

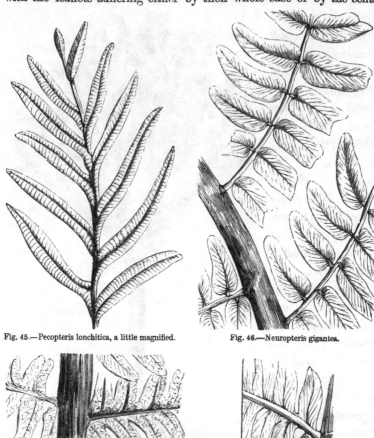

Fig. 45.—Pecopteris lonchitica, a little magnified.

Fig. 46.—Neuropteris gigantea.

Fig. 47.—Lonchopteris Bricii.

Fig. 48.—Odontopteris Brardii.

only; the midrib running through to the point. *Neuropteris* has leaves divided like Pecopteris, but the midrib does not reach the apex of the leaflets, but divides right and left into veins. *Odontopteris* has pinnatifid leaves, like the last, but its leaflets adhere by their whole base to the stalk. *Lonchopteris* has the leaves several times pinnatifid, the leaflets more or less united to one another, and the

Fig. 49.—Sphenopteris artemisiæfolia, magnified.

veins reticulated. Among the numerous species of the period was *Sphenopteris artemisiæfolia* (Fig. 49), of which a magnified leaf is represented above. Sphenopteris has twice or thrice pinnatifid leaves, the leaflets narrow at the base, and the veins generally arranged as if they radiated from the base; the leaflets frequently wedge-shaped. The seas of this epoch included an immense number of zoophytes, nearly four hundred species of mollusks, and a few crustaceans and fishes. Among the fishes the genus *Psammoides* and *coccasteus*, whose massive teeth inserted in the palate are suitable for grinding; and the *Holoplychius* and *Megalichthys*. The mollusks are chiefly brachiopodes of great size.

K

The Productus attained here great development. *Productus Martini* (Fig. 50), *P. semi-reticulatus* and *P. giganteus*, being the most remarkable. Some great Spirifers were there, as *Spirifer*

Fig. 50.—Productus Martini. One-third nat. size.   Fig. 51.—Bellerophon costatus. Half nat. size.   Fig. 52.—Goniatitis evolutus. Nat. size.

trigonalis, S. Glaber. *Terebratula hastata* has been preserved to us with the coloured bands which adorn the shell of the living animal. The *Bellerophons*, Gasteropodes, whose shell rolled round symmetrically upon itself, in some respects resembles the nautilus of our

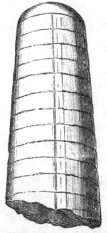

present seas, but without its chambered shell, were then represented by many species, among others by *Bellerophon costatus* (Fig. 51). Again, among the cephalopodes, we find the *Orthoceras* (Fig. 53), with its narrow and erect dwelling, and *Goniatitis evolutus* (Fig. 52), congener of the ammonities, which soon begin to appear in great numbers.

The crustaceans are rare in the rocks of the carboniferous limestone; the Phillipsia are the last of the trilobites, which become extinct in this period. As to the zoophytes, they consist chiefly of Encrinites and polypii. The Encrinites were

Fig. 53.—Orthoceras laterale. represented by the genus *Platycrinus* and *Cyathocrinus*. We also have in these rocks some Bryozoanes.

Among the polypiers of the period, we may include the genus Lithostrotion and Lonsdalea, of which *L. basaltiforme* (Fig. 54), and *Lonsdalea floriformis* (Fig. 55), are the representatives, with *Am-*

X.—Ideal view of marine life in the Carboniferous Period.

plexis coralloides. Among the Bryozoares, the genus *Fenestrella*
and *Polypora.* To these we may add a group
of animals which will play a very important
part, and become very abundant in the beds
of later geological periods, but which may be
traced already in the carboniferous deposits.
We speak of the *Foraminifera* (Fig. 56),
microscopic animals, which can scarcely be
said to have a distinct individual existence,
but clustered either in one body, or divided
into segments, and covered with an ordinary
testaceous envelope, as in Fig. 57, *Fusulina*
cylindrica. These little creatures, which
during the jurassic and cretaceous period
formed enormous masses of rock, begin to
make their appearance in that which now
engages our attention.

Fig. 54.—Lithostrotion
basaltiforme.

The plate opposite (Pl. x.) is an ideal

Fig. 55.—Lonsdalea floriformis.

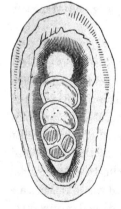

Fig. 56.—Foraminifer of the mountain
limestone, forming the centre of an
oolitic grain. Power 120.

Fig. 57.—Foraminifera of the chalk, obtained by
brushing it in water. Power 120.

aquarium, in which some of the more prominent species which in-

K 2

habited the seas during the carboniferous age are represented. On the right is a tribe of polypiers, with reflections of dazzling white : the species represented are, nearest the edge, the *Lasmocyathus*, the *Chætitus*, and the *Ptlypora*. The mollusk which occupies the extremity of the elongated, conic and sabre-like tube—an *Aploceras*, seems to prepare the way for the ammonite; for if this elongated shell was rounded and turned round its centre it would approximate to the ammonite and nautilus. In the centre of the first plane we have *Bellerophon huilcus*, the *Nautilus Koninckii*, and a *Productus*, with the numerous spines which surround the shell.

On the left are other polypiers : the *Chonetas* at the surface, extended and furnished with small spines, and the *Cyathophyllum*, with straight cylindrical stems, some Encrinites, *Cyathocrinus* and *Platycrinus*, rolled round the trunk of a tree, or with their flexible stem floating in the water. Some fishes, *Amblypterus*, move about in the middle of these creatures, the most part of whom are immoveably attached, like plants, to the rock on which they are rooted.

In addition, this engraving shows us a series of islets, rising above the tranquil sea. One of these is occupied by a forest, in which a distant view is presented of the general form of the grand vegetation of the period.

The rocks formed by deposit from the sea during the era of carboniferous limestone are important, inasmuch as they include coal, though in quantities much smaller than in the succeeding sub-period of the real coal deposit. They consist essentially of a compact limestone, of a greyish, blue, and even black colour. The blow of the hammer produces a somewhat fetid odour, which is owing to the decomposed organic matter—the modified substance of the mollusks and zoophytes —of which it is to so great an extent composed, and whose remains are still easily recognised. In the North of England, and many other parts of the British Islands, the carboniferous limestone forms, as we have seen, lofty mountain masses, to which the term *mountain limestone* is sometimes applied.

In Derbyshire the formation rises into rugged, lofty, and fantastic-shaped mountains, whose summits mingle with the clouds, while its picturesque character appears here, as well as farther north, in the *dales* or valleys, where rich meadows, through which the mountain streams force their way, seem to be closed abruptly by masses of rock, rising like the grey ruins of some ancient tower; while the base is pierced with caverns, and its sides covered with mosses and ferns, for which the limestone has a natural affinity.

The formation is *metalliferous*, and yields rich veins of lead in Derbyshire, Cumberland, and other counties of Great Britain. The rock is found in Russia, in the north of France, and in Belgium, where it furnishes the common marbles, known as Flanders marble, *marbres de Flanders*, and small granite, *petit granite*. These marbles are also quarried in other localities, such as Regneville (La Manche), either for the manufacture of lime or for ornamental stonework. One of the varieties quarried at Regneville, being black, with great yellow veins, is very pretty.

In France, the *carboniferous rock*, with its sandstones and conglomerates, schists and limestones, is largely developed in the Vosges, in the Lyonnais, and in Languedoc, often in contact with syenites and porphyries and other igneous rocks, by which it has been penetrated and overthrown, and even *metamorphosed* in many ways, and its constitution totally changed. In the United States it occupies a grand enough position in the rear of the Alleghanies, and it is found forming considerable ranges in our Australian colonies.

In virtue of their antiquity, as compared with the secondary and tertiary limestone rocks, the carboniferous rocks are generally more marked in character. The valley of the Meuse, from Namur to Chockier, above Liége, is of this formation; and many of our readers will remember with delight the picturesque character of the scenery, especially that of the left bank of the celebrated river.

COAL MEASURES (SUB-PERIOD).

This terrestrial period would be characterised in a remarkable manner, by the abundance and peculiarity of the vegetation which then covered the islands and continents of the whole globe. Upon all points of the earth, as we have said, this flora presented a striking uniformity. In comparing it with the vegetation of the present day, the learned French botanist, M. Brongniart, who has given particular attention to the flora of the coal measures, arrives at the conclusion that it presented considerable analogy with that of the islands of the equatorial and torrid zone, in which the maritime climate and elevated temperature exist in the highest degree. It is believed that the islands were very numerous at the period; that, in short, the dry land formed a vast archipelago upon the one general ocean, of no great depth, the islands being connected together and formed into continents as they gradually emerged from the ocean.

This flora, then, consists of great trees, often, as we have seen in the preceding formation, *herbaceous*, but also of many small plants, which would form a close, thick grass, or sod, half buried in marshes of an almost unlimited extent. M. Brongniart indicates as belonging to the period 500 species belonging to families which we have already seen dawning on the Devonian horizon, but which now attain this prodigious development. The ordinary dicotyledons and monocotyledons—that is, plants having seeds with two lobes in germinating, and plants having one seed lobe—are almost entirely absent; the cryptogamic, or flowerless plants, predominate; in particular, the ferns abound, of the lycopods and equisetacea; forms insulated and actually destroyed in these same families. A few dicotyledonous gymnosperms, or naked-seeded plants forming a genera of conifers, have completely disappeared, not only from the existing flora, but from the period under consideration: there is no trace of them in the subsequent flora, except in their fossilized state. Such is a general view of the features most characteristic of the coal period, and in general of the transition epoch. It differs altogether and absolutely from that of to-day: the climatric condition of these remote ages of the globe, however, enables us to comprehend the characteristics which distinguish its vegetation. Continual rains and an intense heat, a soft light, veiled by permanent fogs, engendered this special vegetation, of which we search in vain for anything analogous in our days. The nearest approach to the climate and vegetation proper to the geological phase which occupies us would probably be found by transporting ourselves by our thoughts to certain islands, or to the littoral of the Pacific ocean—the island of Chloë, for example, where it rains during 300 days in the year, and where the sun is shut out by permanent fogs: where the arborescent ferns become forests; in their shade grow herbaceous ferns, which rise three feet and upwards above a soil nearly all marsh; which gives shelter also to a mass of cryptogamic plants, greatly resembling, in its main features, the flora of the coal measures. This flora was, we have said, uniform and poor in its botanic genera, compared to the abundance and variety of the existing flora; but the few families which existed then included many more species than they now present in the same countries. The fossil ferns of the coal series in Europe, for instance, comprehends about 200 species, while all Europe can now produce only fifty. The gymnosperms, which now muster only twenty-five species in Europe, numbered more than 120.

It will simplify the classification of the flora of the carboniferous epoch if we give a tabular arrangement adopted by the best authorities:—

Dr. Lindley.	Brongniart.	
I. Thallogens .	1. Cryptogamous Amphigens, or Cellular Cryptogames	Lichens, sea-weeds, fungi.
II. Acrogens .	2. Cryptogamous Acrogens	Mosses, equisetums, ferns, Lycopodes, Lepidodendrons.
III. Gymnogens .	Dicotyledonous Gymnosperms	Conifers and Cycades.
IV. Exogens . .	Dicotyledonous Angiosperms	Compositæ, Leguminosæ, Umbelliferæ, Cruciferæ, heaths. All European except conifers.
V. Endogens .	Monocotyledons	Palms, lilies, aloes, rushes, grasses.

Calamites are found in striated, jointed, cylindrical or compressed fragments, with channels furrowed in their sides, and sometimes surrounded by a bituminous coating, the remains of a cortical integument.

They were originally hollow, but the cavity is usually filled up with the substance into which they themselves have been converted. They were separable at their articulations, and when broken across at that part they show a number of striæ, originating in the furrows of the sides and turning inwards towards the centre of the stem. It is not known whether this structure was connected with an imperfect diaphragm stretched across the hollow of the stem at each joint, or merely represented the ends or woody plates of which the solid part of the stem is composed. Their extremities have been discovered to taper gradually to a point, as represented in *C. cannæformis* (Fig. 58), or to end abruptly, the intervals becoming shorter and smaller. The obtuse point is now found to be the root of the tree. Brongniart classes the Calamites with the Equisetaceæ; later botanists consider that they belong to an extinct family of plants. The

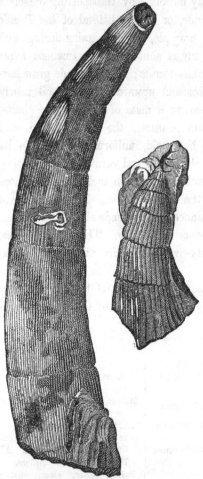

Fig. 58.—Calamites cannæformis.   One-third nat. size.

Stigmaria is the commonest of all plants in the coal formation; not a mine is opened, nor a heap of shale thrown out, but there occur fragments of its stem, marked externally with small cavities in the centre of slight tubercles, irregularly arranged. From the tubercles arise long ribbon-shaped bodies, said to have been traced to the length of twenty feet. The arms of this plant is a mass of cellular matter, having in its centre a hollow cylinder composed of spiral vessels destitute of medullary processes. This is still of questionable origin.

In the family of the Sigillarias we have already presented the bark of *S. lævigata*, at page 124: we here give a piece of the bark of *S. reniformis*, one-third the natural size.

In the family of the Asterophyllites, the leaf of *Asterophyllites foliosa* (Fig. 60), and the foliage of *Annularia orifolia* (Fig. 61) are remarkable. In addition to these we present, in Fig. 63, a

Fig. 59.—Sigillaria reniformis.

restoration of one of these Asterophyllites, the *Sphenophyllum*, after M. Eugene Deslongchamps. This herbaceous tree, like the Calamites, would present the appearance of an immense asparagus, twenty-five to thirty feet high : it is represented here with its branches and *fronds*, which bear some resemblance to the leaves of the gincko. The bourgeon, as represented in

Fig. 60—Asterophyllites foliosa.

the figure, is terminal and not auxiliary, as in some of the Calamites.

During the coal period the vegetable kingdom would seem to have reached its culminating point; if so, the animal kingdom was poorly represented. Some remains have been found, both in America and

Germany, consisting of portions of the skeleton and the impression of footsteps of a reptile, which has received the name of Archegosaurus. In Fig. 62, the head and neck of *Archegosaurus minor*, found in 1847 in the coal basin of Saarbruck, between Strasburg and Treves, is represented. Among the few animals of this period we find

Fig. 61.—Annularia orifolia.

a few fishes, analogous to those of the Devonian rocks. These are the *Holoptychius* and *Megalichthys*, having jaw-bones armed with enormous teeth. Some winged insects would probably join this slender cortége of living beings. It is then correct to say that the immense forests and marshy plains, crowded with trees, shrubs, and herbaceous vegetables, which formed on the innumerable isles of the period a thick and tufted carpet, were almost devoid of animals.

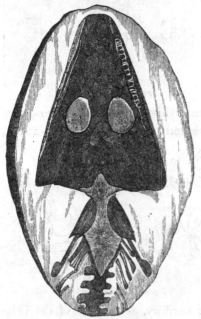

Fig. 62.—Head and neck of Archegosaurus minor.

On the opposite page (Pl. XI.), M. Riou has attempted, under the directions of M. Deslongchamps, to reproduce the aspect of nature during the period. A marsh and

XI.—Ideal view of a marshy forest of the Coal Period.

forest of the coal period are here represented with a short and

Fig. 63.—Sphenophyllum restored.

thick vegetation, a sort of grass, composed of herbaceous ferns and mares-tail. Several trees of forest height raise their heads above this lacustrine vegetation: Here are the indications of the species represented :—

On the left are seen the naked trunk of a *Lepidodendron* and a *Sigillaria*, an arborescent fern rising between the two trunks. At the foot of these great trees an herbaceous fern and a *Stigmaria* appear, whose long ramification of roots, provided with reproductive spores, extend to the water. On the right the naked trunk of another *Sigillaria*, a tree whose foliage is altogether unknown, a *Sphenophyllum* and a *conifer.* It is difficult to describe with precision the species of this family, whose imprints are, nevertheless, very abundant in the coal formation.

In front of this group we see two trunks broken and overthrown. These are a *Lepidodendron* and *Sigillaria*, mingling with a heap of vegetable débris in course of decomposition, from which a rich humus will be formed, upon which a new generation of plants will soon develop themselves. Some herbaceous ferns and buds of *Calamite* rise out of the marshy water.   .

A few fishes, belonging to the period, swim on the surface of the water, and the aquatic reptile *Archegosaurus* shows its long and pointed head—the only part of the animal which is known. A *Stigmaria* extends its roots into the water, and the pretty Asterophyllites rise above it in the first plane, with finely-cut stems.

A forest, composed of *Lepidodendrons* and *Calamites*, forms the background to the picture.

FORMATION OF BEDS OF COAL.

Coal, as we have said, is only the result of partial decomposition of the plants which covered the earth during a geological period of immense duration. No one now has any doubt that this is its origin. In coal-mines it is not unusual to find small débris of the very plants whose trunks and leaves characterize the coal measures, or carboniferous era. More than once, also, immense trunks of trees have been

met with in the middle of a bank of coals. In the coal-mines of
Treuil, at St. Etienne, for instance, the trees, *Lepidodendrons*, are not
even mixed with the coal, but with the bed of overlying rocks, as

Fig. 64.—Treuil coal-mine, at St. Etienne.

represented in the engraving, which has been reproduced from
M. Ad. Brongniart's design.

In England it is the same: entire trees are found lying across the
coal-beds. Sir Charles Lyell tells us that in Parkfield Colliery, South
Staffordshire, there was discovered in 1854, upon a surface of some
hundreds of yards, a bed of coal which has furnished more than
seventy-three trunks furnished with roots, some of them measuring

more than eight feet in circumference : their roots formed part of a
bed of coal more than eighty feet thick, resting on an argillaceous bed
two inches thick, under which was a second forest superposed on a
band of coal from two to five feet thick.   Underneath was a third
forest, with great trunks of *Lepidodendrons*, of *Calamites* and other
trees.

In the Bay of Fundy, in Nova Scotia, Sir Charles Lyell found in a
coal-field, upwards of twelve hundred feet thick, sixty-eight different
levels, presenting evident traces of many ground-plans of forests,
where the trunks of the trees were still furnished with roots.

We have endeavoured to establish the true geological origin of coal
in order that no doubts may exist in the minds of our readers on a
subject so important.   In order to explain the presence of coal in the
bosom of the earth, this is the only possible hypothesis.   These
vegetable débris may result from the burying of plants which may
have been brought from afar and transported by rivers or maritime
currents, forming immense rafts, which may have grounded in different places to be covered subsequently by the earth ; or the trees
may have grown on the spot where they perished, and where they are
now found.   Let us examine each of these theories :

Can the coal-beds result from the transport by water, and burying
underground, of immense rafts formed of the trunks of trees ?   The
hypothesis has against it the enormous height which must be conceded
to the raft in order to give the coal-beds the thickness they actually
attain in the collieries.   If we take into consideration the specific
weight of wood, and its contents in carbon, we find that the coal at
the depôts is only about seven-hundredth parts of the volume of the
primitive wood and other vegetable materials from which it is formed ;
and if we take account of the numerous voids from loose packing of
the raft, as compared with the compactness of coal, this may fairly be
reduced to five-hundredths. A bed of coal, for instance, sixteen inches
thick would have required a raft three hundred and eight feet high
for its formation.   Such a raft, we need hardly say, could neither float
on our rivers, nor on the greater part of our seas ; their accumulations

could never arrange themselves with the regularity of the stratified
coal-beds, not to mention their equal thickness over many hundreds of
yards that is observable in the more extensive coal deposits: and how
regularly they succeed each other by superposition, separated by banks
of sandstone or earth-beds. Again, admitting the possibility of an
accumulation, slowly and gradually, of vegetable débris as it reaches
the mouth of a river, would they not be buried, then, with great quantities of mud and earth? Now, in the most part of coal-beds, the proportion of earthy matter does not exceed fifteen per cent. If we bear
in mind, finally, the remarkable parallelism existing in the stratification
of the coal-beds, and the fine preservation in which the imprints of the
most delicate vegetable forms are discovered, it will, we think, be demonstrated that these formations have taken place in perfect tranquillity. We are, then, forced to the conclusion that coal results from
the fossilization which has taken place where the plants lived and
died.

In order to comprehend entirely the phenomena of the transformation of the forest and herbaceous plants which inhabited the
marshes of the ancient world into coal, there is a last consideration to
be presented. During the coal period, the terrestrial crust, then
scarcely consolidated, formed only a very elastic crust, by reason of its
immense extent, and from its reposing upon the yielding interior fluid
mass. This elastic crust would be agitated by the alternate movement of elevation and depression of the internal liquid mass under
the impulse of solar and lunar attraction to which they would be
subject, as our seas are now; giving birth to a sort of subterranean
tide, operating at intervals, more or less distant, upon weak parts in
the crust. It would perhaps be through one of the openings thus
produced that the vegetable masses would find themselves submerged, and the shrubs and herbs, after having thus covered for a time
the surface of the earth, would finish by being buried under the waters.
After this submersion, new forests may have developed themselves in
the same place. Through another weak point the second forests have
been depressed in their turn, and drowned in the overflowing waters.

It is probably by the repetition of this double phenomena—this burying of whole regions of forest, and the development upon the same
site of new masses, that the enormous heaps of plants, half decomposed, which constitute the coal measures have accumulated in a long
series of ages.

But has coal been produced from the great vegetables only, for
example, from the great forest-trees of the period, such as the Lepidodendrons, Sigillarias, Calamites and Sphenophyllums ?   This is scarcely
probable, for many coal deposits contain no vestiges of the great trees
of the period, but only of ferns and other herbaceous plants of small
size.   It is, therefore, presumable that the grand vegetation has been
almost unconnected with the production of coal, or at least that they
have taken merely an accessory part in the fossilization.   In all probability there existed in the coal period, as in our days, two distinct
kinds of vegetation :   one formed of lofty forest-trees ; the other,
herbaceous, and aquatic, developing itself on marshy plains.   It is the
latter class of vegetation, probably, which has furnished the great mass
of matter for coal, in the same manner as the same class of plants
have, during the historic times and up to the present day, supplied the
peat turf, which may be considered as a sort of contemporaneous
coal.

To what modifications have the vegetables of the ancient world been
subjected to attain the carbonized state, charged with bitumen, which
constitutes coal ?   The submerged plants would at first be a light
spongy mass, in all respects resembling the peat moss of our moors
and marshes.   In continuing under water these vegetable masses are
subjected to partial decomposition—a sort of fermentation, the chemical
phases of which are not very easily defined.   What we may affirm,
however, is that the decomposition and fermentation of the peat moss
of the ancient world was accompanied by the production of much carburated hydrogen, either gaseous or liquid.   And such is the origin of
the carburated hydrogen with which the coal is impregnated.   This,
with the tar oils which penetrated the bituminous schists, this
emission of bicarbonated hydrogen gas would probably continue until

after the turf-beds were buried under the soil. The mere weight of
the superincumbent mass, continued at an increasing ratio during a
long series of ages, would give to coal the very considerable density
which distinguishes it in its state of aggregation.

The heat emanating from the interior fires of the globe, and which
is presumed still to make itself felt at the surface, would also exercise
a great influence upon the final result. It is to these causes, that is to
say, to the pressure and to the heat, be it more or less, of the great
central terrestrial fires, that we may attribute the differences which
exist in the mineral characters of the coal, in proportion as they are
raised from the base of the coal measures towards the upper deposits.
The inferior beds are drier, and more compact than the upper, because
their mineralization, so to speak, has been completed under the influence of a higher temperature, and at the same time under a greater
pressure.

An experiment which was attempted for the first time in 1833, at
Saint-Bel, and afterwards by M. Cagniard de la Tour, and completed
at Saint Etienne in 1858, is quite a demonstration of the process by
which coal was formed : these gentlemen made the attempt to produce very compact coal artificially, by subjecting wood and other
vegetable matters to the double influence of heat and pressure.

The apparatus employed for this experiment by M. Baroulier, at
Saint Etienne, permitted him to expose the vegetable matter enveloped in moist clay, strongly compressed, at the temperature long
maintained of from 200° to 300° Centigrade. This apparatus, without being absolutely closed, offered considerable obstacles to the escape
of gases or vapours of the kind generated by decomposing organic
matter operating in the midst of humidity, and under a pressure
which opposed itself to the dissolution of the elements of which it was
composed. By placing in these conditions saw-dust of various kinds
of wood, products were obtained which resembled in many respects,
sometimes the brilliant sparkling, and at others the dull culmy coal ;
these differences, moreover, varying with the conditions of the experiment and the nature of the wood employed : thus, to all appearance,

L

explaining the striated appearance of coal composed alternately of
sparkling and dull heavy veins.

When the stems and leaves of herbaceous ferns are compressed
between beds of clay or puzzolana, they are decomposed by the pressure, and form themselves into compact carbonized blocks, arranged in
layers, having impressions of the leaves of plants such as blocks of
recent coal-beds frequently present. These experiments, which have
been conducted by Dr. Tyndall, leave us no room to doubt that coal
has been formed at the expense of the vegetables of the ancient world.

Fig. 65.—Stratification of the coal-beds.

Passing from these speculations to the actual coal fields :

This formation is composed of a succession of beds, more or less
thick, composed of divers gritstones, termed *grauwacke*, of clay and of
schists, sometimes bituminous and inflammable—in short, *coal*. These
three rocks form among themselves *strata*, which may alternate up to
150 times. Carbonate of iron may be considered a constituent of
this rock : its extensive dissemination in connection with coal, in
some parts of Great Britain, has been of immense advantage to the
iron works of that country, in many parts of which the high-blast-
furnaces for the manufacture of iron rise by hundreds alongside of the
coal-pits from which they are fed. In France, it may be noted, this

strong carbonated iron only occurs in nodules, much interrupted, so that it becomes necessary in that country to find other minerals to supply the wants of the foundries which had been established, taking for their base the coal measures of England. Fig. 65 gives an idea of the ordinary arrangement of the coal-beds, in which the coal is seen enclosed between two horizontal and parallel beds of schistose clay, mixed with nodules of carbonate of iron, a disposition very common in English collieries; the coal basin of Aveyron, in France, presenting an analogous arrangement.

The frequent presence of the carbonate of iron in or near the coal formations is a happy circumstance for mineralogical industry. When the miner finds in the same spot the iron mineral and the combustible, the necessary arrangements for working them can be established at comparatively small cost. The extent of the coal measures, in various parts of the world, may be briefly and loosely stated as follows:—

ESTIMATED COAL MEASURES OF THE WORLD.

	Square miles.
North America	310,500
Great Britain	6,200
France	1,550
Belgium	775
Rhenish Prussia and Saarebruck	1,550
Westphalia	590
Bohemia	620
Saxony	66
The Asturias, in Spain	310
Russia	160
Islands of the Pacific and Indian Ocean	Unknown.

The American continent, then, contains more extensive coal-fields than Europe; it contains, very nearly, two square miles of coal-fields for every five miles of its surface; but, it must be added, that these immense fields of coal have not hitherto been productive in proportion to their extent. The productiveness of the coal-fields may be stated as follows; for the year 1864:—

								Tons.
British Islands	.	.	.	.	.	.	.	80,000,000
United States	.	.	.	.	.	.	.	10,000,000
Belgium	.	.	.	.	.	.	.	8,000,000
France	.	.	.	.	.	.	.	6,000,000
Germany	.	.	.	.	.	.	.	6,000,000
Austria	.	.	.	.	.	.	.	900,000
Spain	.	.	.	.	.	.	.	500,000

We thus see that the United States only produce one-eleventh part of the whole of Europe, and one-eighth of Great Britain.

The coal measures of England and Scotland cover a large area, and attempts have been made to estimate the quantity of fuel they contain; but all such attempts must be a mere approximation to the truth when it is considered that in the coal-fields of South Wales, ascertained by actual measurement to attain the extraordinary thickness of 12,000 feet, the colliers reckon on having a hundred different beds of coal, varying from six inches to more than ten feet; when, moreover, we have the published opinion of Sir R. Murchison that the whole of the Permian series of rocks have for their *floor* coal-fields more or less powerful. So far as ascertained, however, it may be stated that Mr. Phillips estimates the coal-bearing strata of the north of England at 3,000 feet; but these coal-fields contain, along with many beds of the mineral in a more or less pure state, interstratified beds of sandstones, shales and limestone, the real coal seams, to the number of twenty or thirty, not exceeding sixty feet in all. In Somersetshire and the west of England and South Wales, where the whole series are from 10,000 to 12,000 feet thick, the hundred seams of coal will probably be in a similar proportion. The Scottish coal measures present a thickness of 3,000 feet of coal-bearing strata, with the same intercalation of other carboniferous rocks. In short, when it is considered that the whole of the Transition rocks which we have been describing were deposited in the sea-basins until they were filled up; that these basins emerged from the deep by upheaval or some other operation of nature; that they were again submerged, to pass through another course of upheaval and subsidence, filling-up, denudation and elevation; and this repeated many times and at vast intervals—this continued alternation of coal-beds, limestone, slate and sandstone is probably what might be expected to result.

The coal-basin of Belgium and of the north of France forms a zone, nearly continuous from Liége, Namur, Charleroi, and Mons, to Valenciennes, Douai, and Bethune. The beds there number from 50 to 110, and their thickness ranges between ten inches and six feet. Some coal-basins which are disposed beneath the secondary formations in the centre and south of France present beds less numerous, but thicker and less regularly stratified. The two basins of the Saône-and-Loire, the principal mines of which are Creuzat, Blauzy, Montchanin, and Epinac, only contain ten beds; but some of these attain 100, and even 120 feet in thickness, as at Montchanin. The coal-basin of the Loire is that which contains the greatest total thickness of coal-beds: there the beds are twenty-five in number. After those of the north, of the Saône-and-Loire, and of the Loire, the principal basin in France is that of the Allier, where very powerful beds occur at Commentry and Bezenet; the basin of Brassac, which commences at the confluence of

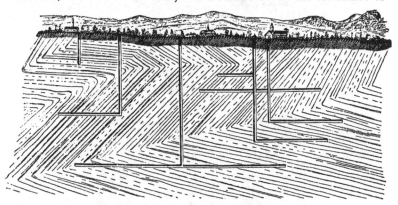

Fig. 66.—Contortions of coal-beds.

the Allier and the Alagnon; the basin of the Aveyron, known by the collieries of Decazeville and Aubin; the basin of the Gard, and of Grand-Combe. Besides these principal basins, however, a great number, little less important, occur, which yield annually to France from six to seven millions of tons of coal.

The beds of coal are rarely found in the position in which their transformation took place, which would be horizontal. They are generally much twisted and contorted, with numerous dislocations. They are found broken up and distorted by faults, and even folded back on themselves into zigzag form, as represented in the engraving (p. 149), which is the direction assumed by all the coal-beds of the basins of Belgium and the north of France: foldings which permit of vertical pits, in which the coal can be extracted by traversing the same beds many times.

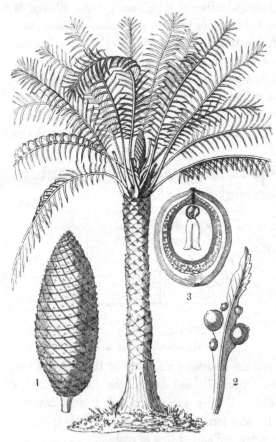

Fig. 67.—Cycas circinalis.

XII.—Ideal landscape of the Permian Period.

PERMIAN PERIOD.

THE earth continuing to cool, fractures would be produced in the
consolidated crust. Through the large openings thus left, the fluid
or viscous matter shut in by the consolidated beds would slowly force
its way to the surface, impelled by the central gases and aided by
the solar and lunar attraction before alluded to. Syenite and porphyry are the substances thus forced to the surface, and they form
there a series of cone-shaped domes or eminences, in form something
between a filbert and a lady's thimble resting on its broadest part. It
is to be noted, however, that porphyry does not now appear on the surface
for the first time. Before the carboniferous age it had surged to the
surface, and its débris is sometimes found in the lower conglomerates of
the system, rounded by the action of the waves.

The porphyritic and syenitic mountains which were thus formed at
the close of the transition period would, in their semi-fluid state, be
infinitely too high in temperature to admit as yet of vegetation. They
would rise in naked arid downs on the surface of the earth, covered
in other parts with the rich vegetation of the coal period.

On the opposite page an ideal view of the earth during the Permian
period is presented (PL. XII.). In the background, on the right, is seen
a series of syenitic and porphyritic domes, recently thrown up; while

a mass of steam and vapour rises in columns from the water, resulting
from the still-smoking and scarcely-consolidated matter.   In the foreground, on the right, rise groups of tree-ferns, Lepidodendrons, and
Walchias, of the preceding period. At the edge of the sea, left exposed
by the retiring tide, are Mollusks and Zoophytes of the period—*Productus, Spirifers* and *Encrinites;* pretty little plants —the*Asterophyllites*
of the carboniferous age are growing at the water's edge, not far from
the shore.   Having attained a certain height in the cooler atmosphere,
the columns of steam would become condensed and finally fall in torrents of rain.   The evaporation of water in such vast masses being
necessarily accompanied by an enormous disengagement of electric
fluid, this gloomy picture of the primitive world is lit up by brilliant
flashes of lightning, accompanied by the reverberating noise of thunder.

During the Permian period the vegetation would be nearly the
same as that already described as belonging to the coal period, with
some few species and even genera not hitherto observed. M. Ad.
Brongniart has described some forms, which he places intermediate
between the Carboniferous and Permian age, but still belonging to the
Equisetaceæ—ferns and palms, with some conifers ; and Professor King
has published a valuable memoir of the Permian rocks of England, as
compared with those of Thuringia, in the proceedings of the Palæon-
tographic Society, of which we append an analysis :—

NORTH OF ENGLAND.	THURINGIA.	MINERALS.
1. Lower sandstone and sand of various colours	1. Rothliegendes	1. White limestone with gypsum and white salt.
2. Marl slates . . .	2. Mergel - Schiefer or Kupferschiefer	2. Red and green grits with copper ore.
3. Compact limestones .	3. Lower Zeichstein	3. Magnesian limestones.
4. Fossiliferous limestone .	4. Dolomite or Upper Zeichstein	4. Marl stones.
5. Brecciated and imperfectly brecciated limestone . . .	5. Rauchwacke . .	5. Conglomerates.
6. Crystalline or concrete and non - crystalline limestone . .	6. Stinkstem . .	6. Oolitic.

At the base of the system lies a band of *lower sandstone*, of various colours, separating the magnesian limestone from the coal in Yorkshire and Durham; sometimes associated with red marl and gypsum, but with the same obscure relations in all these beds which usually attend the close of one series and the commencement of another; the imbedded plants being, in many cases, identical with those of the carboniferous series. In Thuringia the *Rothliegende*, or *red-lyer*, a great deposit of red sandstone conglomerate, associated with porphyry, basaltic trap, and amygdaloid, lies at the base of the system. Among the fossils of this age are the silicified trunks of tree-ferns, called *Psaronius*, their bark surrounded by dense masses of air-roots, which double or quadruple the diameter of the stem; in this respect bearing a strong resemblance to the arborescent ferns of New Zealand.

The marl slate consists of hard calcareous shales, marl slates, and thin bedded limestone, the latter nearly thirty feet thick in Durham, and yielding many fine specimens of Ganoid and Placoïd fishes— *Palæoniscus, Pygopterus, Cælacanthus,* and *Platysomus*—genera which all belonged to the coal measures, and which Mr. King thinks probably lived at no great distance from the shore; but the Permian species of the marl slates of England are identical with those of the copper slate of Thuringia. Agassiz was the first to point out a remarkable peculiarity in the form of the fishes before and after this age : nearly the whole of the nine thousand species now living have the tail-fins either single or equally divided, or "homocercal." In the Palæoniscus and most of the fossil fishes of the Permian and earlier periods of the earth's history, the tails of fishes have the upper fin much longer than the lower, and the vertebral column prolonged into the upper lobe, as in the sharks. This Agassiz calls "heterocercal." The compact limestone is rich in bryozoares. The fossiliferous limestone of this period, Mr. King considers, is a deep-water formation, from the numerous bryozoa which it contains. Some of them, as *Fenestrella retiformis*, found in the Permian rocks of England and Germany, sometimes measuring eight inches in width.

Many species of Mollusks, and especially Brachiopodes, appear in the Permian seas of this age, *Spirifera* and *Productus* being the most characteristic.

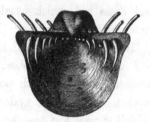 

Fig. 68.—Productus horridus.   Half natural size.

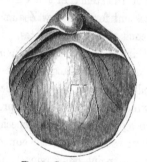 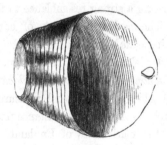

Fig. 69.—Spirifer trigonalis.   Fig. 70.—Cyrtoceras depressum.

Another genera, resembling the Productidæ, now occur, which have not been observed in strata newer than the Permian.   The *Stropholosia* are abundant in the yellow magnesian limestone, accompanied by *Spirifer undulatus*.   *S. Schlotheimii* is widely disseminated both in England, Germany, and Russia, with *Lingula Crednerii,* and other palæozoic brachiopodes.   Here also we note the first appearance of the oysters, but still in small numbers.   The *Fenestrella* represent the bryozoare mollusks.

The brecciated limestone and the concretionary masses overlying it, although Mr. King has attempted to separate them, are considered by Professor Sedgwick as forms of the same rocks.   They contain no foreign elements, but seem to be composed of fragments of the underlying limestone. Some of the angular masses at Tynemouth cliff are two feet in diameter, angular, and none of them water-worn.

The crystalline formation is seen upon the coast of Durham and Yorkshire, between the Wear and the Tees, and Mr. King thinks that the character of the shells and the absence of corals indicate their deposit in shallow water.

Among the ferns characteristic of the period may be mentioned *Sphenopteris dichotoma* and *S. Artemisiæfolia: Pecopteris lonchitica* and *Neuropteris gigantea* are figured on page 128. "If we attempt," says Lyell, "to draw a line between the secondary and primary fossiliferous strata, it must be drawn through the middle of what was once called the New Red. The lower half of this group will rank as palæozoic, or primary, while its upper members will form the base of the secondary or mesozoic series."

Among the *Equiseta*, Colonel Von Gutbier found *Calamites gigas* and sixty other fossil trees, most of them ferns of the Permian formation of Saxony, forty of which have not been found elsewhere. Among these are several species of *Walchia*, a genus of conifers, of which an example is given in the margin. They bear some resemblance to the *Araucarias*, which have been introduced from North America into our pleasure-grounds during the last half-century, in their stems, leaves, and cones.

Fig. 71.—Walchia Schlotheimii.

Among the genera enumerated by Colonel Von Gutbier are some fruit called *Cardescarpon*, and the Asterophyllites and Annularia, so characteristic of the carboniferous age. The Lepido-

dendrons are also common in the Permian rocks of Saxony, Russia, and Thuringia; also the *Nöggerathies*, a family of large trees, intermediate between the Cycades and

Fig. 72.—Trigonocarpum Nöggerathi.

Conifers. The fruit of one of these is represented in the margin.

PERMIAN ROCKS.

We now come to inquire into the physiognomy of the earth in the Permian period. Of what do the beds consist? what the extent, the mineralogical constitution of the rocks deposited by the seas of the period? To furnish a reply, geologists divide them into three ascending orders of stratification.

1. New red sandstone; 2. Magnesian limestone; 3. Permian, or sandstone of the Vosges.

The *new red sandstone,* which attains a thickness of from three hundred to six hundred feet, is found over great part of Germany, in the Vosges, and in England. Its fossil remains are few and rare: they include silicified trunks of conifers, some impressions of ferns, and calamites.

In England, the new red sandstone, surmounted by mottled marl, occupies, in the midland counties, a very considerable extent of country. Ravines, both narrow and deep, intersect table lands of considerable elevation, sometimes so precipitous and so red as to look like a wall of new bricks; then, here and there, in the midst of the sandstone, spotted and mossy as it is when clothed with lichens, rises some grand mass of chalk, upright as a Druid altar, giving appearances of the most remote antiquity to the rocks which surround them. Such is in general the style of the landscape. The aspect of this formation is not, however, so bold and striking as that of the primitive rocks.

The *magnesian limestone* or zechstein, so called by the Germans in consequence of the numerous metalliferous deposits met with in its diverse beds, presents in France only a few insignificant fragments; but in Germany and England it attains the thickness of a hundred and fifty yards. It is composed of magnesian limestone and bituminous clay, the last black and fetid. The subordinate rocks consist of marl, gypsum, and inflammable bituminous schists, which are met with in great proportions in the district of Mansfeld in Thuringia, among the minerals of argentiferous grey copper and lead largely worked in that country. The cupriferous schists are remarkable for the numbers of fossil fishes which they contain, whence they are called *Kupferschiefer* in Thuringia.

The mountain limestone in England has attained considerable development in Derbyshire and Yorkshire, where its accumulation of dolomite forms an excellent building-stone, the Houses of Parliament being built of it. As we proceed northwards the Permian mountains attain a great height, but they present few fossil remains: the primitive types of life seem to disappear, and those which survive are modified: we have reached a period of decay—at least, of transition. Nevertheless, the Permian rocks present impressions of feet, which seem to indicate that, in the midst of this decline of other animal forms, the race of reptiles was increasing: we find also on the flat surface of some of the new red sandstone rocks in England traces of ancient sea waves, which have left their marks, proving that tides existed. Upon other blocks of stone are small hollows, evidently the impressions left by heavy rain-drops in the soft mud, which has since hardened into stone, such as is represented at page 22. Sometimes these depressions have the lips more marked on one side than upon the other, as if the water had forced its way in that direction, by which we know that the skies of that age were covered with clouds; but we are without indications of the wind, or of the point of the horizon whence it blew.

The Permian formation, which has been used to designate the period, derives its name from the province of *Perm* in Eastern Russia. This rock, known in France as the sandstone of the Vosges, is habitually of a red colour; all the northern parts of the Vosges are composed of it, and here it attains a thickness of a hundred to a hundred and fifty yards, forming numerous flat summits, which testify to an ancient plateau cut up and intersected by water-courses: it contains only a few rare vegetable remains.

France, Germany and England present only some out-croppings of trifling extent as compared to its immense extension in the Eastern part of European Russia, where it is composed of a powerful alternation of limestone, of clay and sandstone, containing some Productus, some ferns, and even reptiles and fishes analogous to those of the zechstein of Western Europe; even gypsum and rock-salt, cropping out of the soil, is worked on a large scale. In France the Permian rock is sometimes confounded by the resemblance of its schists to those of the upper schists of the coal-fields, and thus deceive the miners who labour in the beds, leading them to imagine they are near to coal when it is altogether absent.

The extent of ocean which covered the globe in the Permian period would still be immense as compared with the seas of the present day. The chain of the Vosges, stretching across Rhenish Bavaria, the Grand Duchy of Baden, as far as Saxony and Silesia, would be under water. They would communicate with the ocean, which would cover all the midland and western counties of England and part of Russia. In other parts of Europe the continent has varied very little since the primitive Devonian and carboniferous ages. In France, the central plateau would form a great island, which extended towards the south, probably as far as the foot of the Pyrenees: another island would consist of the mass of Brittany. In Russia, the continent would have extended itself considerably toward the east: finally, it is probable that, at the end of the carboniferous period, the Belgian continent would stretch from the department of the Pas-de-Calais and Du Nord, in France, and would extend up to and beyond the Rhine.

In England, the Silurian archipelago, now filled up and occupied by deposits of the Devonian and carboniferous systems, would be covered with carboniferous vegetation; dry land would now extend, almost without interruption, from Cape Wrath to the Land's End; but, on its eastern shore, the great mass of the country now lying less than three degrees west of Greenwich would, in a general sense, be under water, or form islets rising out of the sea. Alphonse Esquiros thus eloquently closes the chapter of his work in which he treats of this formation in England: "We have seen seas, vast watery deserts, become populated; we have seen the birth of the first land and its increase; ages succeeding each other, and nature in its progress advancing among ruins; the ancient inhabitants of the sea, or at least their spoils, have been raised to the summits of lofty mountains. In the midst of these vast cemeteries of the primitive world we have met with the remains of millions of beings; entire species sacrificed to the development of life. Here terminates the first mass of facts constituting the infancy of the British Islands. But great changes are still to produce themselves on this portion of the earth's surface."

Having thus described the *transition epoch*, it may be useful, before entering on the *secondary epoch*, to glance backwards at the facts which we have had under consideration.

In this period plants and animals appear for the first time upon the globe, now getting cool. We have said the seas of the epoch were dominated by the fishes known as *Ganoïds*, so called from γάνος, *to glitter*, from the polish of the scales which cover their bodies, sometimes in a very complicated and fantastic manner, the *Trilobites* are curious crustaceans, which appear and disappear in the epoch; an immense quantity of mollusks, cephalopodes and brachiopodes; the *Encrinites*, the product of animals of a most curious organization—a species of mineral flowers, which form the most graceful ornaments of our Palæontological collections.

But, among all these beings, those which rule, those which were truly the kings of organic life, as it existed in the transition seas, were the fishes and

Fig. 73.—Lithostrotion.

Fig. 74.—Rhyncholites, upper, side, and internal views. 1, Side view (muschelkalk of Luneville); 2, upper view (same locality); 3, upper view (lias of Lyme Regis); 4, calcareous point of an under mandible, internal view, from Luneville. (Buckland.)

Ganoïds, which have left no animated being behind them of similar organization. Furnished with a sort of resisting cuirass, they seem to have received from nature, the means of protection to insure their existence and permit them to triumph over all the causes of destruction which threatened them in the seas of the ancient world.

In the transition epoch the living creation was in its infancy. No mammifera roamed the forests; no bird had yet displayed its wings. No mammifers therefore, no maternal instinct, none of the soft affections which are, with animals, like the precursors of intelligence. No birds, therefore, no songs in the air. Fishes, mollusks, and crustaceans silently occupied their bed in the depths of the sea. The immoveable zoophyte lived there, the obscure, and almost unconscious life of these imperfect beings. On the land we only find a few muddy reptiles, of small size—forerunners of those monstrous Saurians which will appear in the secondary epoch.

The vegetation of the epoch belongs chiefly to plants of an inferior organization. With a few vegetables of a higher order, that is to say, Dicotyledons, it was the cryptogamia, the ferns, the lycopods, and the equisetacea, then at their maximum of development, which formed the great mass of the vegetation of the epoch.

Let us also consider, in this short analysis, that during the epoch under consideration what we designate *climate* did not exist. The same animals and the same plants covered the globe from the equator to the poles; since we find in the transition formations of Spitzbergen and Melville Island the same fossils which we meet with in similar rocks in the torrid zone : we must conclude, then, that the temperature of this epoch was uniform all over the globe, and that the heat of the earth itself was sufficient to annul and render inappreciable the calorific influence of the sun.

During this period the progressive cooling of the earth occasioned frequent ruptures and dislocations of the terrestrial crust; and between these an opening was made for the passage of the rocks called *igneous,* such as granite, afterwards porphyry, and then syenite, which surged slowly across the fissures in a viscous state, and formed mountains of granite, of porphyry, and syenite, or simply of clefts, which were slowly filled with oxides and metallic sulphates, forming what are now designated veins. The great mountain range of Ben Nevis offers a striking example of the first of these occurrences: through the granite base a distinct natural section can be traced of porphyry, ejected through the granite, and of syenite through the porphyry. These geological commotions, which called forth, not over the whole extent of the earth, but only in certain places, great movements of the soil, would appear to have been more frequent towards the close of this epoch, and especially at the moment which formed the passage, as it were, between the Permian and the Triasic periods; but it has been a part of our plan to consider apart, in a former chapter, the phenomena of eruptions, and the character of the rocks called eruptive.

The convulsions and subversions by which the surface of the earth was agitated did not extend, let it be noted, to the whole circumference; the effects were restrained and local. It would, then, be wrong to assert, with many geologists, that the dislocations of the soil which accompany the agitations of the surface extended to both hemispheres, destroying all living creatures in each. The fauna and flora of the Permian period does not differ essentially from the fauna and flora of the preceding age, which shows that no general revolution occurred to disturb the entire globe between these two epochs. Here, as in all analogous cases, then, it is useless to recur to any general cataclysm to explain the passage from one epoch to another. Have we not, almost in our own day, seen certain species of animals die out and disappear, without the least geological revolution? Without speaking of the Beaver, which was abundant two centuries ago on the banks of the Rhone, and in the Cevennes, which still burrowed in the small river Beivre in the middle ages, which lived in our own rivers within the memory of man almost, and which is still found in America and other countries; we may cite many examples of animals which have become extinct in times by no means remote from ours. Such are the *Dinoris* and the *Griornis*, colossal birds of New Zealand and Madagascar, and the *Dodo*, which lived in the Isle of France in

M

1626.   The *Ursus squalus*, the *Cervus Megaceros*, the *Bos primo-genus*, were all contemporary with man, and have been long extinct. We no longer know the gigantic wood-stag, figured by the Romans on their monuments, and which they had brought from England for the fine quality of its flesh.   The Erymanthean boar, so widely disseminated in the ancient historic period, no longer exists among our living races, any more than the crocodiles *lacunosus* and *laciniatus*,

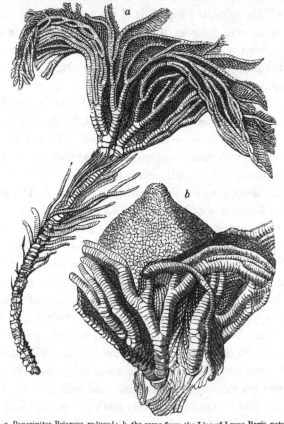

Fig. 75.—*a*, Pancrinites Briareus, reduced ; *b*, the same from the Lias of Lyme Regis, natural size.

found by Geoffroy St. Hilaire in the catacombs of ancient Egypt.   Many races of animals figured on the mosaics of Palestrina, and engraved or painted along with species actually living, are no longer found among living animals in our days. Must we imagine a series of geological revolutions to account for the disappearance of animals which have evidently become extinct in a natural way? What has come to pass in our days, it is reasonable to conclude may have taken place in the times anterior to the appearance of man.

Fig. 76.—Terebellaria ramosissima.

SECONDARY EPOCH.

During the *transitionary epoch* our globe would appear to have been appropriated only to beings which lived in the waters—above all, to the crustaceans and fishes: during the *secondary epoch* reptiles seem to have been its chief inhabitants. Beings of this class and of astonishing dimensions would seem to have multiplied themselves most singularly: they seem to have been kings of the earth. At the same time, however, while the animal kingdom thus developed itself, the vegetation has lost much of its vigour.

Geologists have agreed among themselves to divide the epoch into three periods: 1. the *Triasic;* 2. the *Jurassic;* 3. the *Cretaceous* period. A division which it is convenient to adopt.

The Triasic, or New Red Period.

This period has received the name of Triasic because the rocks of which it is composed, which are more fully developed in Germany than either in England or France, were formerly called the Trias or Triple Group by German writers because it was divided into three groups, as follows :—

English.	French.	German.
Gypseous shales and sandstone . . . }	Marne irisée . .	Keuper.
Wanting . . {	Muschelkalk or Calcaire Coquilier . . . }	Muschelkalk.
Sandstone and quartzose conglomerate . . }	Gres bigarré . .	Bunter sandstone.

Of these three groups geologists now form only two, the shelly, or new red sub-period, which embraces the two first, and the saliferous sub-period.

New Red Sandstone.

In this new phase of the revolutions of the globe the animated beings on its surface differ much from those which belonged to the transition epoch. The curious crustaceans which we have described under the name of *Trilobites* have disappeared; the molluscous cephalopodes and brachiopodes are here small in numbers, as are the ganoïd and placoïd fishes, whose reign also seems to have terminated during this period. But that of the *Ammonites* now begins, and assumes in the period a prodigious development. Vegetation is subjected to analogous changes. The cryptogamic plants, which reached their maximum in the transition epoch, become now less numerous, while the conifers take a corresponding extension. Some kinds of terrestrial animals have disappeared, but they are replaced by genera as numerous as they are new : for the first time the turtle appears in the bosom of the sea and on the banks of lakes. The Saurian reptiles assume great development; they prepare the way for the enormous Saurians, which appear themselves in the following period, whose skeleton presents such proportions, whose form is so incomprehensible as to strike with astonishment all who contemplate their gigantic and, so to speak, threatening, remains.

The seas of this sub-period, which is named after the innumerable masses of shells included in the rocks which it represents, included, besides great numbers of mollusks, Saurian reptiles of twelve different genera, some turtles, and six new genera of fishes clothed with the cuirass. Let us pause at the mollusks which peopled the Triasic seas.

Among the shells characteristic of the conchylian period, we may mention *Natica Gadlardoti, Rostellaria antica, Lima lineata, Avicula socialis, Terebratula communis, Mytilus edulilformis, Myophoria Goldfussii, Possidonia minuta,* and *Ce-ratites nodosus.* The *Ceratites,* of which a species is here represented, formed a genus closely approximating to the *Am-monites,* which seem to have occupied a position so important in the ancient seas, but which seem to have no existence in those of our era, either in species or genera. This Ammonite is found in the muschelkalk of Germany, a forma-tion which has no equivalent in Eng-land, but which is a compact greyish limestone underlying the saliferous rocks

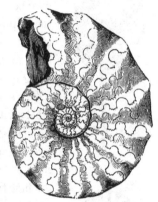

Fig. 77.—Ceratites nodosus (muschelkalk).

in Germany, and includes beds of dolomite with gypsum and rock-salt.

The Mytilus or Mussels, which properly belonged to this age, are acephalous or headless mollusks, with elongated triangular shells, hooked at the point and termination: they are common in our seas. The *Lima myophoria, Possidonia* and *Avicula* are acephalous mol-lusks of the same period. The two genera *Natica* and *Rostellaria* belong to the Gasteropodes, and are abundant in the muschelkalk in France, Germany and Poland.

Among the Echinoderms belonging to this period may be men-tioned *Encrinus moniliformis* and *lilliformis* (Fig. 78), whose remains, constituting in some localities whole beds of the soil, show the slow pro-

gress with which this zoophyte formed its beds of limestone rock in
the clear seas of the period. To these may be added among the mollusks, *Avicula subcostata*, *Myophoria pesauseris* and *Patella lineata*.

In the muschelkalk are found the skull and
teeth of *Placodus gigas*, a reptile which was
originally placed by Agassiz among the class of
fishes, but better specimens have satisfied Professor Owen that they belonged to a Saurian reptile.

It may be added, that the presence of a few
genera, proper to the transition epoch, which
disappear for ever during this sub-period, and
the appearance for the first time of some other
animals belonging properly to the Jurassic period,
give to the conchylian fauna the appearance of
being one of passage from one age to the other.

The seas, then, presented a few reptiles, probably
inhabitants of the banks of rivers, as *Phytosaurus*,
Capilosaurus, &c., and sundry fishes, as *Sphæ-
rodus* and *Pernodus*. In this sub-period we
shall see nothing of the land turtles, which, for
the first time, appear in this epoch of the world's
history; but it is to be noted with some care
that in this age a gigantic reptile appears, on
which the opinions of geologists were long at

Fig. 78.—Encrinus lilliformis. variance. In the clay rocks of the conchylian
period imprints of the foot of some animal were
discovered which very much resembled the impression that might
be made in soft clay by the outstretched fingers and thumb of the
human hand. These traces were made by a species of reptile furnished with four feet, the two fore-feet being much broader than the
hind ones. The head, pelvis, and scapula only of this strange-looking
animal have been found, but these are considered to have belonged to
a gigantic batrachian. It is thought that the head was not naked,
but protected by a bony cushion; that its jaws were armed with conical

teeth, of great strength and of a complicated structure. This curious and uncouth-looking creature, of which the engraving (Fig. 79) is a restoration, has been named the Cheirotherium, or Labyrinthodon.

Another reptile of great dimensions—which would seem to have

Fig. 79.—Labyrinthodon restored. One-twentieth natural size.

been intended to prepare the way for the appearance of the enormous Saurians which present themselves in the Jurassic period—was the *Nothosaurus*, a species of crocodile, of which a restoration has been attempted in PL. XIII.

It was long supposed, from other impressions which appear in the same rocks, that birds made their appearance in the period which occupies us: the flags on which these imprints occur show the traces of an animal of great size, presenting the impression of three toes, like some of the Struthionidæ, or ostriches, accompanied by rain-drops. But the opinion which has attributed these impressions to birds is no longer entertained. No remains of the skeletons of birds have been met with in rocks of the period, and the impressions in question are all that can be alleged in support of the hypothesis. We may perhaps attribute the traces to some unknown reptile with more probability.

M. Ad. Brongniart places the commencement of dicotyledonous gymnosperm plants in this age. The characteristics of this flora consist in numerous ferns, constituting genera now extinct, such as *Anopteris* and *Crematopteris*. The true *Equisetum* are rare in it. The Calamites, or, rather, the *Calamodendrons* abound. The gymnosperms are represented by the genus of Conifers, *Voltzia* and *Haidingera*, of which both species and individuals are very numerous in the formation of the period.

Among the species which characterize this strata we may mention *Neuropteris elegans, Calamites arenaceus, Voltzia heterophylla,* and *Haidingera speciosa.* The Haidingera, belonging to the tribe of *Abietineae,* were plants with broad leaves, analogous to those of our *Dammara,* growing close together, and nearly imbricated, or lying over each other, as in the *Araucarias.* The fruit, which are cones with rounded scales, are imbricated, and have only a single seed, thus bearing out the strong resemblance which has been traced between these fossil plants and the Dammara.

The *Voltzias,* which seem to have formed the greater part of the forests of the period, were a genus of Cupressineae, now extinct, which are well characterized among the fossilized conifers of the period. The alternate spiral leaves, forming five to eight rows sessile, that is, sitting close to the branch and drooping, have much in them analogous to the cryptomerias. Their fruit was an oblong cone, with scales loosely imbricated, cuneiform, or wedge-shaped, and commonly of from three to five obtuse lobes. In Fig. 80 we have a part of the stem, a branch with leaves and cone. In his 'Botanic Geography,' M. Lecoq thus describes the vegetation of the ancient world in the first period of the triasic age:—"While the variegated sandstones and mottled clays were being slowly deposited in regular beds by the waters, magnificent ferns still exhibited their light and elegantly-carved leaves. Divers *Protopteris* and majestic *Neuropteris* associated themselves in extensive forests, where vegetated also the *Crematopteris*

Fig. 80.—Branch and cone of Voltzia restored.

typica of Schimper, the *Anomopteris Mongeotii* of Brongn., and the pretty *Trichomanites myriophyllum* of Brongniart. The conifers of this epoch attain a very considerable development, and would form graceful forests of green trees. Elegant monocotyledons representing the forms of equatorial countries would show themselves for the first time, the *Yuccites Vogesiacus* of Schimp. would form groups at once thickly serried and of great extent.

"A family hitherto doubtful appears under the elegant form of *Nilsonia Hogardi*, Schimp.; *Ctenis Hogardi*, Brongn. It is still seen in the *Zamites Vogesiacus*, Schimp.; and the group of the Cycades share at once in the organization of the conifers and the elegance of the palms which now decorate the earth, while revealing in these forms its vast fecundity.

"Of the herbaceous plants which formed the undergrowth of the forests, or which luxuriated in its tepid marshes, the most remarkable is the *Œtheophyllum speciosum*, Schimp. Their organization approximates to the Lycopodiaceæ and Thypaceæ, the *Œtheophyllum stipulara*, Brongn., and the curious *Schizoneura paradoxa*, Schimp. Thus we can trace the commencement of the reign of the Dicotyledons with naked seeds, which afterwards become so widely disseminated. A few Angiosperms, composed principally of two families, the conifers and Cycadaceæ, still represented in the existing vegetation. The first, very abundant at first, associated themselves with the cellular Cryptogames, which still abound, although they are decreasing, then with the Cycadaceæ, which present themselves slowly, but will soon be observed to take a large part in harmonizing the vegetable kingdom."

The accompanying engraving (PL. XIII.) gives an idealized picture of the plants and animals of the period. The reader must imagine himself transported to the shores of the conchylian sea at a moment when its waves are agitated by a violent but passing storm. The reflux of the tide exposes some of the aquatic animals of the period. Some fine Encrinites are seen, with their long, flexible stems, and a few Mytulus and Terebratulas. The reptile which occupies the rocks, and prepares to throw itself on its prey, is the *Nothosaurus*. Not far from it are other reptiles, its congeners, but of a smaller species. Upon the down on the shore is a fine group of the trees of the period,

XIII.—Ideal landscape of the Conchylian Sub-period.

that is, of *Haidingeras*, with large trunks, with drooping branches and inclined foliage, of which the cedars of our age give some idea. The elegant *Voltzias* are seen in the second plane of this curtain of verdure. The reptiles which lived in these primitive forests, and which would give to it so strange a character, are represented by the *Labyrinthodon*, which descends towards the sea on the right, leaving upon the sandy shore those curious traces which have been so strangely preserved to our days, as if they were intended to answer the interrogations of science by their wonderfully-preserved vestiges.

CONCHYLIAN ROCKS.

The rocks of this period were composed of—

I. *Variegated sandstone*, which contains many vegetable, but few animal remains, although we constantly find imprints of the steps of the Labyrinthodon.

II. Beds of compact limestone, often greyish, sometimes black, alternating with marl and clay, which are sprinkled with such numbers of shells that the name of shelly limestone (*muschelkalk*) has been given to the formation by the Germans.

The conchylian formation shows itself in France, in the Pyrenees, around the central plateau in the Var, and upon both slopes of the Vosges. It is represented in Germany, in Belgium, in Switzerland, in Sardinia, in Spain, in Poland, in the Tyrol, in Bohemia, in Moravia, and in Russia. M. D'Orbigny states from his own inquiries that it covers vast surfaces in the mountainous regions of Bolivia, in South America. It is recognized in the United States, in Columbia, in the Great Antilles, and in Mexico.

The conchylian rocks in France are reduced to the variegated sandstone, except around the Vosges, in the Var, and the Black Forest, where it is accompanied by the muschelkalk. In Germany it furnishes building-stone of excellent quality : many great edifices, in particular the cathedrals, so much admired on the Rhine, such, for example, as those of Strasburg and Fribourg, are constructed of this stone: the

sombre tints of the stone singularly relieving the grandeur and
majesty of the Gothic architecture. Whole cities in Germany are
built of the brownisn-red stones drawn from its mottled sandstone
quarries. In England, in Scotland, and in Ireland this formation
extends from north to south through the whole length of the country.
" The old land," says Professor Ramsay, " consisted in great part of
what we now know as Wales and the adjacent counties of Herefordshire,
Monmouthshire and Shropshire, of part of Devon and Cornwall, Cumberland, and probably the Pennine chain and all the mountainous parts
of Scotland. Around old Wales, on three sides of Cumberland, and
probably all round and over great part of Devon and Cornwall the new
red sandstone was deposited: part at least of this oldest of the
secondary rocks was formed of the waste of the older palæozoic strata
that had then risen above the surface of the water. The new red
sandstone consists in its lower members of beds of red sandstone and
conglomerate, more than 1,000 feet thick, and above them are red
and green marls, chiefly red, which in Germany are called the keuper
strata, and in England the new red marl. These formations range
from the mouth of the Mersey, round the borders of Wales, to the
estuary of the Severn, eastward into Warwickshire, and thence
northward into Yorkshire and Northumberland, along the eastern
border of the magnesian limestone. They also form the bottom of
the valley of the Eden, and skirt Cumberland on the west. In the

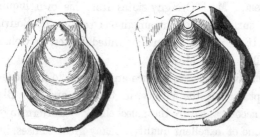

Fig. 81.—Pecten orbicularis.

centre of England the unequal hardness of its sub-divisions sometimes giving rise to minor escarpments, overlooking plains and undulating grounds of softer strata."

"Different members of the group rest in England, in some region or other," says Lyell, "on almost every member of the palæozoic series, on Cambrian, Silurian, Devonian, carboniferous and Permian, and there is evidence everywhere of disturbance, contortion, partial upheaval into land, and vast denudations which the older rocks underwent before and during the deposition of the new red sandstone group."

Among the most abundant of the shells belonging to the upper trias in all the countries where it has been examined are the avicula, cardiums and pectens, one of which is given in Fig. 81.

SALIFEROUS SUB-PERIOD.

The formation which characterizes the saliferous period is of moderate extent, and derives its name from the salt deposits of the period.

What is the origin of the great deposits of marine salt which occur in this formation, alternating constantly with thin beds of clay or marl ? We can only attribute them to the evaporation of vast quantities of sea-water fortuitously introduced into the depressions, cavities or gulfs, which the sandy downs afterwards separated from the great body of the sea. In PL. XIV. an attempt is made to represent the natural fact that during this period banks of immense extent existed, on which the very considerable masses of rock-salt were deposited which are now found in the rocks of the period. On the right is the sea, with a down of considerable extent, separating it from a tranquil basin of smooth water. At intervals, and from various causes, the sea, clearing the down, enters and fills the basin. We may even suppose that a gulf exists here which at one time communicated with the sea : the winds having raised this sandy down, the gulf is transformed by degrees into a basin closed on all sides. However that may be, it is pretty certain that if the waters of the sea were once shut up in this basin, with an argillaceous bottom without any opening, evaporation from the effects of solar heat would take place, and a bed of marine

salt would be the result of this evaporation, mixed with other mineral
salts, such as chloride of sodium, sulphate, magnesium and chloride of
potassium, which form sea-water.   This bed of salt left by the evaporation of the water would soon receive an argillaceous covering from the
mud suspended in the miry waters of the basin, thus forming a first
alternate bed of marine salt and of clay or marl.   The sea making
fresh breaches across the barriers, the same process takes place with
a similar result, until the basin is filled up.   By the regular and
tranquil repetition of this phenomenon, continued during a long
succession of ages, this abundant deposit of rock-salt has been formed
which occupies a position so important in the rocks of secondary formations.   On the opposite page (PL. XIV.) is an ideal representation of the
earth during the saliferous age, which explains at the same time the
origin of the rock-salt in this formation.   A theoretic section in the
foreground shows the salt beds as formed by the geological machinery
we have described.   The beds are enclosed obliquely, in consequence
of a movement of the soil subsequent to their deposit.

This hypothesis probably leaves the resulting stratification too much
to the chapter of accidents to account for the phenomena under consideration without other accessories ; which, however, are not wanting.

There is in the delta of the Indus a singular region, called the
Runn of Cutch, which extends over an area of 7,000 square miles,
which is neither land nor sea, but is under water during the monsoons,
and in the dry season is encrusted here and there with salt, about an
inch thick—the result of evaporation.   Dry land has been largely
increased here during the present century by subsidence of the waters
and upheavals by earthquakes. "That successive layers of salt may have
been thrown down one upon the other on many thousand square miles
in such a region is undeniable," says Lyell.   " The supply of brine
from the ocean is as inexhaustible as the supply of heat from the sun.
The only assumption required to enable us to explain the great thickness of salt in such an area is the continuance for an indefinite period
of a subsiding movement, the country preserving all the time a general
approach to horizontality."   The observations of Mr. Darwin on the

XIV.—Ideal landscape of Saliferous Sub-period.

atols of the Pacific prove that such a continuous subsidence is probable. Hugh Miller, after ably discussing various spots of earth where, as in the Runn of Cutch, evaporation and deposit take place, adds: "If we suppose that, instead of a barrier of lava, sand-bars were raised by the surf on a flat arenaceous coast, during a slow and equable sinking of the surface, the waters of the outer gulf might occasionally topple over the bar and supply fresh brine when the first stock had been exhausted by evaporation."

There is little to be said of the animals which belong to this period. They were nearly the same with those of the conchylian age.

In the saliferous age the islands and continents presented few mountains ; they were intersected here and there by large lakes, with flat and uniform banks. The vegetation on their shores was very abundant, and we possess its remains in great numbers. The saliferous flora consisted of ferns, equisetaceæ, cycades, conifers, and a few plants, which M. Ad. Brongniart classes among the dubious monocotyledons. Among the ferns may be quoted many species of *Sphenopteris* and *Pecopteris*. Among them, *Pecopteris Stuttgartiensis*, a tree with channeled trunk, which rises to a considerable height without throwing out branches, and terminates in a crown of leaves finely cut and with long petioles; the *Equisetites columnaris*, the great equisetum analogous to the horse-tails of our age, but of infinitely larger dimensions; its long fluted trunk surmounted by an elongated fructification dominating over all the other trees of the marshy soil.

The *Pterophyllum Jœgeri* and *Munsteri* represented the cycades, the *Taxodites Munsterianus* represented the conifers, and, finally, the trunk of the calamites was covered with a creeping plant, having elliptic leaves, with a recurving nervature borne upon its long petioles, and the fruit disposed in bunches: this is the *Presteria antiqua*, a doubtful monocotyledon, according to Brongniart, but M. Unger places it in the family of *Smilax*, of which it will thus be the first representative. The same botanist classes with the canes a marshy plant very common in this

N

period, the *Palæoxyris Munsteri*, which Brongniart classes with the
Presteria, among his doubtful monocotyledons.

The vegetation of the triasic period is thus characterised by Lecoq,
in his 'Botanical Geography:'—" The cellular *Cryptogamæ* predominate
in this as they do in the carboniferous epoch, but the species have
changed, and many of the genera also are different; the *Chladephlebis*,
the *Sphenopteris*, the *Coniopteris*, and *Pecopteris* predominate over the
others in the number of species. The Equisetaceæ are more developed
than in any other formation. One of the finest species, the *Calamites
arenaceus* of Brongniart, must have formed great forests. The fluted
trunks resemble immense columns, terminating at the summit in leafy
branches, disposed in graceful verticillated tufts, foreshadowing the
elegant forms of our *Equisetum sylvaticum*, besides growing alongside
of curious Equisetum and singular Equisetites, a species of which,
E. columnaris, raised its herbaceous stem to a great height with its
sterile articulations.

" What a singular aspect these ancient rocks would present, if we
add to them the forest-trees *Pterophyllum* and the *Zamites* of the fine
family of cycadaceæ, and the conifers, which seem to have made their
appearance at the same time in the humid soil!

" It is during this epoch, while yet under the reign of the dicotyledonous angiosperms, that we discover the first true monocotyledons.
The *Presteria antiqua*, with its long petals, drooping and creeping
round the old trunks, its bunches of bright-coloured berries like the
Smilax of our own age, to which family it appears to have belonged.
Besides, the triasic marshes gave birth to tufts of *Palæoxyris Munsteri*,
a cane-like graminæ, which, in all probability, cheered the otherwise
gloomy shore.

" During this long period the earth preserved its primitive vegetation; new forms are slowly introduced, and they multiply slowly. But
if our present types of vegetation are deficient in these distant epochs,
we ought to recognize also that the plants which in our days represent
the vegetation of the primitive world are often shorn of their grandeur.
Our Equisetaceæ and Lycopodiaceæ are but poor representatives of the

Lepidodendrons; the Calamites and Asterophyllites had already run their race before the epoch of which we write."

The principal features of triasic vegetation are represented in PL. XIV.: on the cliff, on the left of the ideal landscape, the graceful stems and lofty trees are groups of *Calamites arenaceus*: below are the great "horse-tails" of the epoch, *Equisetum columnaris*, a slender tapering species, of soft and pulpy consistence, which, rising erect, would give a peculiar physiognomy to the solitary shore.

SALIFEROUS ROCKS.

These rocks consist of a vast number of argillaceous and marly beds, irregularly coloured, but chiefly red, with a dash of yellow, black and green. These are the colours which in earlier days gave the name of *variegated marl* to the series. These beds of red marl often alternate with sandstones which are also variegated in colour. As subordinate rocks we find in this formation some deposits of a poor pyritic coal and of gypsum. But what especially characterises the rocks are the formidable beds of rock-salt which are included. These saliferous beds, often twenty-five to forty feet thick, alternate with beds of clay, the whole attaining a thickness of 160 yards. In Germany, in Wurtemberg, in France, at Vic and at Dieuze, in the Meurthe, the rock-salt of the saliferous formation have become important branches of industry. In the Jura, salt is extracted from the water, charged with chlorides, which issue from this formation.

Some of these deposits are placed very deep in the soil, and cannot be reached without very considerable labour. The salt-mines of Vielizka, in Poland, for example, can be procured on the surface, or by galleries of little depth, because the deposit belongs to the tertiary period; but the deposits of salt in the triasic age are placed so much deeper as to be only approached by a regular process of mining by galleries, and the ordinary mode of reaching the salt is by digging pits, which are afterwards filled with water. This water, charged with marine salt, is then pumped up into

troughs, where it is evaporated, and the crystallized mineral obtained
by deposit.

The saliferous formation presents itself in Europe on many points,
and it is not difficult to follow its traces.   In France it appears in the
department of the Indre, of the Cher, of the Allier, of the Nièvre, of
Saône-and-Loire; upon the western slopes of the Jura its outliers crop
out near Poligny and Salins upon the western slopes of the Vosges; in
the Doubs it shows itself; then it skirts the conchylian area in the
Haute-Marne; in the Vosges it assumes large proportions in the
Meurthe at Luneville and Dieuze; in the Moselle it extends northward
to Bouzonville, and on the Rhine, to the east of Luxembourg, as far as
Dockendorf.   Some traces of it show themselves upon the eastern
slopes of the Vosges on the lower Rhine.

It shows itself again in Switzerland and in Germany, in the canton of
Basle, in the Argovie, in the Grand Duchy of Wurtemberg, and in the
Tyrol, where it gives its name to the city of Saltzbourg.

In the British Islands the saliferous deposits commence in the
eastern parts of Devonshire, and a band, more or less regular, is traced
into Somersetshire, through Gloucestershire, Worcestershire, Warwick,
Leicestershire, Nottingham, up to the banks of the Tees in Yorkshire,
with a bed independent of all the others in Cheshire, which extends
into Lancashire.   " At Nantwich, in the upper trias of Cheshire," says
Sir Charles Lyell, " two beds of salt, in great part unmixed with
earthy matter, attain the thickness of 90 or 100 feet.   The upper
surface of the highest bed is very uneven, forming cones and irregular
figures.   Between the two masses there intervenes a bed of indurated
clay, traversed by veins of salt.   The highest bed thins off towards the
south-west, losing fifteen feet of its thickness in the course of a mile,
according to Mr. Ormerod.   The horizontal extent of these beds is not
exactly known, but the area containing saliferous clay and sandstones
is supposed to exceed 150 miles in diameter, while the total thickness
of the trias in the same region is estimated by Mr. Ormerod at 1700
feet.   Ripple-marked sandstones and the footprints of animals are
observed at so many levels that we may safely assume the whole area

to have undergone a slow and gradual depression during the formation of the new red sandstone."

Not to mention the importance of salt as a source of health, it is in Great Britain, and, indeed, all over the world where the saliferous rocks exist, a most important branch of industry.   The quantity of the mineral produced in England from all sources is between five and six hundred thousand tons annually, and the population engaged in producing the mineral from sources supposed to be inexhaustible, upwards of twelve thousand.

If the saliferous formation is poor in organic remains in France, it is by no means so on the other side of the Alps.   In the Tyrol, and in the remarkable beds of Saint Cassian, Aussec, and Hallstadt, the rocks are stony, with an immense number of marine fossils : among them, Cephalopodes, Ceratites, and Ammonites of peculiar form, having the lobes shaped and cut like the finest needle.   The Orthoceras, which we have seen abounding in the Silurian period, and continued during the deposit of the Devonian and carboniferous periods, appears here for the last time.   We still find here a great number of Gasteropodes and of Lamellibranches of the most varied form, the latter representing the family of Productus.   Sea Urchins—Polypiers of elegant form—seem to have occupied, on the other side of the Alps, the same seas which in France and Germany seem to have been nearly

Fig. 82.—Producta Martini.   Fig. 83.—Patella vulgata.

destitute of animals.   Some beds are literally formed of accumulated shells belonging to the *Avicula*.

In following the grand mountainous slopes of the Alps and Carpathians we discover the saliferous rocks by this remarkable accumulation of Aviculæ. The same facies presents itself under the identical conditions in Syria, in India, in New Caledonia, in New Zealand, and in Australia. It is not the least curious part of this period, that it presents on one side the site of the Alps, which were not yet raised, an immense accumulation of sediment, charged with gypsum, rock-salt, &c., without organic remains, while beyond, a region presents itself equally remarkable for the extraordinary accumulation of the remains of marine mollusks. Among these were *Myophoria lineata,* which is often confounded with Trigonias, *Patella lineata,* and *Stellispongia variabilis.*

France, at this period, was still the skeleton of what it has become. A map of the country represents the primitive rocks occupying the site of the Alps, the Cevennes, and the Puy-de-Dôme, the country round Nantes, and the highlands of Brittany. The transition rocks reach the foot of the Pyrenees, the Contentin, the Vosges, and the Eifel Mountains. Some bands of coal stretch away from Valenciennes to the Rhine, and on the north of the Vosges; these mountains themselves being chiefly triasic.

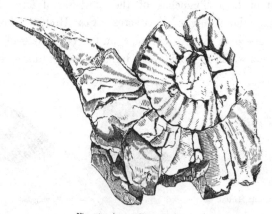

Fig. 84.—Ammonites rostratus.

JURASSIC PERIOD.

THIS period, one of the most important in the physical history of the globe, has received its name from the Jura mountains in France, the Jura range being composed of the rocks deposited by the seas of the period. In the term Jurassic, the formations designated as the "Oolite" and "Lias" are included, both being found in the Jura mountains. The Jurassic period presents a very striking assemblage of characteristics, both in its vegetation and in the animal remains which it contains: many genera of animals belonging to the preceding age have disappeared: new genera have replaced them, comprising a very specially organised group, containing not less than four thousand species.

The Jurassic period is subdivided into two sub-periods : that of the *Lias* and the *Oolite*.

THE LIAS

Is an English provincial name given to an argillaceous limestone, with marl and clay, which forms the base of the Jurassic formation, and passes almost imperceptibly into the lower oolite in some places, where the marlstone of the lias partakes of the mineral character, as well as the fossil remains of the lower oolite: and it is sometimes treated as belonging to that formation. "Nevertheless," says Sir Charles Lyell, "it may be traced throughout a great part of Europe as a separate and independent group of considerable thickness, varying from 500 to 1000 feet, containing many peculiar fossils, and having a very uniform lithological aspect." The rocks which represent the liasic period form the base of the system, and have a mean thickness of about 300 feet. In the inferior part we find sand, quartzose sandstones, which are called the sandstone of the lias, and comprehend the greater part of *Guadersandstein*, or building-stone of the Germans; above it comes some compact limestone, argilliferous, bluish and yellowish; finally the formation terminates in the marlstones, sometimes sandy, and occasionally bituminous.

The lias is generally divided into three formations: 1. the lower; 2. the middle; and 3. the upper; but these have been again subdivided —the first into six zones, each marked by its own group of fossils; the second into three zones; and the third consisting of clay, shale and thin beds of limestone, over which come sandstones. For the purposes of description we shall divide the lias into four groups :—

1. *Lower lias*, which is well developed in the Alps of Lombardy and the Tyrol, in Luxembourg and in France, characterized by abundant remains of *Avicula conterta*.

2. *Graphite limestone*, or lias below *Gryphæa arcuata*, with sandy banks at the base. Independently of the *Gryphæa arcuata*, we find *Ammonites Bucklandi*.

3. The *middle lias*, formed by numerous banks, enclosing, among other fossils, *Pentacrinites, Belemnites paxillosus, Gryphæa obliqua, Saurians*, &c.

4. The *upper lias*, consisting of marlstone, surmounted by a bed of oolitic ironstone generally worked. These marls form powerful beds: the fossils are *Ammonites bifrons, Brachyphyllum, Belemnites tripartitus*. The bed of ironstone is generally remarkably rich in fossils.

In France the lias abounds in the Calvados, in Burgundy, Lorraine, Normandy and the Lyonnais. In the Vosges and Luxembourg M. E. de

Beaumont states that the lias containing *Gryphæa arcuata* and *Plagiostoma gigantea*, and some other fossils, becomes arenaceous: and around the Hartz mountains, in Westphalia and Bavaria, in its lower parts the formation is sandy, and is sometimes a good building-stone.

"In England the lias constitutes," says Professor Ramsay, "a well-defined belt of strata running continuously from Lyme Regis, on the south-west, through the whole of England, to Yorkshire on the north-east, and

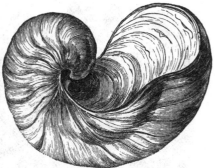

Fig. 85.—Gryphæa incurva.

is an extensive series of alternating beds of clay, shale, and limestone, with occasional layers of jet. The unequal hardness of the clays and limestones of the liasic strata causes some of its members to stand out in distinct minor escarpments, often facing the west and north-west. The marlstone forms the most prominent of these, and overlooks the broad meadows of the lower lias clay, which forms much of the centre of England." In Scotland there are few traces of the lias. Zoophytes, mollusks and fishes of a peculiar organization, but, above all, reptiles of extraordinary size and structure gave to the sea of the liasic period an interest and features quite peculiar. Well might Cuvier exclaim, when the drawings of the Plesiosaurus were sent to him, "Truly this is altogether the most monstrous animal that has yet been dug out of the ruins of a former world." In the whole of the English lias there are about 243 genera, and 467 species of zoophytes and mollusks. The whole series of rocks have been divided into zones characterized by particular ammonites which are found to be limited to them.

Among the zoophytes belonging to the lias we may cite *Asteria lombricalis* and *Palæocoma Fustembergii*, which constitute a genera not dissimilar to the star-fishes, of which its radiated form reminds us.

The *Pentacrinus fasciculosus* is another zoophyte of this epoch, which ornaments many collections by its elegant form. It belongs to the order of Crinoidæ, which is represented in the present epoch by the

Fig. 86.—Pentacrinus fasciculosus. Half natural size.

Medusæ, one of the rare and delicate zoophytes of our seas. This fine zoophyte is represented in Fig. 86.

The oysters made their appearance in the last period, but only in a small number of species ; they increased greatly in number in the liasic seas.

The *Ammonites*, a curious genera of mollusks, which we have seen in small numbers in the preceding age of the trias, become quite special in the secondary epoch, and disappear altogether before our age. They are characteristic of the Jurassic period, and, as we have already said, each zone is characterized by its peculiar species. The name is taken from the resemblance of the shell to the ram's-horn ornaments which decorated the front of the temple of Jupiter Ammon and the bas-reliefs of the statues of this pagan deity. They were cephalopode mollusks with circular shells, winding in

Fig. 87.—Lituites articulatus Convoluted so that volutions touch on all inner parts.

spirals on the same plane, and divided into a series of chambers. The animal only occupied the outer cavities of the shell; all the others were void. A tube is-suing from the first tra-versed all the cavities, as is seen in the sec-tion of *Lituites arti-culatus*, in Fig. 87. This tube enabled the animal to rise to the surface, or sink to the bottom, for the Am-monite could at plea-sure fill the chambers or expel the water, thus rendering it lighter or heavier as occasion

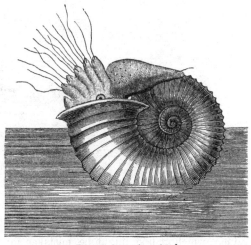

Fig. 88.—Ammonite restored.

required. The nautilus of our seas is provided with the same curious organization, and reminds us forcibly of the Ammonites of geological times.

Shells are the only traces which remain of the Ammonites. We
have no exact knowledge of the animal which occupied and built them.
The attempt at its restoration, as exhibited in Fig. 88, will probably
convey a fair idea of the living Ammonite. We assume that it resembled the nautilus of our days. Like a little sculler, the Ammonite
floated on the surface of the waters; like the nautilus, the shell was an
animated skiff. What a curious aspect these primitive seas must have
presented, covered by myriads of these mollusks of all sizes, rowing
about in eager pursuit of their prey !

The Ammonites presented themselves in the Jurassic age in a great
variety of forms, and of all sizes; some of them of great beauty.
Ammonites bifrons, *A. nodatianus*, *A. bisculatus* and *A. margari-
tatus*, come under this description.

The *Belemnites* were also molluscous cephalopodes of a very
curious organization, which would probably appear in great numbers,
and for the first time in the Jurassic seas. Of this mollusk we only

Fig. 89.—Belemnite restored.

possess a fossilized bone, analogous to that of the cuttle-fish and the
calmar, which at first might be mistaken for a petrified wand. This
simple and mute ruin is very far from giving us any exact idea of
what the animal was to which the name of Belemnite was given.
The little bone was long, slender and dart-like, whence their name.
When first discovered no better name could be found for them than

"Ladies' fingers." They were at last judged, from their surroundings, to be the bony processes of some primitive cuttle-fish. The only part left to us is the terminal part of the Belemnite, probably once surrounded with flesh. Unlike the Ammonite, which floated on the surface and sunk to the bottom at pleasure, the Belemnite, it has been thought, could swim or creep at the bottom of the sea.

In the preceding page (Fig. 89) is the living Belemnite, restored according to Professor Buckland and the celebrated palæontologist, Professor Owen, in which the terminal part of the animal is marked in a darker tint to indicate the place where the bone comes which alone represents in our days this fossilized being. A pretty exact idea of this mollusk may be arrived at by looking at the cuttle-fish. Like the cuttle-fish, the Belemnite secreted a black

Fig. 90.—Belemnites canaliculatus.

liquid, a sort of ink or sepia ; and the fossil has actually been found with the ink dried up, and one geologist has executed a drawing with the ink of this sepia, aged probably many thousands of years.

Among -the Belemnites characteristic of the liasic period, may be cited *B. acutus, pistiliformis* and *sulcatus*, among the acephalous mollusks. The seas of the period contained a great number of the fishes called *ganoïds ;* that is to say, with hard and glittering scales. *Lepidotus gigas* was a fish of great size belonging to this age. A smaller fish was the *Tetragonolepis,* or *Œchmodus Buchii.* The *Acrodius nobilis,* of which the teeth have been recovered, was a fish of which no other part of the skeleton has been preserved. Neither are we better informed as to *Hybodus reticulatus.* The osseous spines which form the anterior part of the dorsal fin of this ancient fish had long been an object of curiosity to geologists under the general name of *Ichthyodorulites,* before it was established that it was a fragment of the fin of the *Hybodus.* The teeth of *Acrodius nobilis* were long known popularly as the " fossil teeth," and the Ichthyodorulites were supposed by some naturalists to be the jaw of some animal, by others, weapons like those of the living *Balistes* or *Silurus;* but Agassiz has shown them to be neither the one nor the other, but bony spines on the fin, like those of the living genera of *Cestracions* and *Chimæras*, in both of which the concave face is armed with small spines like the *Hybodus*. The spines were simply embedded in the flesh, and attached by strong muscles. "They served," says Dr. Buckland, "as in the *Chimæra*, to raise and depress the fin, their action resembling that of a moveable mast lowering backward."

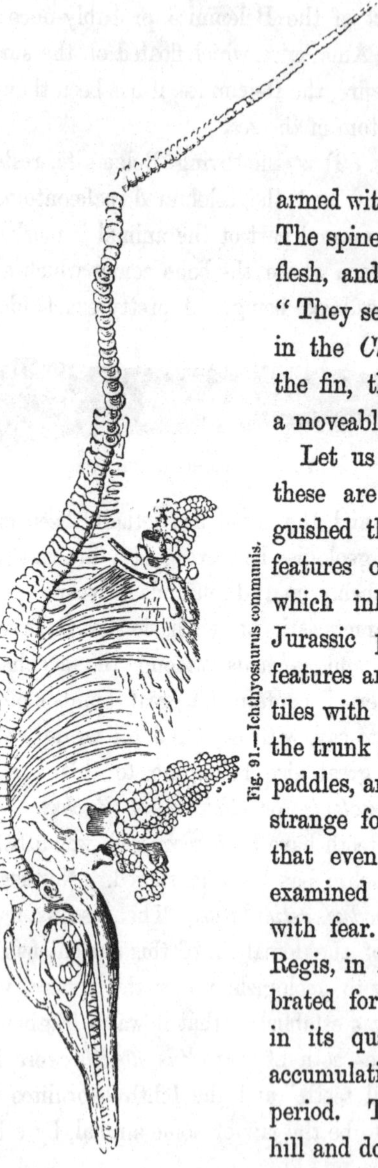

Let us hasten to say, however, that these are not the beings that distinguished the age, and were the salient features of the generation of animals which inhabited the earth during the Jurassic period. These distinguishing features are found in the enormous reptiles with lizard's head, crocodile's teeth, the trunk and tail of a quadruped, whale's paddles, and serpent-like head; and this strange form on such a gigantic scale that even its inanimate remains are examined with a curiosity not unmixed with fear. The country round Lyme Regis, in Dorsetshire, was already celebrated for the curious fossils discovered in its quarries, formed in the muddy accumulations of the quiet sea of the liasic period. The country is hilly—up one hill and down another, is a pretty correct

Fig. 91.—Ichthyosaurus communis.

description of the walk from Bridport to Lyme Regis, where slept the sleep of stones the most frightful creatures the living world has ever imagined. The quarries of Lyme Regis were the cemetery of the Ichthyosaurii, the sepulchre which concealed these dragons of the ancient seas.

In 1811 a simple country girl, who made her precarious living by picking up fossils for which the neighbourhood was already famous, was pursuing her avocation, hammer in hand, when she perceived some bones projecting a little from the cliff. On examination she found it was part of a large skeleton: she cleared away the rubbish and traced the whole creature on the block of stone. She hired workmen to dig out the block of lias in which it was embedded. In this manner was the first of these monsters brought to light: "a monster some thirty feet long, with jaws nearly a fathom in length, and huge saucer eyes; which have since been found so perfect, that the petrified lenses have been split off, as a writer in 'All the Year Round' assures us, and used as magnifiers."

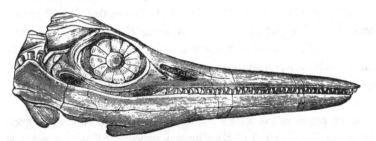

Fig. 92—Head of Ichthyosaurus platydon.

In the structure of the head the lizard form has been particularly observed. (In Fig. 92, the head of *I. platydon* is represented.) As in the Saurians, the nostrils approach the eyes; but, on the other hand, the form and arrangement of the teeth approach to the crocodiles: they are conical, but not set deep or separate in their sockets; they are only ranged in a long and continuous trench hollowed in the bones of the jaw. These jaws have an enormous opening; for in some instances they have been found armed with a hundred and sixty teeth. Let us

add, that teeth lost through the voracity of the animal, or in contests
with other animals, could be replaced many times ; for above every tooth
there is the germ of a new one, and we have described a sufficiently
formidable battery of teeth.

The eyes of this marine Colossus were unusually large; in volume they
frequently exceed the human head, and their structure was one of its
most remarkable peculiarities. Before the orbit of the eye there exists
a series of thin, circular, bony plates, which surround the opening of
the pupil. This apparatus, which exists in the eyes of some birds,
and in those of the turtle and lizard, can be used so as to increase or
diminish the curvature of the transparent cornea, and thus increase or
diminish the magnifying power, according to the requirements of
the animal; performing the office, in short, of a telescope or microscope
at pleasure. The eyes of the Ichthyosaurus were, then, an optical apparatus of wonderful power and of singular perfection. They gave the
animal the power of seeing its prey far and near, and of pursuing it in
the darkness and in the bosom of the deep. The curious osseous plate
we have described furnishes, besides, in its vast ocular globe the power
necessary to support the pressure of a considerable weight of water,
as well as the violence of the waves, when the animal, coming to the
surface to breathe, raised its head above the surface. This magnificent
specimen of the fish lizard, or Ichthyosaurus, as it was named by Dr.
Ure, now forms part of the treasures of the British Museum.

At no period in the earth's history have reptiles occupied so important a place as they did in the Jurassic period. All nature seems to
have been prepared for their appearance. They are as complicated in
their structure as the mammifera which succeeded them. They lived
in the open sea, but they seem to have sought the shore from time to
time: they crawled along the beach, covered with a soft skin, probably
not unlike some of our Cetaceæ, for the Ichthyosaurus may be considered as the whale of its time. Its voracity would be prodigious,
for Dr. Buckland describes a specimen which had under its sides, in
the place where the stomach might be supposed to be placed, the
skeleton of a smaller one—a proof that this monster, not contented

with preying on his weaker neighbours, was in the habit of devouring its own species. In the same waters with him lived the Plesiosaurus, with long neck and form more fabulous than any monster the mind of man had conceived: yet these potentates of the seas are warmed by the same sun and upon the same banks, in the midst of a vegetation not unlike that which the climate of Africa now produces. These strange and gigantic Saurians seem to have disappeared during the following geological period, for, although they have been discovered as low as the trias in Germany, and as high as the chalk in England, they probably only appeared as stragglers in both these epochs; and the Saurians which succeeded them are, so to speak, only the shadows of their powerful congeners of the ancient world.

Confining ourselves to well-established facts, we shall consider in some detail the best known of these fossil reptiles.

The extraordinary creature which bears the name of *Ichthyosaurus*, from the Greek words Ιχθυς σαυρος, fish-lizard, presents certain dispositions and organic arrangements which are met with dispersed

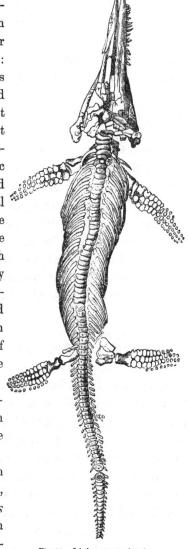

Fig. 93.—Ichthyosaurus platydon

O

in certain classes of animals now living, but they never seem to be again reunited in any one.   It possesses the muzzle of a porpoise, the head of a lizard, the jaws and teeth of a crocodile, the vertebræ of a fish, the sternum of the Ornithorhynchus, the paddles of a whale, and the trunk and tail of an ordinary quadruped.

Bayle appears to have given the best idea of the Ichthyosaurus by describing it as the whale of the Saurians—the Cetacea of the primitive seas.   It was, in short, an animal exclusively marine: on shore it would be an immovable, inert mass: its paddles and fish-like vertebræ, the length of the tail and other parts of its structure, prove that its habits were aquatic, as the remains of fishes and reptiles, and the form of its teeth, show that it was carnivorous.   Like the whale, also, the Ichthyosaurus respired atmospheric air; so that it was under the necessity of coming frequently to the surface of the water, like that inhabitant of the deep.   We can even believe with Bayle that, like the whale, it was provided with vents or blowers, through which it ejected the water it had swallowed in columns into the air.

The dimensions of the Ichthyosaurus varied with the species, of which five are known and described.   These are *I. communis, I. platydon, I. intermedius, I. tenuirostris,* and *I. Cuvierii,* the largest being over thirty feet in length.

The short thick neck of the Ichthyosaurus supported a voluminous head, which continued backwards from behind the eyes in a vertebral column of more than a hundred vertebræ.   The animal being created, like the whale, for rapid movement through the water, its vertebræ had none of the invariable solidity of the lizard or crocodile, but rather the structure and lightness of that of fishes.   The section of these vertebræ presents two hollow cones connected by their summit at the centre of the vertebræ, which would permit of the utmost flexibility of movement. The sides extended themselves in all the length of the vertebral column from the head to the pelvis.   The bones of the os sternum, or that part of the frame which supported the paddles, present the same combinations with those of the sternum in the Ornithorhynchus of New Holland, an animal which presents the singular combination of a mammiferous furred quadruped with the mouth of a duck and palmated feet; which dived to the bottom of the water in search of its food, and

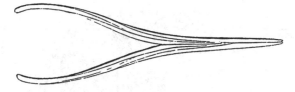

Fig. 94.—Lower jaw of Ichthyosaurus. (Dr. Buckland.)

returned to the surface to breathe the air. In this phenomenon of living nature the Creator seems to have repeated in our days the organic arrangements which appear once before in the Ichthyosaurus.

In order that the animal should be able to move with rapidity in the water, both the anterior and posterior members consisted of fins or paddles, the hands or anterior fins being much the most powerful, being made up of about a hundred bones, of polygonal form, disposed in series representing the phalanges of the fingers. The hand, jointed at the arm, has a great resemblance to the paddles, without distinct fingers, of the porpoise and the whale. A specimen of the posterior fin of *I. communis*, discovered at Barrow-on-Suir, in Leicestershire, in 1840, by Sir Philip Egerton, exhibited on its posterior margin the remains of cartilaginous rays, which bifurcated as they approached the edge, like those in the fins of a fish. "It had previously been supposed," says Professor Owen, "that the locomotive organs were enveloped, while living, in a smooth integument, like that of the turtle and porpoise, which has no other support than is afforded by the bones and ligaments within; but it now appears that the fin was much larger, expanding far beyond the osseous frame-work and deviating widely in its fish-like rays from the ordinary reptilian type." The Professor believes that, besides the fore-paddles, these stiff-necked Saurians were furnished with a tail-fin without radiating bones, and purely tegumentary, expanding vertically to assist them in turning, and not horizontally as in the whale. It is obvious that the Ichthyosaurus was an animal powerfully armed for offence and defence. We cannot say with certainty whether

o 2

the skin was smooth, like that of the whale or lizard, or covered with scales, like the great reptiles of our own age. Nevertheless, as the scales

Fig. 95.—Skeleton of Ichthyosaurus.
Containing teeth and bones of fish in the coprolite form. One-fifteenth natural size.

of the fishes and the cuirass and horny armour of the reptiles of the lias are preserved, and as no such defensive scales have been found belonging to the Ichthyosaurus, it is probable that the skin was smooth.

It is curious to see the exact knowledge which we possess of the antediluvian animals, their habits, and their economy.   Fig. 95

Fig. 96.—Coprolite, enclosing bones of small Ichthyosaurus.

represents the skeleton of an Ichthyosaurus found in the lias of Lyme Regis, which still retained in its abdominal cavity the coprolites, that is to say, the residue of digestion.   The softer parts of the intestinal tube have disappeared, but the *fœces* themselves are preserved, and their examination informs us as to the alimentary regimen of this animal

which has perished from the earth many thousands, perhaps millions of years.   Mary Anning, a simple English girl, to whom we owe most of the discoveries made in the neighbourhood of Lyme Regis, her native place, had in her collection an enormous coprolite of the Ichthyosaurus. This coprolite (Fig. 96) contained some bones and scales of fishes, and of divers reptiles, well enough preserved to have their species identified.   It only remains to be added that, among these bones, those of the Ichthyosaurus were often found, especially the bones of young individuals. The presence of these remains of animals of the same species in the digestive canal of the Ichthyosaurus proves, as we have already had occasion to remark, that this great Saurian was a most voracious monster, since it habitually devoured individuals of its own race. The structure of the jaw leads us to believe that the animal swallowed its prey without dividing it. Its stomach and intestinal tube must, then, have formed a sort of pouch of great volume, filling entirely the abdominal cavity, and responding in extent to the great development of the teeth and jaws.

The perfection in which the contents of the intestinal cavity has been preserved in the fossilized coprolites furnishes indirect proofs that the intestinal tube of the Ichthyosaurus resembles closely that of the shark and the dog-fish—fishes essentially voracious and destructive, in which the intestinal muscle is spiral and contorted, an arrangement which is exactly indicated in some of the coprolites of the Ichthyosaurus, as is evident from the impressions of the intestinal muscle left on the coprolite of which Fig. 97 is a copy.

Fig. 97.—Coprolite of Ichthyosaurus.

It will be readily comprehended that the coprolites, being muffled up as it were in phosphate of lime, a mineral which is indestructible in its nature, would be well preserved. On the coast near Lyme Regis, coprolites are abundant in the liasic formation, and have been found in heaps disseminated over many miles of coast.

Some readers will be surprised that we should dwell on a subject so obscure, and apparently so uninteresting, as the structure of the intestines of extinct reptiles. It will relieve us from this reproach to find, in beings so remote from our times, the same system of organs which belong to modern animals, and to trace them even in organs so perishable as the intestinal tube, composed entirely of soft parts non-adherent to the skeleton. Is it not establishing, by the similitude of the organic apparatus, the continuity of a chain which was to all appearance broken? Is it not establishing the unity and visible continuity of design in the works of creation, through the many ages which have glided away? What an admirable privilege of science, which is able, by an examination of the simplest parts in the organization of beings which lived ages ago, to give to our minds such solid teachings and such true enjoyments! "When we discover," says Dr. Buckland, " in the body of an Ichthyosaurus the food which it has engulphed an instant before its death, when the intervals between its sides present themselves still filled with the remains of fishes which it had swallowed some ten thousand years ago, or a time even twice as great, all these immense intervals vanish, time disappears, and we find ourselves, so to speak, thrown into immediate contact with events which

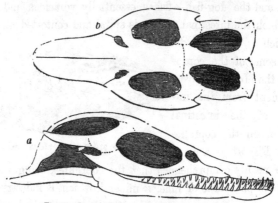

Fig. 98.—Skull of Plesiosaurus restored. (Conybeare.)
a, profile; *b*, seen from above.

took place in epochs immeasurably distant, as if we occupied ourselves with the affairs of the previous day."

The name of *Plesiosaurus*, from the Greek words πλησίος, near, and σαῦρος, lizard, reminds us that this animal also is neighbour by its organization to the Saurians, and consequently to the Ichthyosaurus.

The Plesiosaurus presents in its structure and organs the most curious assemblage we have met with among the organic vestiges·

of the ancient world. One author has compared it to a serpent hid under the carapace of a turtle. Let us remark, however, that there is here no carapace. The Plesiosaurus has the head of a lizard, the teeth of a crocodile, a neck of commensurate length and resembling the body of a serpent, the sides of a chameleon, a trunk and a tail whose proportions are those of an ordinary quadruped, and, finally, the fins of a whale. Let us throw a glance at the remains of this strange animal which the earth has revealed, and which science has restored to us.

The head of the Plesiosaurus presents a combination of the characters proper to the Ichthyosaurus, to the crocodile and the lizard. Its neck includes a greater number of vertebræ than either the camel, the giraffe, or even the swan, which of all the feathered race has the longest neck in comparison to the rest of the body. And in birds it is to be remarked that, contrary to what obtains in the mammifera, where the vertebræ of the neck are always seven, the vertebræ in birds increases with the length of the neck.

The trunk is cylindrical and rounded, like that of all the great marine turtles. It was doubtless uncovered, having neither the scales nor carapace in which some authors have clothed it, for no vestiges of such coverings have been found near either of the skeletons which have been discovered. The dorsal vertebræ attach

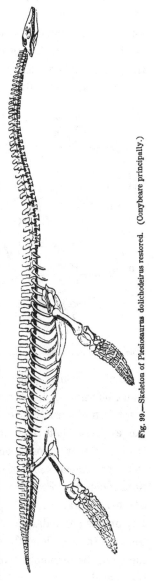

Fig. 99.—Skeleton of Plesiosaurus dolichodeirus restored. (Conybeare principally.)

themselves one to the other by plain surfaces like those of terrestrial quadrupeds, which deprives the whole of its vertebral column of much of its flexibility. Each pair surrounds the body as with a girdle formed of five pieces, as in the chameleon and iguana; from thence, no doubt, as with the chameleon, great facilities existed for the contraction and dilatation of the pulmonary organs.

The breast, the pelvis, and the bones of the anterior and posterior

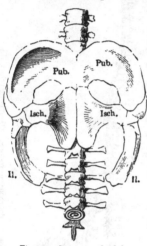

Fig. 100.—Sternum and pelvis.
Pub., pubis; Isch., ischium; Il., ilium.

extremities concurred in furnishing apparatus which permitted the Plesiosaurus, like the Ichthyosaurus and all the living Cetacea, to descend into the water and return to the surface at pleasure. Professor Owen, in his 'Report on British Reptiles,' characterizes them as air-breathing and cold-blooded animals; the proof that they respired atmospheric air immediately, being found in the position and structure of the nasal passages, and the bony mechanism of the thoracic duct and abdominal cavity. In the first, the size and position of the external nostrils, combined with the structure of the paddles,

point at a striking analogy between the extinct Saurians and the Cetaceans, offering, as the Professor observes, "a beautiful example of the adaptation of structure to the peculiar exigencies of species." While the evidence that they were cold-blooded animals is found in the flexible or unanchylosed condition of the osseous pieces of the occiput and other cranial bones of the lower jaw, and of the vertebral column; from which the Professor draws the conclusion that the heart was adapted for transmitting a part only of the blood through the respiratory organs; the absence of the ball and socket articulations of the bones of the vertebræ, the position of the nostrils near the summit of the head, the numerous short and flat digital bones, which must have been enveloped in a simple undivided integumentary sheath, forming

in both fore and hind extremities a fin closely resembling that of the
living Cetacea. The tail is relatively much shorter than in the

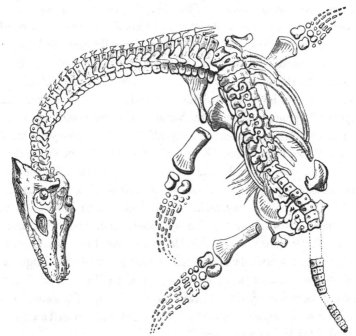

Fig. 101.—Remains of Plesiosaurus macrocephalus. One-twelfth natural size.

Ichthyosaurus, and there is an obvious reason for the curtailment in
the length and flexibility of the neck, which renders a like length and
flexibility in the tail unnecessary : at the same time the paddles are
larger and more powerful, as if to compensate for absence of power in
the tail : the latter, in short, was more calculated to act the part of a
rudder, in directing the course of the animal through the water, than a
propelling power.

Such were the strange combinations of form and structure in the
Plesiosaurus and Ichthyosaurus, genera of animals whose remains
have, after an interment extending to unknown thousands of years,
been revealed to light and submitted to examination—nay, rebuilt,
bone by bone, until we have the complete skeleton before us, and its

habits described as if they had been observed in life.   Conybeare thus
speaks of the supposed habits of this extinct form which he had built
up from scanty materials: "That the Plesiosaurus is aquatic was
evident from the form of its paddles; that it was marine is equally
so, from the remains with which it is universally associated; that it
may have occasionally visited the shore, the resemblance of its extremities to the turtle may lead us to conjecture; its motion, however, must have been very awkward on land; its long neck must have
impeded its progress through the water, presenting a striking contrast
to the organization which so admirably fits the Ichthyosaurus for
cutting through the waves.   May it not, therefore, be concluded that it
swam on or near the surface, arching back its long neck like the swan,
and occasionally darting it down at the fish which happened to float
within its reach ?   It may, perhaps, have lurked in shallow water along
the coasts, concealed among the sea-weeds, and, raising its nostrils
to the surface from a considerable depth, may have found a secure
retreat from the assaults of dangerous enemies, while the length and
flexibility of its neck may have compensated for the want of strength
in its jaws, and incapacity for swift motion through the water, by the
suddenness and agility of the attack they enabled it to make on every
animal fitted to become its prey."

  The Plesiosaurus was first described by the Rev. W. D. Conybeare
and Sir Henry de la Beche, in the 'Geological Society's Transactions'
for 1821, and a restoration of *P. dolichodeirus*, the most plentiful of
these fossils appeared in the same work for 1824.   The first specimen
was discovered, as the Ichthyosaurus had been, in the lias of Lyme
Regis; since then other individuals and species have been found in the
same geological formation in various parts of England, Ireland, France,
and Germany, and with such variations in their structure that
Professor Owen has felt himself justified in recording sixteen distinct
species, of which we have represented *P. dolichodeirus* (Fig. 99), as
restored by Conybeare, and *P. macrocephalus* (Fig. 101), with its skeleton, as moulded from the rock of Lyme Regis, which has been placed
in the Gallery of Palæontology of the Museum of Natural History.

XV.—Ideal scene of the Lias with Ichthyosaurus and Plesiosaurus.

The Plesiosaurus was scarcely so large as the Ichthyosaurus. The specimen of *I. platydon* in the British Museum probably belonged to an animal four-and-twenty feet in length, and some are said to indicate thirty feet, while there are species of Plesiosaurus measuring eighteen and twenty. On the opposite page (PL. XV.) an attempt is made to represent these grand reptiles of the lias in their native element, and as they lived.

Cuvier says of the Plesiosaurus, " that it presents the most monstrous assemblage of characteristics that has been met with among the races of the ancient world." It is not necessary to take this expression literally; there are no monsters in nature; the laws of organization are never positively infringed; and it is more accordant with the general perfection of creation to see in an organization so special, in a structure which differs so notably from that of the animals of our days, the simple augmentation of a type, and some times also the beginning and successive perfecting of these beings. We shall see, in examining the curious series of animals of the ancient world, that the organization and physiological functions go on improving unceasingly, and each of the extinct genera which preceded the appearance of man, present for each organ, modifications which always tend towards greater perfection. The fins of the fishes of Devonian seas become the paddles of the Ichthyosaurii and of the Plesiosaurii; these, in their turn, become the membranous foot of the Pterodactyle, and, finally, the wing of the bird. Afterwards come the articulated fore-foot of the terrestrial mammalia, which, after attaining remarkable perfection in the hand of the ape, becomes, finally, the arm and hand of man; an instrument of wonderful delicacy and power, belonging to an enlightened being gifted with the divine attribute of reason! Let us, then, dismiss this idea of monstrosity, which can only mislead us, and only consider the antediluvian beings as digressions. Let us look on them, not with disgust; let us learn, on the contrary, to read in the plan traced for their organization, the work of the Creator of all things, as well as the plan of creation.

These reflections will assist us to appreciate in its true light one of the most singular of the inhabitants of the ancient world, the *Ptero-dactylus*, from πτερόν, winged, δάχτυλος, fingers, discovered in 1828, which made Cuvier retract what he had said of the Ichthyosaurus, and award to it the palm of monstrosity. Half vampire, half woodcock, with crocodile's teeth along its tapering bill, and scale-armour over its lizard-like body, "qualified," says Dr. Buckland, "like Milton's fiend, for all services and all elements, the creature was a fit companion for

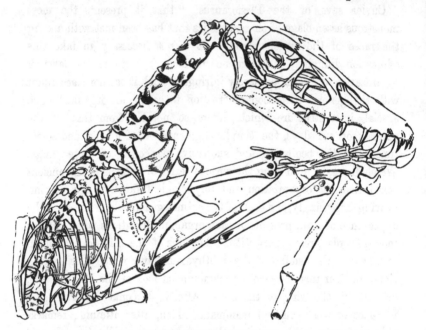

Fig. 102.—Pterodactylus crassirostris.

the kindred reptiles that swarmed in the seas or crawled on the shores of a turbulent planet :—

> The fiend
> O'er bog, or steep, through strait, rough, dense, or rare,
> With head, hands, wings, or feet, pursues his way,
> And sinks, or swims, or wades, or creeps, or flies.

" With flocks of such-like creatures flying in the air, and shoals of

Ichthyosauri and Plesiosauri swarming in the ocean, and gigantic crocodiles and tortoises crawling on the shores of primeval seas, lakes, and rivers, air, land, and sea must have been strangely tenanted in these infant days of our land."

The strange structure of this animal gave rise to most contradictory opinions from naturalists. One made it a bird, another a bat, and others a flying reptile. Cuvier was the first to detect the truth, and prove, from its organization, that the animal was a Saurian. "Behold," he says, "an animal which in its osteology, from its teeth to the end of its claws, presents all the characters of the Saurians; nor can we doubt that their characteristics existed in its integuments and softer parts, in its scales, its circulation, its generative organs:

it was at the same time provided with the means of flight, but when stationary it could not have made much use of its anterior extremities, even if it did not keep them always folded as birds fold their wings. It might, it is true, use its small anterior fingers to suspend itself from the branches of trees, but when at rest it must have been generally resting on its hind

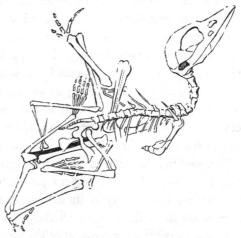

Fig. 103.—Pterodactylus brevirostris.

feet, like the birds again, and like them it must have carried its neck half-erect and curved backwards, so that its enormous head should not interrupt its equilibrium." This diversity of opinion need not very much surprise us after all, for it had the form of a bird in its head and the length of its neck, of the bat in the structure and proportion of its wings, and the reptile in the size of its head and in its beak armed with at least sixty pointed teeth.

Dr. Buckland describes eight distinct species, varying in size from that of a snipe to a cormorant. Of these *P. crassirostris* (Fig. 131) and *P. brevirostris*, were both discovered in the lias of Solenhofen. *P. macronyx* belongs to the lias of Lyme Regis.

The long rows of teeth which armed the jaw, the small number of cervical vertebræ, the narrowness of the sides, the form of the pelvis, separate the Pterodactyli from the birds. A brief comparison between the structure of the head, and the wing of the bats and the corresponding parts in the Pterodactylus, equally distinguishes the latter from the bats, or flying mammifera. Its jaw, provided with conical teeth analogous to those of the Saurians, its narrow sides, the form of the pelvis, the numbers and proportions of the bones of its fingers, closely resemble the reptiles. The Pterodactylus was, then, a reptile provided with a wing resembling bats, and formed, as with this mammifer, by a membrane which connected the body with an excessively elongated joint attached to the fourth finger, which thus become expansers to the membranous wing. The Pterodactylus was, as we have seen, an animal of small volume; the largest specimens did not exceed the size of the swan, and the smallest about the size of the snipe. On the other hand, its head was enormously disproportioned to the rest of the body. We cannot admit, therefore, that this animal could fly, and, like a bird, beat the air. The membranous appendage which connected its long finger to the body was rather a parachute than a wing. It served to moderate the velocity of its descent when it dropped on its prey from a height. Essentially a climber, it could only raise itself by climbing up tall trees or rocks, after the manner of lizards, and throw itself thence to the ground, or upon the lower branches, displaying its parachute to break the fall.

The ordinary position of the Pterodactylus was upon its two hind feet. It held itself firmly upright, the wings folded, and walking on its two hind feet. Habitually it perched on trees: it could climb along rocks and cliffs, aiding itself with claws and feet: it is even probable, Dr. Buckland thinks, that it had the power of swimming, so common with reptiles, and possessed especially by the vampire bat of the island of Bonin. It is believed that the smaller species lived upon insects, and that the larger ones preyed upon fish, upon which it could throw itself after the manner of the sea-gull.

The most startling feature in the organization of this animal is the strange combination of two powerful wings attached to the body of a reptile. The imagination of the poets had long dwelt on such a combination; the *Dragon* was a creation of their fancy, and it had long played a great part in the pagan mythology. The Dragon, or flying reptile, breathing fire and poisoning the air with his fiery breath, had, according to the fable, disputed with man the possession of the earth. Gods and demigods claimed, among their most famous exploits, the glory of having vanquished this powerful and redoubtable monster. From pagan fictions the Dragon passed into the poesy of the Greeks and Latins, and later still into that of the Renaissance, and to modern times. What a part did not the Dragon play in the verses of Tasso and Ariosto! Consecrated by the superstition of the earlier peoples, transferred from pagan mythology to Greek and Roman poesy, and finally into the poetic fictions of the middle ages, the Dragon, according to a very apposite thought of Lacepède, always has and ever will belong to the supernatural.

The Pterodactylus is an animal which might respond to this type of religion, and of ancient poesy; but we see the Dragon greatly curtailed in the poor, climbing, and leaping reptile which lived in Jurassic times. Among the animals of our epoch only a single reptile is found provided with wings, or digital appendages analogous to the membranous wings of the bats, and which can be compared to the Pterodactylus. This is called the *Dragon*, or Draconidæ, a family of Saurians, which has been described by Daudin, distinguished by having their first six ribs false, instead of hooping round the abdomen, extending in nearly a straight line, and sustaining a production of skin which forms a wing analogous to that of the Pterodactylus. Independent of the four feet this wing sustains the animal like a parachute as it leaps from branch to branch, but the creature has no power to beat the air with it as birds do when flying. This reptile lives in the forests of the most burning African countries, and in some isles of the Indian Ocean, about Sumatra and Java. The only known species is that figured below, which comes from the East Indies.

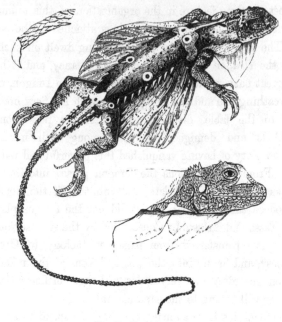

Fig. 104.—Draco fimbriatus.

What a strange population was that which occupied the earth at this stage of its history, when the waters were filled with creatures so extraordinary as those whose history we have traced! Plesiosaurii and Ichthyosaurii filled the seas, upon the surface of which floated innumerable ammonites in light skiffs, some of them of the size of a good-sized cart-wheel, while gigantic turtles and crocodiles crawled on the banks of rivers and lakes. No mammifera, no birds had yet appeared; nothing broke the silence of the air, if we except the breathing of the terrestrial reptiles and the flight of winged insects.

The earth has cooled progressively up to the Jurassic period, the rains have lost their continuity and abundance, the pressure of the atmosphere has sensibly diminished. All these circumstances second the appearance and the multiplication of innumerable species of animals, whose singular forms then show themselves on the earth. We can scarcely imagine the prodigious quantity of mollusks and zoophytes whose remains lie buried in the Jurassic rocks, forming themselves entire beds of immense height and extent.

The same circumstances concurred to favour the production of plants. If the shores and seas of the period received from the formidable being we have described such a terrible aspect, the vegetation which covered the land had also its peculiar character and appearance. Nothing that we know ŏf in the existing scenery of the globe surpasses the rich vegetation which decorated the few continents of the period. A temperature still of great elevation, a humid atmosphere, and, we have no reason to doubt, a brilliant sun, were provocative of a luxuriant vegetation, such as some of the tropical islands, with their burning temperature and maritime climate, can alone give us an idea, while it recalls some of the Jurassic types of vegetation. The elegant Voltzias of the trias had disappeared, but the palms remained, whose slender and delicate·stems rose erect in the air with their graceful panicles; the gigantic rushes also remained, and though the tree-ferns had lost the enormous dimensions of the carboniferous age, they had lost none of their fine and delicately-cut leaves.

Alongside these vegetable families, which remained a legacy from the preceding age, an entire family—the Cycadaceæ—present themselves for the first time. They soon become numerous in genera, such as Zamites, the Pterophyllum, and the Nilsonias. Among the numerous species which characterize this age, we may cite the following, arranging them in families :—

FERNS.	CYCADEÆ.	CONIFERA.
Odontopteris cycades.	Zamites distans.	Taxodites.
Taumopteris Munsteri.	Zamites heterophyllus.	Pinites.
Camptopteris crenata.	Zamites gracilis.	
	Pterophyllum dubium.	
	Nilsonia contigua.	
	Nilsonia elegantissima.	
	Nilsonia Sternbergii.	

P

The *Zamites*, trees of elegant appearance, seem to be forerunners of the palms, which will make their appearance in the following epoch, closely resembling the existing Zamias, which are trees of tropical America, and especially of the West India islands; they were so numerous in species and in individuals that they seem to have formed by themselves alone one half of the forests during the period which engages our attention. The number of their fossilized species exceed in number the living species. The trunk of the Zamites, simple and covered with scars left by the old leaves, supports a thick crown of leaves more than six feet in length, disposed in a fan-like shape, parting from a common centre.

The *Pterophyllum* were great trees, of considerable elevation, covered with large pinnated leaves from top to bottom. Their leaves, thin and membranaceous, were furnished with leaflets truncated at the summit and traversed by fine nervures, not convergent, but abutting on the terminal truncated edge.

The *Nilsonia*, finally, were Cycadaceæ resembling the Pterophyllum, but with thick and coriaceous leaves, and short leaflets contiguous to, and in part attached to the base; they were obtuse or nearly truncated at the summit, and would present nervures arched or confluent towards that summit.

The essential characters of vegetation during the lias sub-period were, 1. The great predominance of the Cycadaceæ, thus continuing the development which commenced in the previous period, expanding into numerous genera belonging both to this family and that of the *Zamites* and *Nilsonias*; 2. The existence among the ferns of many genera with reticulated veins or nervures, and under forms of little variation, which scarcely show themselves in the more ancient formations.

On the opposite page (PL. XVI.) is an ideal landscape of the liasic period; the trees and shrubs characteristic of the age are the elegant Pterophyllum, which appears in the extreme left of the picture, and the Zamites, which are recognizable by their thick and low trunk and fanlike tuft of foliage. The great horses-tail, or Equiseteæ of this epoch

XVI.—Ideal landscape of the Liasic Period.

mingle with the great tree-ferns and the cyprus, a conifer congenerous to those of our age. Among animals we see the Pterodactylus specially represented. One of these reptiles is seen in a state of repose, resting on its hind feet. The other is represented, not flying after the manner of a bird, but throwing itself from a rock in order to seize upon a winged insect, the dragon-fly (*Libellulæ*).

Fig. 105.—Millepora alcicornis.

OOLITIC SUB-PERIOD.

This period is so named because many of the limestones entering into the composition of the formations it represents originate almost entirely in an aggregation of rounded concretionary grains of singular appearance, resembling the roe or eggs of certain fishes, each of which is a nucleus of sand, round which concentric layers of limestone have accumulated; hence the name, from ὠόν, egg, λιθος, stone.

The Oolite is usually subdivided into three sections, the *Lower*, *Middle*, and *Upper* Oolite. These rocks form in England a band some thirty miles broad, ranging across the country from Yorkshire, in the north-east, to Dorset, in the south-west, but with a great diversity of mineral character, which has led to a further subdivision of the formation derived from particular beds in the central and south-western counties :—

Lower: 1, inferior oolite; 2, fullers' earth; 3, Stonesfield slate; 4, Cornbrash, or Forest oolite.

Middle: 1, Oxford clay; 2, coral rag.

Upper: 1, Kimmeridge clay; 2, Portland stone; 3, Purbeck stone.

The alternations of clay and limestone to the lias and oolite formations give some marked features in the outline both of France and England: broad valleys, separated from each other by limestone ranges of hills more or less elevated. In France, the Jura mountains are composed of them; in England, the slopes of this formation are more gentle—the valleys are intersected by brooks, and clothed with a rich vegetation: it forms what is called a tame landscape, as compared with the more savage grandeur of the primitive rocks—it pleases more than it surprises. It yields materials, also, more useful than some of the older formations, numerous quarries being met with which furnish excellent building materials, especially at Bath, where the stone is at first soft and easily fashioned, becoming hard on exposure to the air.

The annexed section will give some idea of the configuration which
the stratification assumes, such as may be ob-
served in proceeding from the north-east to
the south-west, from the banks of the Ouse
to Carmarthenshire.

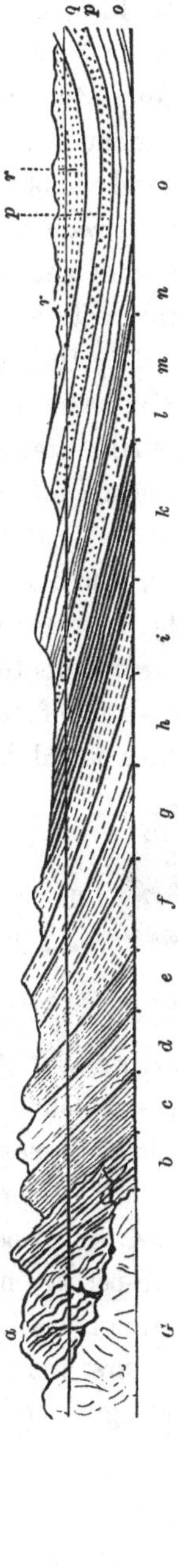

LOWER OOLITE FAUNA.

The most salient and characteristic feature
of this age is undoubtedly the appearance on
the earth of animals belonging to the class of
mammifera. But the organization, quite special, of the first of the mammalia will certainly
be matter of astonishment to the reader, and
must satisfy him that nature proceeded in the
creation of animals by successive steps, by
transitions which, in a manner almost imperceptible, connects the beings of one age with
others more complicated in their organization.
The first mammifers which appear upon the
earth, for example, did not enjoy all the organic
attributes belonging to the more perfect creations of the class. In this great class the
young are brought forth living, and not from
eggs, like birds, reptiles, and fishes. But the
first mammifers which God placed on earth
were not so organized ; they belonged to that
order of animals quite special, and never numerous, which only deposits a gelatinous mass
containing at once the egg, and the young
animal ; in short, marsupial animals. The
mother nurses this mass during a certain time
in a sort of pouch in the neighbourhood of the
abdomen. After a sojourn more or less prolonged in this pouch, and under the influ-

Fig. 106.—General view of the succession of British strata, with the elevations they reach above the level of the sea.

a, Granitic rocks ; *a*, gneiss ; *b*, mica-schist ; *c*, Skiddaw or Cumbrian slates ; *d*, Snowdon rocks ; *e*, Plynlymmon rocks ; *f*, Silurian rocks ; *g*, old red sandstone ; *h*, carboniferous limestone ; *i*, millstone grit ; *k*, coal measures ; *l*, magnesian limestone ; *m*, new red sandstone ; *n*, lias ; *o* lower, middle, and upper oolite ; *p*, greensand ; *q*, chalk ; *r*, the tertiary strata.

ence of maternal warmth, the perfect animal bursts its bonds, and enters into life and light; a sort of middle course between oviparous generation, in which the animals are hatched from eggs after exclusion from the mother's body, like birds, and viviparous, in which the animals are brought forth alive, as in the mammalia.

In classical works of natural history the animals under consideration are ranged as *mammiferous didelphæ.* They are brought forth in an imperfect state, and during their embryonic life the connection with the visceral cavity is maintained by the bones called *marsupial,* which are attached by their extremities to the pelvis, whence they are called *marsupial mammalia.* The opossums, kangaroos, and Ornithorhynchus are the actual representatives of this group.

The name of *Thylacotherium,* or *Amphitherium,* or *Phascolotherium,* is given to the first of these marsupial mammifers which made their appearance, whose remains have been discovered in the lower oolite, and in its most recent stage, namely, that called the *great oolite.* Fig. 107 represents the jaw of the first of these animals, and Fig. 108 the other—both of the natural size. These jaw-bones represent all that has been found belonging to these early marsupial animals, and

Fig. 107.—Jaw of Thylacotherium Prevosti.   Fig. 108.—Jaw of Phascolotherium.

Baron Cuvier and Professor Owen have both decided as to their origin. The first was found in the Stonesfield quarries. The Phascolotherium, also a Stonesfield fossil, was the ornament of Mr. Broderip's collection. The animals, which lived on the land during the lower oolitic period, would be nearly the same with those of the lias. The insects were perhaps more numerous.

The marine fauna included reptiles, fishes, mollusks, and zoophytes. Among the first were Pterodactylus, and a great Saurian, the Teleosaurus, belonging to a family which made its appearance in this

age, and which reappears in the following epoch. Among the fishes,
the Ganoïds and Ophiopsis predominate. Among the Ammonites,
Ammonites Humphrysianus, A. bullatus
(Fig. 109), *A. Brongniarti, Nautilus line-
atus,* and many other representatives of
the cephalopodous mollusks. Among the
Brachiopodes are *Terebratula digona*
(Fig. 110) and *spinosa.* Among the Gasteropodes, *Pleurotomaria conoidea* is

Fig. 109.—Ammo-
nites.   Fig. 110.—Tere-
bratula digona.

remarkable for its elegant shape and markings, and very unlike any
of the living *Pleurotoma* as represented by *P. Babylonia.* *Ostrea
Marshii* and *Lima proboscidea,* which belong to the Acephala, are

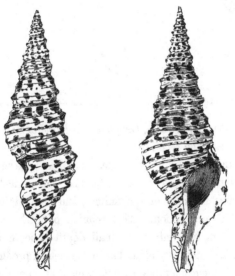

Fig. 111.—Pleurotoma Babylonia.

fossil mollusks of this epoch, to which also belong *Entalophora cellari-
oides, Eschara Ranviliana, Bidiastopora cervi-cornus;* elegant and
characteristic molluscous Bryozoares. We give a representation of
two living species, as exhibiting the form of these curious products.

The Echinoderms and Polypiers appear in great numbers in the deposits of the lower oolite: *Apiocrinus elegans, Hyboclypus gibberulus, Dysaster Endesii* represent the first; *Montlivaltia caryo-*

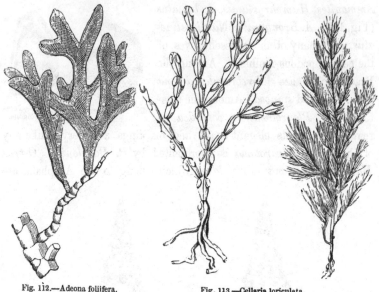

Fig. 112.—Adeona foliifera.   Fig. 113.—Cellaria loriculata.

phyllata, Anabacia orbulites, Cryptocænia bacceformis, and *Eunomia radiata* represent the second.

This last and most remarkable species of zoophyte presents itself in great masses many yards in circumference, and indicates an accumulation requiring myriads of the animal, and necessitates a long series of ages for its production. This reunion of little creatures living under the waters, but only at a small depth beneath the surface, as Mr. Darwin has demonstrated, has nevertheless produced banks, or rather islets, of considerable extent, which at one time constituted veritable reefs rising out of the ocean. These reefs were principally constructed in the Jurassic period, and their extreme abundance is one of the characteristics of this geological age. The same phenomenon continues in our day, but by a new race of zoophytes, which carry on their operations, preparing a new continent, probably, in the *atols* of the Pacific Ocean.

The flora of the epoch was very rich. The Ferns continue to figure there, but the size and bearing were sensibly inferior to what they had been in the preceding period. Among them Otopteris, distinguished for its simply pinnated leaves, whose leaflets are auriculate at the base: of

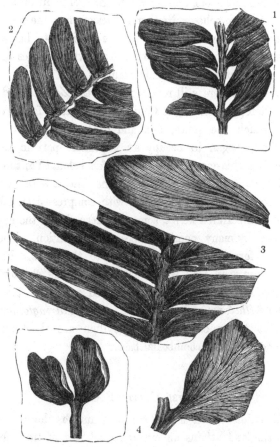

Fig. 114.—1, Otopteris dubia; 2, Otopteris obtusa; 3, Otopteris acuminata; 4, Otopteris cuneata.

the five species, 1, *O. dubia;* and 2, *O. obtusa;* and 3, *O. acuminata;* and 4, *O. cuneatea*, are from the oolite. In addition to these we may name *Conipteris Murrayana, Pecopteris Desnoyersii, Pachypteris lanceolata,* and *Phlebopteris Phillipsii ;* and among the Lycopodiaceæ, *Lycopodus fulcatus.*

The vegetation of this epoch has a peculiar physiognomy, from the presence of the family of the Pandanaceæ, or screw-pines, so remarkable for their aerial roots, and for the magnificent tuft of leaves which terminates their branches. Neither the leaves nor the roots of these plants have, however, been found in the fossil state, but we possess specimens of their large and spherical fruit, which leaves no room for doubt as to the nature of the entire plant.

The Cycadaceæ were still represented by the *Zamias,* and by many species of Pterophyllum. The Coniferæ, that grand family of modern times, to which the pines, firs, and other trees of our northern forests belong, began to occupy an important part in the world's vegetation from this epoch. The first conifers belonged to the genera of *Thuites, Taxites,* and *Brachyphyllum.* The *Thuites* were true *Thuyas,* evergreen trees of the present epoch, with compressed branches, small imbricated and serrated leaves, somewhat resembling the cypress, but distinguished by many points of special organization. The *Taxites* have been referred, with some doubts, to the yews. Finally, the *Brachyphyllum* were trees which, according to the characteristics of their vegetation, seem to have approached nearly to two existing genera, the *Arthotaxis* of Tasmania, and the *Weddringtonias* of South Africa. The leaves of the Brachyphyllum are short and fleshy, inserted by a large and rhomboidal base.

Lower Oolite Rocks.

The formation which actually represents the lower oolite, and which in England attains an average thickness of from five to six hundred feet, forms a very complex system of stratification, which includes the two formations of *Bajocien* and *Bathonian,* adopted by M. D'Orbigny and his followers. The lowest beds of the *inferior oolite* occur in Normandy, in the Lower Alps (Basses-Alps), in the neighbourhood of Lyons. They are remarkable near Bayeux for the

variety and beauty of their fossils : the rocks are composed principally
of limestones—yellowish, brown, or red, charged with hydrate of iron,
often oolitic, and reposing on calcareous sandy gravel. These deposits
are surmounted by alternate layers of clay and marl, blue or yellow—
the well-known *fullers' earth;* so called because it is employed to
extract the grease employed in the manufacture of woollen fabrics.
The second stratum of the inferior oolite, which attains a thickness of
one hundred and fifty to two hundred feet on the coast of Normandy,
and is well developed in the neighbourhood of Caen and in the Jura,
has been divided into four series of beds, in an ascending scale :—

1. The *Great Oolite,* which consists principally in a very characteristic oolitic limestone of fine grain, white, soft, and well developed
at Bath, and also at Caen in Normandy. At the level of the great
oolite the Stonesfield beds occur, in which were found the bones of the
marsupial mammifera, to which we have already alluded, and along
with them bones of reptiles, principally Pterodactyli, together with
some finely-preserved fossils of plants, fruits, and insects.

2. *Bradford Clay,* which is a bluish marl, containing many fine
encrinites, but which have only a local existence, such fossils appearing to be almost entirely confined to limestone formations. " In this
case, however," says Lyell, " it appears that the solid upper surface of
the great oolite had supported for a time a thick submarine forest of
these beautiful zoophytes, until the clear and still water was invaded
by a current charged with mud, which threw down the encrinites, and
broke most of their stems short off near the point of attachment, the
stumps still remaining in their original position."

3. *Forest Marble,* which consists of an argillaceous shelly limestone,
abounding in marine fossils, and sandy and quartzose marls, quarried
in the forest of Wichwood, in Wiltshire.

4. The *Cornbrash* (wheatlands), formed of the broken limestone,
or calcareous sandstone, which covers the fields appropriated to the
cultivation of certain cereals ; hence its name.

The lower oolite ranges over great part of England, but "attains
its maximum in the neighbourhood of Cheltenham, where it can be

subdivided into three parts.   Passing north, the two lower divisions, each more or less characterized by its own fossils, disappear, and the ragstone north-east of Cheltenham lies directly on the lias, apparently as conformably as if it formed its true and immediate successor.   In Dorsetshire, on the coast, the series is again perfect, though thin.   Near Chipping Norton, in Oxfordshire, the inferior oolite disappears altogether, and the great oolite, having first overlapped the fullers' earth, passes across the inferior oolite, and in its turn seems to lie on the upper lias with a regularity as perfect as if no formation in the neighbourhood came between them.   In Yorkshire the changed type of the inferior oolite, the prevalence of sands, land-plants, and beds of coal leave no doubt of the presence of terrestrial surface on which the plants grew; and all these phenomena lead to the conclusion that various and considerable oscillations of level took place in the British

Fig. 115.—Meandrina dædalæa.
a, entire figure reduced; b, portion, natural size.

area during the deposition of the strata, both of the inferior oolite and the formations which immediately succeed it."

The inferior oolite here alluded to is a thin bed of calcareous free-

XVII.—Ideal landscape of the Lower Oolite.

stone, resting on and sometimes replaced by yellow sand, which
repose on the lias. Fullers' earth lies between this and the great
oolite, at the base of which lie the Stonesfield slates, a shelly limestone,
flagstones some six feet thick, rich in organic remains, which ranges
over Oxfordshire and towards the north-east. At Colley Weston, in
Northamptonshire, fossils of *Pecopteris polypodioides* are found. In
the great oolite formation, near Bath, are many corallines, among
which *Eunomia radiata* is conspicuous. The fossil is not unlike the
brain coral of the tropical seas. The work of this coralline seems to
have been suddenly stopped by " an invasion," says Lyell, " of argillaceous matter, which probably put a sudden stop to the growth of
Bradford Encrinites, and led to their preservation as marine strata."
The Cornbrash passes down to the forest marble, sometimes, as at
Bradford, near Bath, in masses of clay. Rippled slabs of fossil oolite,
used as a roofing slate, may be traced over a broad band of country in
Wiltshire and Gloucestershire, separated from each other by seams
of clay, in which the undulating ridges of the sand are preserved,
and even the footmarks of small crustaceans are still visible.

On the opposite page (PL. XVII.) is represented an ideal landscape
during the period of the lower oolite. On the shore are types of the
vegetation of the period. The *Zamites*, with its large trunk covered
with fan-like leaves, resembled in form and bearing the Zamias of
tropical regions ; a *Pterophyllum*, with its stem covered from base to
summit with its finely-cut feathery leaves; conifers closely resembling our
cypress, and an arborescent fern. What distinguishes this sub-period
from that of the lias is a group of magnificent trees, *Pandanus*, remarkable for their aerial roots, their long leaves, and globular fruit.

Upon one of the trees of this group the artist has placed the
Phascolotherium, not very unlike to our opossum. It was the first
of the mammalia which gave animation to the ancient world. The
artist has here enlarged the dimensions of the animal in order to seize
the form ; let the reader reduce it in his thoughts to one-sixth, for it
was not larger than an ordinary-sized cat.

A crocodile and the fleshless skeleton of the Ichthyosaurus remind us that reptiles still occupied an important place in the animal creation. A few insects, especially dragon-flies, fly about in the air. Ammonites float on the surface of the waves, and the terrible Plesiosaurus, like a gigantic swan, swims about in the sea. The circular reef of coral, the work of ancient polypi, foreshadow the atols of the great ocean, for it was during the Jurassic period that the polypiers of the ancient world were most active in the production of coral reefs and islets.

MIDDLE OOLITE.

The terrestrial flora of this age was composed of ferns, cycadaceæ and conifers. The first represented by the *Pachypteris microphylla*, the second by *Zamites Moreana*. *Brachyphyllum Moreanum* and *majus* appear to have been the conifers characteristic of the period : fruits have also been found in the rocks of the period, which appear to belong to the palms, but this point is still obscure and doubtful.

Numerous vestiges of the fauna which animated the period are also revealed in the rocks of the period. Certain hemipterous insects appear on the earth for the first time, and the bees among the Homenoptera; butterflies among the Lepidoptera, and dragon-flies among the Neuroptera. In the bosom of the ocean, or upon its banks, roamed the *Ichthyosaurus, Pterodactylus crassirostris*, the *Pleurorostrus* and *Geosaurus;* the two latter beings very imperfectly known.

Another reptile congenerous to the Pterodactylus lived in this epoch. It was the *Ramphorynchus*, and was distinguished from the former by a long tail. The imprints which this curious animal has left upon the sandstone of the period indicate at once the impression of its feet and the linear furrow left by its tail. Like the Pterodactylus, the Ramphorynchus, which was about the size of a crow, could not precisely fly, but, aided by the natural parachute formed by the membrane connecting the fingers and the body, it could throw itself from a height upon its prey. Fig. 116 represents this animal restored. The foot-

XVIII.—Ideal landscape of the Middle Oolitic Period.

prints in the soil are those which always accompany the remains of
the Ramphorynchus in the oolitic rocks, and they show the imprints
at once of the anterior and posterior feet and the tail.

Another family of reptiles appear in the middle oolite, of which we

Fig. 116.—Ramphorynchus restored. One quarter natural size.

have had a glimpse in the great oolite of the preceding section. This
is the *Teleosaurus*, which the recent investigations of M. E. Deslongchamps on this family of fossils permits us to reconstruct. The
Teleosaurus permits us to form a pretty exact idea of these crocodiles
of the ancient seas—these cuirassed reptiles, which the German
geologist Cotta describes as "the great barons of the kingdom of
Neptune, armed to the teeth and clothed in an impenetrable panoply ;
the true filibusters of the primitive seas."

The Teleosaurus has an anatomical resemblance to the Gavials of
India. They inhabited the banks of rivers, perhaps the sea itself:
they were longer, more slender, and more active than the living species ;
they were about thirty feet in length, of which the head may be from
three to four feet, with their enormous jaws well defended beyond the
ears, sometimes with an opening of six feet, through which they can
engulph, in the depths of their enormous palate, animals of the size of
an ox.

On the opposite page (PL. XVIII.) the *Teleosaurus cadomensis* is
represented, after the sketch of M. E. Deslongchamps, carrying from the

sea in its mouth a *Geotheutis*, a species of calmar of the oolitic epoch. This creature has the curious peculiarity of being cuirassed both on back and belly. In order to show this peculiarity, a living individual is represented on the shore, and a dead one is floating on its back in shallow water, leaving the ventral cuirass exposed.

Behind *Teleosaurus cadomensis* in the engraving, another Saurian, the *Hyleosaurus*, is represented, which makes its appearance in the cretaceous epoch. We have here adopted the restoration which has been so ably executed by Mr. Waterhouse Hawkins, at the Crystal Palace, Sydenham.

Besides the numerous fishes with which the oolitic seas swarmed, they contained some Crustaceans, Cirropodes, and various genera of Mollusks and Zoophytes. *Eryon arctiformis*, represented in Fig. 117, belongs to the class of Crustaceans, of which the spiny

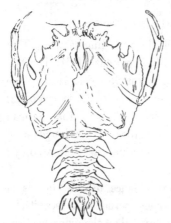

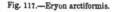

Fig. 117.—Eryon arctiformis.   Fig. 118.—Perfect Ammonite.

lobster is the type. Among the Mollusks were some Ammonites, Belemnites and Oysters, of which 123 species have been described, twenty-three of which are common to the Oxford clay and the Cornbrash which lies at the top of the preceding section, and 106 limited to the Oxford clay. Of these we may mention *Ammonites refractus, Jason* and *cordatus, Ostrea dilatata, Terebratula diphya, Diceras arietana,*

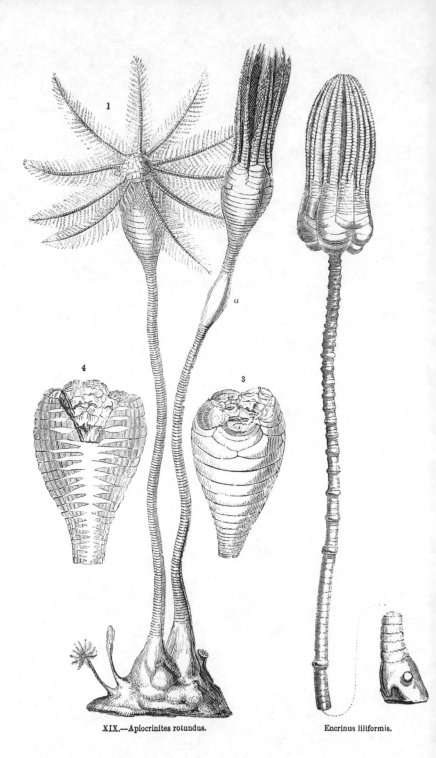

XIX.—Apiocrinites rotundus.　　　Encrinus liliformis.

Belemnites hastatus, and *Puzosianus*. In some of the finely-laminated clays the ammonites are very perfect, but somewhat compressed, with a lateral lobe, as in Fig. 118. Similar elongations belong to some Belemnites discovered by Dr. Mantell in the Oxford clay.

Among the Echinoderms, *Cidaris glandifirus* and *Apiocrinus Roissyanus* and *rotundus*, the graceful *Saccocoma pectinata, Millericrinus no͘dotianus, Comatula costata,* and *Hemicidaris crenularis* may be named. *Apiocrinites rotundis*, figured in PL. XIX., is a restoration reduced: 1, being expanded; *a,* closed; 3, a cross section of the upper extremity of the pear-shaped head; 4, a vertical section showing the enlargement of the alimentary canal, with the hollow lenticular spaces which descend through the axis of the column, forming the joints, and giving elasticity and flexure to the whole stem, without risk of dislocation. *A. rotundis* is found at Bradford in Wiltshire, Abbotsbury in Dorset, at Soissons and Rochelle. This species—known as the Bradford Pear Encrinite—is only found in the fossil state, and in strata above the lias.

The Polypiers of this epoch occur in great abundance, but it is chiefly during the oolitic period that they seem to have been formed. We have already remarked that these aggregations of Polypiers are often met with at great depth in the soil. These small calcareous structures have been formed in the ancient seas, and the same phenomena is extending the terrestrial surface in our days in the seas of Oceania, where reefs and atols of coral are rising by slow and imperceptible steps, but with no less certainty. Although their mode of production must always remain to some extent a mystery, the investigations of M. Lamaroux, Mr. Charles Darwin, and M. D'Orbigny have gone a long way towards explaining their operations; for the zoophyte in action is an aggregation of these minute polypi. Describing what he believes to be a sea-pen, *Virgularia Patagonia*, Mr. Darwin says: " The zoophyte consists of a thin, straight, fleshy stem, with alternate rows of polypi on each side, and surrounding an elastic stony axis. The stem at one extremity is truncate, but terminates at the other in a vermiform fleshy appendage. The stony axis which gives strength to the stem may become at this extremity a mere vessel, filled with granular matter. At low water hundreds of these zoophytes may be seen projecting with the truncated end upwards, a few inches above the surface of the muddy sand. When touched or pulled, they suddenly draw themselves in so as nearly or quite to disappear. By this action the highly-elastic axis must be bent at the lower extremity, where it is naturally slightly curved; and I imagine it is by this elasticity alone that the zoophyte is enabled to rise again from the mud. Each polypus, though closely united to its brethren, has a distinct mouth, body and tentacula. Of these polypi in a large specimen there must be many thousands. Yet they act by one movement. They have also one central axis, connected with a system of obscure circulation." Such is the brief account given by a very acute observer of these singular beings. They manipulate the calcareous matter held in suspension in the oceanic waters and produce the wonderful structure we have now under consideration; and these calcareous banks have been in course of formation during all geological ages. They just reach the level of the waters, for the polypi perish as soon as they are so far above the surface that neither the waves nor the flow of the tides can reach them. In the oolitic rocks these banks are frequently found from twelve to fifteen feet thick, and many leagues in length, and preserving for the most part the relative positions which they occupied in the sea while in course of formation.

The rocks which now represent the middle oolitic period are usually divided into the *Oxford clay*, the lower member of which is an arenaceous limestone, known as the *Kelloway clay*, which in Wiltshire and other parts of the south-west of England attains a thickness of eight or ten feet, with the impression of numerous ammonites, and other shells. In Yorkshire it reaches the thickness of thirty feet, around Scarborough; it forms well-developed beds of bluish-black marl in the department of Calvados, in France. It is the base of the argillaceous clay which forms the soil of the valley of the Ange, renowned for its rich pasturages and magnificent cattle, like the soil formed out of fossilized mollusks. The same beds form the base of the oddly-shaped but fine rocks of La Manche, which are popularly known as the *Vaches Noires*, black cows—a locality celebrated also for its fine ammonites transformed into pyrites.

The *Oxford clay* constitutes the base of the hills in the neighbourhood of Oxford, forming a bed of clay sometimes not less than five hundred feet thick. It is found well-developed in France, at Trouville, in the department of the Calvados, and at Neuvisy, in the department of the Ardennes, where it attains a thickness of about three hundred feet. It is bluish, sometimes whitish limestone, often argillaceous and bluish marl. The *coral rag* takes its name from this fact—that the limestone of which it is chiefly composed consists of an aggregation of numerous fragments of corals, and of polypiers entire or in a mass, and sometimes of enormous masses of polypiers; in short, beds of petrified coral, not unlike that in progress in the Pacific Ocean, supposing them to be covered up for ages and fossilized. This coralline stratum extends through the chalky hills of Berkshire and North Wilts, and it occurs again near Scarborough. In France it is found in the departments of the Meuse, of the Yonne, of the Ain, of the Charente-inférieure. In the Alps the *Dueras limestone* is regarded by most geologists as coeval with the English coral rag.

UPPER OOLITE.

Some marsupial mammifers have left their remains in the upper oolite as in the lower. They belong to the genera of *Spalacotherium*. Besides the Plesiosaurus and Teleosaurus, there still lived in the maritime regions a crocodile, the *Macrorhynchus;* the monstrous *Pœcilopleuron*, among the enormous griffins with sharp cutting teeth, one of the most formidable animals of this epoch; the *Hylœosaurus, Cetiosaurus, Stenosaurus* and *Streptospondyles*, and among the turtles, the *Emys* and *Platemys*. As in the lower oolite, so also in the upper, insects similar to those by which we are surrounded pursued their flight in the meadows or hovered on the surface of the water. Of these animals, however, too little is known for us to give any very precise indication on the subject of their special organization.

The most remarkable fact which occurs in this period is the appearance of the first bird. Hitherto the Mammalia, and these only imperfectly-organised species, namely, the marsupials, have alone appeared. It is interesting to witness birds appearing immediately after. In the quarries of lithographic stone at Solenhofen, the remains of a bird, with feet and feathers, have been found, but without the head. These curious remains are represented in Fig. 119, in the position in which they were discovered. It is usually designated the bird of Solenhofen.

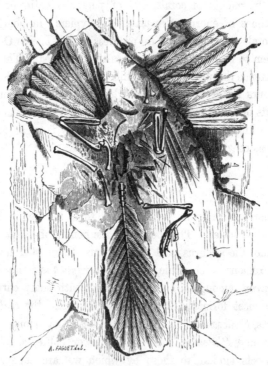

Fig. 119.—Bird of Solenhofen (Archæopteryx).

The oolitic seas of this section contained fishes belonging to the genera of *Asterocanthus, Strophodes, Lepidotus,* and *Microdon*. The cephalopodous mollusks were not numerous, the predominating genera belonged to the lamellibranches and to the gasteropodes, which lived near the shore. The reef-making madrepores or corallines were more numerous. A few zoophytes in the fossil state testify to the existence of these extraordinary animals. The fossils characteristic of the fauna of the period include *Ammonites decipiens* and *giganteas, Natica elegans* and *hemispherica, Ostrea deltoidea* and *virgula, Trigonia gibbosa, Pholadomya multicostata* and *acuticostata, Terebratula subsella,* and *Hemicidaris Purbeckensis*. Some fishes, turtles, paludines, physas, unios, planorbis, and the little

Fig. 120.—Shell of Planorbis corneus.

crustacean bivalves, the Cyprises, constituted the fresh-water fauna of the period. The forms of some of these are represented in the margin.

The terrestrial flora of the period consisted of Ferns, Cycadaceæ, and Conifers; in the ponds and swamps some Zosteras. The *Zosteras* are monocotyledonous plants of the family of the Naïdaceæ, which grow in the sandy mud of maritime regions, forming there, with their long,

Fig. 121.—Shell of Physa fontinalis.

narrow, and ribbon-like leaves, vast prairies of the most beautiful green. At the low tide these masses of verdure appear somewhat exposed.

They would form a retreat for a great number of marine animals, and afford nourishment to others.

On the opposite page an ideal landscape of the period (PL. XX.) represents some of the features of the upper oolite, especially the vegetation of the Jurassic period. The *Sphenophyllum* among the tree-ferns are predominant in this vegetation; some *Pandanas*, a few *Zamites*, and many *Conifers*, but we perceive no palms. A coral islet rises out of the sea, having somewhat of the form of the *atols* of Oceania, indicating the importance these formations assumed in the Jurassic period. The animals represented are the *Crocodileimus* of Jourdan, the *Ramphorynchus*, with the imprints which characterise its footsteps, and some of the invertebrated animals of the period, as the Asterias, Comatulas, Hemicidaris, Pteroceros. Aloft in the air floats the bird of Solenhofen, the *Archæopteryx*, which has been reconstructed from the skeleton, with the exception of the head, which remains undiscovered.

The rocks which represent the upper oolite are usually divided into two series: 1. The *Kimmeridge clay*; 2. The *Portland sandstone and sand*, which is sometimes subdivided into the *lower* and *upper Purbeck beds*.

The *Kimmeridge clay* is specially composed of numerous blue or yellowish argillaceous beds, which pass into the state of clay and bituminous schists, sometimes forming an impure coal, several hundred feet in thickness. These beds are well developed at Kimmeridge, in Wiltshire, whence its name. In some parts of Wiltshire the beds of bituminous matter have a peaty appearance, but there is an absence of the impressions of plants which usually accompany the formation, derived from the decomposition of vegetables. These rocks, with their characteristic fossils, *Cardium striatulum* and *Ostrea deltoidea*, are found throughout England, and in some parts of Scotland: in France, at Tonnese, near Yonne, at Havre, at Honfleur, at Mauvage; in the department of the Meuse it is so rich in fossils of *Ostrea deltoidea*

XX.—Ideal landscape of the Upper Oolitic Period.

and *virgula* that, "near Clermont in Argonne, a few leagues from St. Menehould," says Lyell, "where these indurated marls crop out beneath the gault, I have seen them (*Gryphea virgula*) on decomposing leave the surface of every ploughed field literally strewed over with this fossil oyster."

The second section of this series consists of the oolitic limestone of Portland, and in the cliffs of the Purbeck beds in the little peninsula of Dorset, formed of alternate beds of marine and fresh-water formation, yielding fossil remains which enable us to reconstruct the fresh-water fauna and flora. The lacustrine deposits consist principally of *Cypris* in petrified limestone.

The Isle, or rather peninsula of Portland, lying off the Dorset coast, rises considerably above the sea level, presenting on the side of the port a bold line of cliffs, connected with the mainland by the Chesil bank, an extraordinary formation, consisting of shingles and pebbles loosely piled on the blue clay and stretching ten miles westward along the coast. The quarries occupy the north part of the island. The story told of this remarkable island is an epitome of the revolutions the surface of the earth has undergone. The quarries are situated on the north side of the island. The beds which occupy the summit of the oolite are of a dark-yellowish colour; they are burnt in the neighbourhood for lime. The next bed is of a whiter and more lively colour. It is the stone of which the portico of St. Paul's and many of the houses of London built in Queen Anne's time were built. The building-stone contains fossils exclusively marine. Upon this bank rests a bed of limestone formed in lacustrine waters. Finally, upon this bed rests another bed of bluish substance: this is a bed of very well-preserved vegetable earth or *humus*, quite analogous to our vegetable soil, of the thickness of from fifteen to eighteen inches, of a blackish colour; it contains a strong proportion of ligneous earth; it swarms with the silicified remains of conifers and other plants, analogous to the *Zamias* and *Cycas*: this soil is known as the dirt-bed. The trunks of great numbers of silicified trees and tropical plants are found here erect, and their roots fixed in the soil, and of species differ-

ing from any of our forest-trees. " The ruins of a forest upon the
ruins of a sea," says Esquiros, " the trunks of these trees were petrified while still growing. The region now occupied by the narrow
channel and its environs had been at first a sea in whose bed the
oolitic deposits which now form the Portland stone accumulated : the
bed of the sea gradually rose and emerged from the waves. Upon the
land thus rescued from the deep, plants began to grow : they now
constitute with their ruins the soil of the dirt-bed. This soil, with its
forest of trees, was afterwards plunged again into the waters—not the
bitter waters of the ocean, but in the fresh waters of a lake formed at
the mouth of some great river."

" Time passed on however : an alluvial soil, brought from the interior
by the rolling waters, formed a bed of mud over the dirt-bed : finally, the
whole region was engulphed anew in some grand convulsion or by
some system of succession deposits, until the day when the Isle of
Portland was again revealed to light. " From the facts observed," says
Lyell, " we may infer : 1. That the Portland beds, which are full of
marine shells, were overspread with fluviatile mud, which became dry
land, and were covered with forests throughout a portion of the south
of England, the climate being such as to admit of the growth of *Zamia*
and *Cycas*. 2. This land at length sank down and was submerged
with its forests beneath a body of fresh water, from which sediment
was thrown down containing fluviatile shells. 3. The regular and
uniform preservation of this thin bed of black earth over a distance of
many miles shows that the change from dry land to the state of a fresh-
water lake or estuary was not accompanied by any violent rush of
water, since the black earth and trees must inevitably have been swept
away had any such violent catastrophe taken place."

The soil known as the *dirt-bed* is horizontal in the Isle of Portland ;
but we discover it again not far from there in certain cliffs having
an inclination of 45°, where the trunks continue perfectly parallel
among themselves, affording a fine example of a change in the position
of beds originally horizontal. Fig. 122 represents this species of
geological *humus*. " Each *dirt-bed*," says Sir Charles Lyell, " may, no

doubt, be the memorial of many thousand years or centuries, because we find that two or three feet of vegetable soil is the only monument which many a tropical forest has left of its existence ever since the ground on which it now stands was first covered with its shade."

Fig. 122.—Geological humus. *a.* Fresh-water calcareous slate; *b,* dirt-bed, with roots and stems of trees; *c,* fresh water; *d,* Portland stone.

This bed of vegetable soil is then at the summit of that long and complicated series of beds which constitute the Jurassic period : these ruins, still vegetable, remind us forcibly of the coal-beds, for they are nothing else than a less advanced state of that kind of vegetable fossilization which was perfected on such an immense scale, and during an infinite length of time in the coal period.

The Purbeck beds, which are sometimes subdivided into lower, middle, and upper beds, are fresh-water formations, intimately connected with the upper Portland beds. But there they begin and end, being scarcely recognizable except in Dorsetshire, in the sea-cliffs of which they were first studied. They are finely exposed in Durdlestone Bay, near Swanage, and at Lulworth Cove on the same coast. The *lower beds* consist of a purely fresh-water marl, eighty feet thick, containing shells of *Cyprises, Limnæa,* and some *Serpula* in a bed of marl of brackish water, and some *Cypris*-bearing shales, strangely broken up at the west end of the Isle of Purbeck.

The *middle series* consists of twelve feet of strata known as the "cinder-beds," formed of a vast accumulation of *Ostrea distorta,* resting on fresh-water strata full of *Cypris fasciculata, Planorbis* and *Limnæa,* by which this strata has been identified as far inland as the vale of Wardour in Wiltshire. Above the cinder-beds are shales and limestone, partly of fresh water and partly of brackish origin, in which are fishes, many species of Lepidotus, and the crocodilian reptile,

Macrorhynchus. On this rests a purely marine deposit, with *Pectens, Avicula,* &c. Above, again, are brackish beds with *Cyrena,* overlying which is thirty feet of fresh-water limestone, with *fishes, turtles* and *Cyprides.*

The *upper beds* are purely fresh-water strata, about fifty feet thick, containing. *Paludina, Physa, Limnæa,* all very abundant. In these beds the Purbeck marble, formerly much used in the ornamental architecture of the old English cathedrals, is quarried.

A few words, in explanation of the term *oolite,* applied to this sub-period of the Jurassic formation. In a great number of rocks of this series the elements are neither crystalline nor amorphose—they are, as we have already said, oolitic; that is to say, they have the form of the eggs of certain fishes. The question naturally enough arises, Whence this singular form assumed by certain rocks in their elementary form ? Sometimes it is averred that the grains are fishes' eggs petrified : sometimes, that they are formed of the eggs of aquatic flies or crustaceans in a like condition : again, it is asserted that the grinding action of the sea acting upon the precipitated limestone produces rounded forms analogous to the grains of sand which cover the oolitic regions. These hypotheses may be well founded in some cases. The marine sediments which are deposited in some of the warm bays of Teneriffe are found to take the spheroidal granulated form of the oolite. But these local facts cannot be made to apply to the whole extent of the oolitic formation. We must, therefore, look further for an explanation of the phenomena.

It is stated that if the cascades of Tivoli, for example, can give birth to the oolitic grains, the same thing happens in the quietest basins : in the stalactitic caverns, oolitic grains develop themselves, which afterwards acquire a clammy feeling, from the continued but very slow affluence of the calcareous waters giving birth to a certain species of oolitic rock.

On the other hand, it is stated that tubercles, more or less scanty, develop themselves in the marl in consequence of the concentration of

the calcareous elements, without the possibility of any flow of water
or any other imaginable cause intervening between the oolites and the
various-sized grains; from which it would almost appear that the
oolites are equally the product of concentration.

Finally, from research to research, it is found that the oolite perfectly constituted—that is to say, in concentric beds, as in the Jurassic
limestone,—develops itself in vegetable earth where the assistance of
flowing water is not more admissible than in the preceding instance.

Thus we arrive at the conclusion, that if nature sometimes forms
and finishes her crystals in her workshop in the course of solidification,
she gives birth also to spheroidal forms surrounding various centres,
which are sometimes spontaneous and in other cases the débris of
fossils, or even of simple stony objects. Nevertheless, all mineral
substances are not alike calculated to produce oolitic grains : putting
aside some particular cases, this property is reserved for limestone
and oxyde of iron.

In respect to the distribution of the Jurassic formation on the terrestrial globe, it may be stated that in France the Jura mountains are
almost entirely composed of these rocks, the several series of beds
being all represented in them : this circumstance, in fact, induced Von
Humboldt to name the formation after this range. The upper lias
also predominates in the Pyrenees and in the Alps ; it exists in
Spain, in many parts of Northern Italy, in Russia, notably in the
government of Moscow, and in the Crimea; but it is in Germany
where it occupies the most important place. A thin bed of oolitic
limestone presents at Solenhofen, in Bavaria, a geological repository
of great celebrity, containing fossil plants, fishes, insects, crustaceans,
with some Pterodactyli, admirably preserved : it yielded also the first
of the feathered race. The fine quarries of lithographic stone at
Pappenheim, so celebrated all over Europe, belong to the Jurassic
formation.

It has recently been announced that these rocks have been found in
India; they contribute largely to the formation of the massive Hima-

layas, and to the chain of the Andes in South America; finally, from recent investigations, it is found in New Zealand.

In England the lias constitutes a well-defined belt about thirty miles broad, extending from Dorsetshire, in the south, to Yorkshire, in the north, forming alternate beds of clay, shales and limestone, with layers of jet on the coast near Whitby. It is rich, as we have seen, in ancient life, and that in the strongest forms imaginable. From the unequal hardness of the rocks it comprises, it stands only boldly in some of the minor ranges of hills, adding greatly to the picturesque beauty of the scenery in the centre of the country. In Scotland the formation occupies a very limited space.

A map of the country at the close of the Jurassic period would probably double the extent of dry land in the British Islands. The great basin of the London clay, as well as the Paris basin, were still beds of the primordial ocean, but Devon and Cornwall had long risen from the sea, and it is probable that the Jurassic beds of Devonshire and France were connected by a tongue of land running from Cherbourg to the Jurassic beds of North Devon, and that Boulogne, still an islet, was similarly connected with the Weald.

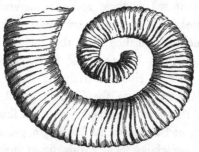

Fig. 123.—Crioceratites Duvalli Tropæum Sowerby.
A disconnected ammonite.

THE CRETACEOUS PERIOD.

THE name *Cretaceous* is given to this epoch in the history of our globe because the rocks deposited by the sea during the period are almost entirely composed of chalk or carbonate of lime.

Chalk, however, does not now appear for the first time as a part of the terrestrial crust; we have already seen limestone intervening among the terrestrial materials in the Silurian period; the Jurassic formation is composed of carbonate of lime in most of its beds, and these beds are enormous in number as well as extent: it appears, therefore, that in the period called *Cretaceous*, chalk was no new substance in the constitution of the globe; and if geologists have been led to give this name to the period, it is because it accords better than any other substance with the characteristics of the period; with the vast accumulations of chalky earth and lime in the basin of Paris, and the complicated beds of greensand, so called, and chalk of the same period in England.

We have already endeavoured to establish the origin of chalk, in speaking of the Silurian and Devonian periods, but it may be useful to recapitulate the facts, even at the risk of repeating ourselves.

We have said that lime was in all probability brought to the surface by thermal waters flowing abundantly through the fissures, dislocations, and fractures in the soil, created by the progressive cooling of the globe; the central nucleus being the grand reservoir and source of the materials which form the solid crust. In the same manner, therefore, as the several eruptive substances, such as granites, porphyries, trachytes, basalts, and lava, have been ejected, so has water in a state of ebullition, and charged with carbonate of lime, and often accompanied with silica, been thrown upon the surface of the soil. We need not repeat the names of the Iceland geysers, the wells of Plombières, and the well-known thermal sources here: through such channels, it is enough to repeat, enormous masses of silica and carbonate of lime, and other substances reached the surface.

But how comes lime in the state of a bicarbonate dissolved in these thermal waters to form rocks? This is what we propose to explain.

During the primary geological periods, water covering nearly the whole terrestrial crust, these thermal waters, as they reached the surface, were discharged into the sea and united themselves to the waves of the vast primordial ocean, and the waters of the sea became sensibly calcareous—they contained, it is believed, from one to two per cent. of lime. The innumerable animals, especially zoophytes and mollusks with solid shells, with which the ancient seas swarmed, seized upon this lime, out of which they built up their mineral dwelling—the shell. In this liquid and decidedly calcareous medium, the foraminifera and polypi of all forms swarmed, an innumerable population. Now what became of the bodies of these creatures after death? They were of all sizes, but chiefly microscopic; that is, so small as to be individually all but invisible to the naked eye. The destructible animal matter disappeared in the bosom of the waters by decomposition, but there remained the indestructible organic matter; that is to say, the carbonate of lime forming their testaceous covering: these were buried in the sandy sea basin, forming calcareous deposits, which, accumulating in thick beds, soon became agglutinated in a unique mass, and formed a series of continuous beds superposed on each other: these, increasing imperceptibly in the course of ages until the ocean bed was filled up with them, form the rocks of the *cretaceous* period we have now under consideration

These statements are not, as the reader might conceive from their nature, a romantic conception invented to please the imagination, or to explain a system—the time is past when geology could be considered as the romance of nature—nor has what we advance the character of an arbitrary conception. One is no doubt struck with surprise on learning for the first time that all the limestone rocks, all the calcareous stones which have been employed in the construction of our dwellings, our cities, our castles and cathedrals throughout the historic period are deposits of the seas of the ancient world, and were originally only an aggregation of the shells of mollusks, or the débris of the testaceous covering of foraminifers and other zoophytes—nay, that they were abstracted from the water itself, and manipulated by these minute creatures, and that this would appear to have been the great object of their creation in such myriads. Whoever will take the trouble to observe, and reflect on what he observes, will find all his doubts vanishing. Examine the chalk with a microscope, he will find that it is composed of a mass of débris of numerous zoophytes, of minute ammonites, of divers kinds of shells, and, above all, of foraminifera so small that their very minuteness seems to have rendered them indestructible. A hundred and fifty of these small beings placed end to end will only occupy the space of about the twelfth part of an inch.

Much of this curious information was unknown, or at least only suspected, when Ehrenberg began his microscopical investigations. From small samples of chalk reduced to powder, placed upon the object-glass, and subjected to the microscope, Ehrenberg prepared the designs which we reproduce from his learned micrographical work, in which some of the elegant forms discovered in the chalk are illustrated, greatly magnified. Fig. 124 represents the chalk of Meudon, in France, in which ammonites and other forms appear. Fig. 125, from the chalk of Gravesend, contains similar objects. Fig. 126 is a sample of chalk from the island of Moën, in Denmark; and Fig. 127, that which is found in the tertiary rocks of Cattolica, in Sicily. In all these samples the ammonites appear, with clusters of round foraminifera and other zoophytes. In two of these engravings (Figs. 124 and

126), the chalk is represented in two modes—in the upper, by transparency or reflection, in the lower, half the mass is exhibited by superficial light.

Observation, then, establishes the truth of the explanation we have given concerning the formation of the chalky or cretaceous rocks, but the question still remains, How were these rocks, originally deposited in the sea, elevated into hills of considerable height, with bold escarpments,

Chalk under the Microscope.

Fig. 124.—Chalk of Meudon.

like those known as the North and South Downs? The answer to this involves other questions which have scarcely got beyond hypothesis.

When the Portland beds had been deposited, the entire oolite series in the south and centre of England and other regions was raised above the sea level, and became dry land. Above these Purbeck beds, as Professor Ramsay tells us, in the district known as the Weald, " we have a series of beds of clays, sandstones, and shelly limestones, indicating by their fossils that they were deposited in an estuary where fresh water and occasionally brackish water and marine conditions prevailed. The Wealden and Purbeck beds indeed represent the delta of an immense river which in size may have rivalled the Ganges or Mississippi, whose waters carried down to its mouth the remains of land plants, small mammalia, and great terrestrial reptiles, and mingled them with the remains of shells, fish, and other forms native to its waters. I do not say that this immense river was formed

Chalk under the Microscope.

Fig. 125.—Gravesend Chalk. (After Ehrenberg.)

by the drainage of what we now call Great Britain—I do not indeed know where this continent lay, but I do know that England formed a part of it, and that in size it must have been larger than Europe, and was probably as large as Asia or the great continent of America." Speaking of the extent of the Wealden clay. Sir Charles Lyell says, " It cannot be accurately laid down, because so much of it is concealed beneath newer marine formations. It has been traced above two hundred miles from west to east; from the coast of Dorset

R

to near Boulogne, in France, and nearly two hundred miles from north-west to south-east; from Surrey and Hampshire, to Beauvais in France;" but he expresses doubts, supposing the formation to have been continuous, if it were contemporaneous, the region having undergone frequent changes, the great estuary having altered its form, and even shifted its place. Speaking of a hypothetical continent, Sir Charles Lyell says, "If it be asked where the continent was placed from the

Chalk under the Microscope

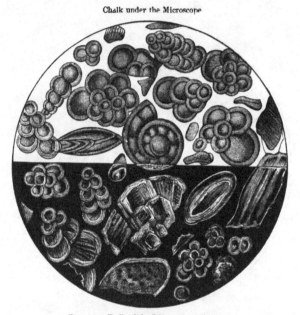

Fig. 126.—Chalk of the Isle of Moën, Denmark.

ruins of which the Wealden strata were derived, and by the drainage of which a great river was fed, we are half tempted to speculate on the former existence of the Atlantis of Plato. The story of the submergence of an ancient continent, however fabulous in history, must have been true again and again as a geological event."

"The proof that the Wealden series were accumulated as a freshwater deposit lies," he adds, "partly in the nature of the strata, but chiefly in the nature of the organic remains. The fish give no

positive proof, but a number of crocodilian reptiles give more con-
clusive proofs, together with the shells, most of them being of fresh-
water origin, such as Paludina, Planorbis, Lymnæa, Physa, and
such like, which are found living in many ponds and rivers of the
present day. Now and then we find bands of marine remains not
mixed with fresh-water deposits, but interstratified with them, showing

Chalk under the Microscope.

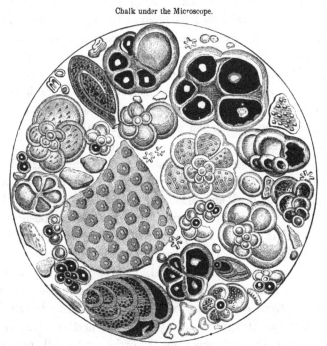

Fig. 127.—Chalk of Cattolica, Sicily.

that at times the mouth and delta of the river had sunk a little, and
that it had been invaded by the sea; then by gradual change it was
lifted up, and became an extensive fresh-water area. The episode
at last comes to an end by the complete submergence of the Wealden
area, and upon these fresh-water strata a set of marine sands and
clays, and upon these again thick beds of pure white earthy limestone
of the cretaceous period were deposited. The lowest of these forma-

R 2

tions is known as the Lower Greensand; then followed the clays of the gault, which were succeeded by the Upper Greensand. Then, resting upon the Upper Greensand comes the vast mass of chalk which in England consists of soft white limestone, containing, in the upper part, numerous bands of interstratified flints, which were mostly sponges originally, since silicified and converted into flint. The strata of chalk where thickest are from 1,000 to 1,200 feet in their thickest part. Their upheaval into land brought this epoch to an end; the conditions which had contributed to its formation ceased in our area, and as the uppermost member of the secondary rocks, it closes the record of Mesozoic life in England."

Let us add, to remove any remaining doubts, that in the basin of a modern European sea—the Baltic—a curious assemblage of phenomena bearing on the question are now in operation. The basin of the Baltic continues unceasingly to rise, and has done so for several centuries, in consequence of the constant deposit which takes place of calcareous shells and testaceous envelopes added to the natural accumulations of sand and mud. The Baltic Sea will certainly be filled up one day by these deposits, and this modern phenomenon, which we find in the act, so to speak, brings directly under our observation an explanation of the manner in which the cretaceous rocks were produced in the ancient world, more especially when taken in connection with another branch of the same phenomenon to which Sir Charles Lyell called attention, in a recent address to the Geological Society. It appears that just as the northern part of the Scandinavian continent is now rising, and while the middle part south of Stockholm remains unmoved, the southern extremity in Scania is sinking, or at least has sunk, within the historic period: from which he argues that there may have been a slow upheaval in one region, while the adjoining one was stationary, or in course of submerging.

After these explanations as to the manner in which the cretaceous rocks were formed, let us examine into the state of animal and vegetable life during this important period in the earth's history.

The vegetable kingdom of this period is the vestibule, as it were, to the vegetation of the present time. Placed at the summit of the secondary epoch, this vegetation is a preparation, and forms a transition, as it were, to the tertiary epoch, which, as we shall see, has a tendency to be confounded with that of our age.

The landscapes of the ancient world still show us some species of vegetables of forms strange and little known, which are now extinct. But during the period whose history we are tracing the vegetable kingdom begins to fashion itself in a manner less mysterious; the palms appear; and in the several species we recognize some which differ little from those of the tropics of our days. The dicotyledons increase slightly in number amid ferns and Cycadaceæ, which have lost much of their importance in numbers and size: we observe an obvious increase in the dicotyledons of our own temperate climate, such as the alder, the wych-elm, the maple, and the walnut-tree—trees of our native land, we salute thee with joy!

"As we retire from the times of the primitive creation," says Lecoq, "and slowly approach those of our own epoch, the sediments seem to withdraw themselves from the polar regions and restrict themselves to the temperate or equatorial zones. The great beds of sand and limestone, which constitute the chalky formation, announces a state of things very different from that of the preceding ages. The seasons are no longer marked by indications of central heat; zones of latitude already show signs of their existence; already the biological conditions of living beings are such as we can comprehend, and the vegetation takes quite a peculiar form.

"Hitherto two classes of vegetation predominated: the cellular *Cryptogams* at first, the dicotyledonous *Gymnosperms* afterwards; and in the epoch which we have reached—the transition epoch of vegetation—the two classes which have reigned heretofore become enfeebled, and a third, the dicotyledonous *Angiosperms*, timidly take possession of the earth. They consist at first of a small number of species, occupying only a small part of the soil, of which it afterwards takes its full share; and in the following periods, as in that of our times, we

shall see that its reign is firmly established : during the cretaceous
period, in short, we witness the birth of the first dicotyledonous *Angiosperms*. Some arborescent ferns still maintain their position, and the
elegant *Protopteris Singeri*, Presl., and *P. Buvigneri*, Brongn., still
unfold their light fronds to the winds of this agitated period. Some
Pecopterii, differing from the Wealdean species, live along with them.
Some *Zamites*, *Cycadæs* and *Zamiostrobus* announce that in the
cretaceous period the temperature was still high. New palms show
themselves ; among others, *Flabellaria chamæropifolia* is especially
remarkable for the majestic crown which it carries.

"The *Conifera* have resisted better than the *Cycadeæ :* they formed
then, as now, great forests, where *Damarites*, *Cunninghamias*,
Araucarias, *Eleoxylons*, *Abietites* and *Pinites* remind us of numerous
forms still existing, but now dispersed all over the earth.

"From this epoch date the *Comptonias*, attributed to the Myricæ ;
Almites Fresii, Nils., which we consider as a Betulaceæ ; *Carpinites
areniaceus*, Gœp., which would be a Cupiliferæ ; the *Salictites*, which are
represented to us by the arborescent willows ; the Acerinæs would have
their *Acerites cretaceæ*, Nilson, and the Juglanditæ, the *Juglandites
elegans*, Gœp. But the most interesting botanical event of this
period is the appearance of the *Crednaria*, with its triple-veined
leaves, of which no less than eight species have been found and
described, but whose place in the systems of classification still remains
uncertain. The *Crednarias*, like the *Salicites*, were certainly trees, as
were, for the most part, the species of this remote epoch."

In the following illustration, the reader has represented two of the
palms belonging to the cretaceous period, restored from the imprints
of the fossil remains left by the trunk and branches in the rocks of
the period. (Fig. 128.)

But if the vegetation of this period exhibits sensible signs of approximation to that of our era, we cannot say the same of the animal
creation. The time has not yet come when mammifera analogous to
those of our epoch gave animation to the forests, plains, and shores of
the ancient world : even the marsupial mammifera, which made their ap-

pearance in the liasic and oolitic formations, no longer exist, so far as

Fig. 128.—Fossil Palms restored.

is known, and no others of the class have taken their place. No climbing opossum with its young one appears among the leaves of the

Zamites! The earth appears still appropriated by reptiles, which alone reveal, by their accidental breathings, the solitudes of the woods and the silence of the valleys. The reptiles, which seem to have swarmed in the seas of the Jurassic period, partook of the crocodilian organization: the reptiles of this period seem more to have resembled the Saurians; namely, reptiles, of which our lizards may be looked upon as the perfection. In this period their remains indicate that they were borne on higher feet; they no longer creep on the earth, and this seems, to be the only approximation creation makes towards the mammalian form.

It is not without surprise that we revert to the immense development, the extraordinary dimensions, which the lizards attain at this epoch. These animals, which in our days rarely exceed a yard or so in length, seem to have attained in the cretaceous period as much as twenty. The marine lizard, which we study under the name of *Mosasaurus*, was then the scourge of the seas, playing the part of the Ichthyosauri in the Jurassic period: for from the age of the lias to that of the chalk, the Ichthyosaurus, the Plesiosaurus and the Teleosaurus were, judging from their organization, the tyrants of the ocean. Well, they seem to have become extinct in the cretaceous period, and given place to the *Mosasaurus*, to whom fell the formidable task of keeping within just limits the exuberant production of the various tribes of fishes and crustaceans, and preventing them from over-populating the seas. This creature was first discovered in the celebrated rocks of St. Peter's Mount at Maestricht, on the banks of the Meuse, the skull alone being about four feet, while the *Iguanodon Mantelli*, discovered by Dr. Mantell in the Wealdean strata, has since been discovered in the Hastings beds in Tilgate forest, measuring, as Professor Owen estimates, between fifty and sixty feet. These enormous Saurians disappear in their turn, to be replaced in the seas of the tertiary epoch by the Cretaceans; by our whales, in short; and henceforth animal life begins to assume more and more the appearance it presents in the actually existing creation.

Seeing the great extent of the seas of the cretaceous period, fishes were necessarily numerous. Pike, or Esocidæ, and the salmons, and the dory, analogous to those of our days, lived in the seas of the period: they fled before the sharks and voracious dog-fishes, which now appear in great numbers, after just showing themselves in the oolitic period.

The seas were still full of polypiers, sea-urchins, crustaceans of various kinds, and genera of mollusks different from those of the Jurassic period; alongside of gigantic lizards are whole piles of animalculæ— those Foraminifera whose remains are now scattered in infinite profusion in the chalk, upon a surface of immense thickness. The calcareous remains of these little beings, incalculable in number, have indeed covered, in all probability, a great part of the terrestrial surface. It will give an idea of the importance of the period in relation to these organic beings, when it is stated that, in the rocks of the period, two hundred and sixty-eight genera, hitherto unknown, and more than five thousand species, have been found; the rocks formed during the period being upwards of four thousand yards. Where is the geologist who will venture to estimate the time occupied in creating and destroying the animated masses of which this is the cemetery? For the purposes of description it will be convenient to divide the formation into lower and upper, according to their antiquity, their peculiar fossils, and sometimes according to their mineral characters.

THE LOWER CRETACEOUS PERIOD.

French classification.	English equivalents.
Etage Néocomien inférieur.	Wealden beds and Hastings sands.
„ Néocomien supérieur.	Lower greensand.
„ Aptien st.	„ „ upper part.
„ Albien.	The Gault.

The upper division of the Wealden clay is, as we have said, of purely fresh-water origin, and is supposed to have been the estuary of some vast river which, like the African Quorra, may have formed a delta of some hundreds of miles, as suggested by Dr. Dunker and Von Meyer. The lower Wealden is a band of sand, sandstone and calciferous grit clay and shale, the argillaceous strata predominating. This part of the Wealden consists of:—

	Feet.
Ashburnham sands—Marled white and red clay and sandstone	330
Ashdown sands—Hard sand, with beds of calc grit	160
Wadhurst clay—Blue and brown shale, with a little calc grit	100
Tunbridge sand – Sandstone and loam	150

The latter having a hard bed of white sand in its upper part, whose steep natural cliffs produce the picturesque scenery of the "High rocks" of that neighbourhood. This Wealden clay has been shown by Dr. Mantell to crop out beneath the lower greensand in various parts of Kent and Sussex, and again in the Isle of Wight, where it reappears at the base of the chalk; the calcareous sandstone and grit in which Dr. Mantell found the remains of the *Iguanodon* and *Hylæosaurus* forming an upper member of the Tunbridge sand. The formation extends over Hanover and Westphalia; the Wealden of these countries, according to Dr. Dunker and Von Meyer, corresponding in their fossils and mineral characters with those of the English series. So that " we can scarcely hesitate," says Lyell, " to refer the whole to one great delta."

The lower greensand is known also as the *Néocomian*, the Latin name of the city of Neufchatel, in Switzerland, where this formation is largely developed, where also it was first recognized and established as a distinct formation. Dr. Fitton, in his excellent monograph of the lower cretaceous formations, gives the following ascending succession of rocks as observable in many parts of Kent:—

	Feet
1. Calcareous stone, called Kentish rag	60 to 80
2. Sand, with green earthy matter	70 to 100
3. Sand, white, yellowish or brown, with limestone, concrete and chert	70

These divisions, which are traceable more or less from the southern part of the Isle of Wight to Hythe in Kent, present considerable variations. At Atherfield, where sixty-three distinct stratum, measuring 843 feet, have been traced, the limestone is wholly wanting, and some fossils range through the whole series, while others are confined to particular divisions : but Dr. Forbes states that when the same conditions are repeated in overlying strata the same species reappear ; but that changes of depth, of the mineral part of the sea bottom, the pressure or absence of lime or peroxide of iron, the occurrence of muddy, sandy or gravelly bottom, are each marked by the absence of certain species, and the predominance of others.

Among the marine fauna of the Néocomien series the following are the principal. Among the *Acephales*, one of the largest and most abundant shells of the lower Néocomien, as displayed in the Atherfield series, is the large *Perna Mullati* (Fig. 129).

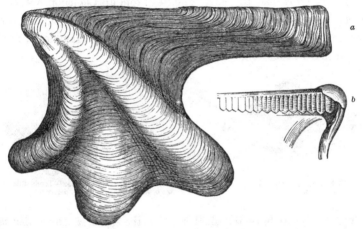

Fig. 129—Perna Mullati. One-quarter natural size.
a, exterior; *b*, part of the upper hinge.

The *Scaphites* have a shell forming a regular spiral scroll upon the

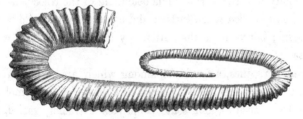

Fig. 130.—Hamites. One-third natural size.

same plane, with contiguous rounds, increasing regularly up to the last evolution, when it detaches itself from the others, and projects a cross more or less elongated.

The *Hamites, Crioceras* and *Ancyloceras,* have club-like terminations at both extremities; they may almost be considered as ammonites with the spiral evolutions dissevered, as in the engraving (Fig. 130). *Ancyloceras Matheronianus* seems to have had spines projecting from the ridge of each of the convolutions.

The *Toxoceras* had the shell also curved, and not spiral.

The *Baculites* had the shell differing from all Cephalopodes, inasmuch as it was elongated, conical, perfectly straight, sometimes very slender, and tapering to a point. Among others, as examples of form, we append the following:—

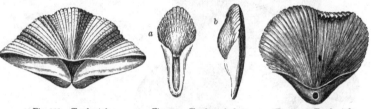

Fig. 131.—Terebratula   Fig. 132.—Terebratula lyra.   Fig. 133.—Terebratula
  canalifera.   *a*, back view; *b*, side view.   deformis.

The *Turrilites* have the shell regular, the spiral *sinister;* that is, turning to the left in an oblique spiral of contiguous whorls. The engraving in the margin will convey the idea of form, although it is the representation of an existing species (Fig. 135).

This analysis of the marine fauna belonging to the Néocomian formation might be carried much further, did space permit, or did it promise to be useful; but without illustration any further description would be valueless.

Numerous reptiles, a few birds, among which are some " Waders," belonging to the genera of *Palæornis* or *Cimoliornis,* new mollusks in considerable quantities, and some zoophytes extremely varied, constitute the rich fauna of the lower chalk. A glance at the more important of these animals, which we only know in a few mutilated fragments, is all our space allows: they are ·true medals of the history of our globe, medals half-effaced by time, which consecrates the memory of departed ages.

In the year 1832 Dr. Mantell added to the wonderful discoveries

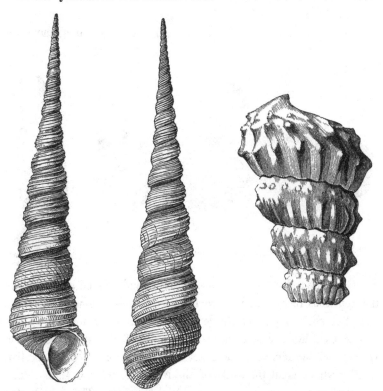

Fig. 134.—Shell of Turritella terebra.

Fig. 135.—Turrilites costatus.

he had made in the same neighbourhood that of the great lizard of the woods, the *Hylæosaurus*, ὕλη, σαῦρος. This discovery was made in the fossil forest of Tilgate, near Cuckfield, and the animal appears to have been about eight yards in length. We have already seen a restoration of this enormous Saurian in PL. IX., page 105. What has been found is reduced to a series of long and pointed bones, which may have formed upon the spine a hard and half-ossified fringe, like the horny spines which run along the back of the reptile which at the present day receives the name of *Iguano*. Some fragments of these great bony plates that have been found mixed with the same débris were probably lodged in the skin of this animal, forming a sort of cuirass for it.

The *Megalosaurus*, the earliest appearance of which is among the

Fig. 136.—Jaw of the Megalosaurus.   Fig. 137.—Tooth of Megalosaurus.

more ancient beds of the Jurassic series, is again found at the base of the cretaceous rocks. It is, as we have seen, an enormous lizard, borne upon feet slightly raised : its length reached about forty-five feet. Cuvier considered that it partook at once of the structure of the Iguana and the Monitors, the latter of which belong to the Lacertian reptiles which haunt the banks of the Nile and tropical India. The Megalosaurus was probably a terrestrial Saurian. The complicated structure and marvellous arrangement of the teeth prove that it was essentially carnivorous. It fed probably on other serpents of moderate size, such as the crocodiles and turtles which are found in the fossil state in the same beds. The jaw represented above in Fig. 136 is the most important fragment we possess of the animal. It is the lower jaw, and supports many teeth : it shows that the head terminated in a straight muzzle, thin and flat on the sides, like that of the *Gavial*, the crocodile of India. The teeth of the Megalosaurus were in perfect accord with the destructive functions evolved in this formidable creature. They partake at once of the knife, the sabre and the saw. Vertical at their junction with the jaw, they assume with the increased age of the animal a backward curve, giving them the form of a gardener's pruning-knife (Fig. 137). After insisting upon some other

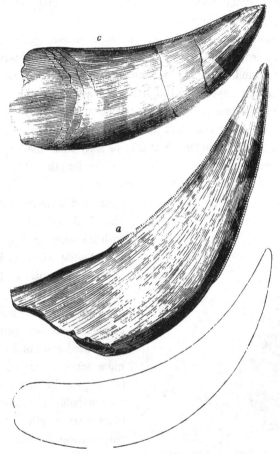

Fig. 138.—*a*, Tooth of Machairodus, imperfect below, natural size; *b*, outline of cast of tooth, perfect, half natural size; *c*, tooth of Megalosaurus, natural size.

particulars respecting these teeth, Buckland says, " With teeth constructed so as to cut with the whole of their concave edge, each movement of the jaws produced the combined effect of a knife and a saw, at the same time that the point made a first incision like that made by the point of a double-cutting sword. The backward curvature taken by the teeth at their full growth render the escape of the prey when once seized impossible. We find here, then, the same arrangements which enable mankind to put in operation many of the instruments which they employ."

The *Iguanodon*, from the Greek word, ὀδούς, tooth, signifying *Iguano*, toothed, was more gigantic still than the Megalosaurus: the most colossal, indeed, of all the Saurians of the ancient world which research has yet exposed to the light of day. Professor Owen and Dr. Mantell are not agreed as to the form of the tail; the former gentleman assigning it a short tail, which would affect Dr. Mantell's estimate of its probable length of fifty or sixty feet: the largest femur bone yet found measures four feet eight in length. The form and disposition of the feet, added to the existence of an osseous horn on the

upper part of the muzzle or snout, almost identifies it as a species with the existing Iguanidæ, the only reptile which is known to be provided with such a horn upon the nose: there is therefore no doubt as to the resemblance between these two beings; but while the largest of living Iguanidæ scarcely exceeds a yard in length, its fossil congener was probably fifteen or sixteen times that length. It is difficult to resist the feeling of astonishment, not to say incredulity, which creeps over one while contemplating the disproportion so striking between this being of the ancient world and its congener of the new.

Fig. 139.—Nasal horn of Iguanodon. Two-thirds natural size.

The Iguanodon carries, as we have said, a horn on its muzzle; the

XXI.—Ideal scene in the Lower Cretaceous Period, with Iguanodon and Megalosaurus.

bone of its thigh, as we have seen, surpasses that of the elephant; the
form of the bone and feet demonstrate that it was formed for terrestrial locomotion; and its dental system shows that it was herbivorous.

The teeth (Fig. 140), which are the most important and charac-

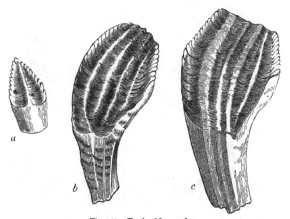

Fig. 140 —Teeth of Iguanodon.
a, young tooth; *b, c*, teeth further advanced, and worn.

teristic organs of the whole animal, are not fixed in distinct sockets
like the crocodiles, but fixed on the internal face of a dental bone; that
is to say, in the interior of the palate, as in the lizards.   The place
thus occupied by the edges of the teeth, their trenchant and saw-like
form, their mode of curvature, the points where they become broader
or narrower which turn them into a species of nippers or scissors—
are all suitable for cutting and tearing the coriaceous and resisting
plants which are also found among the remains buried with the
reptile, a restoration of which is represented in PL. XXI.

The cretaceous seas contained great numbers of fish, among which
some were remarkable for their strange forms.   The *Beryx Lewesiensis* (1) and the *Osmeroïdes Mantelli* (2), (Fig. 141), are restorations
of these two species as they are supposed to have been in life.   The
Odontaspis is a new genera of fishes which may be mentioned.

The seas of the lower cretaceous period were remarkable in a zoological point of view for the great number of species, and the multiplicity of generic forms of molluscous cephalopodes.   The ammonites

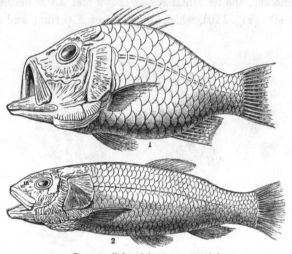

Fig. 141.—Fishes of the cretaceous period.
1, Beryx Lewesiensis; 2, Osmeroides Mantelli.

assume dimensions quite gigantic, and we find among them new species distinguished by their furrowed transverse spaces, as in the *Hamites* (Fig. 130.)   Some of the *Ancyloceras* attained the magnitude of six feet, and other genera, as the *Scaphites*, the *Toxoceras*, the *Crioceras* and other mollusks, unknown till this period, appeared now. Many Echinoderms, or sea-urchins, and zoophytes, have enriched these rocks with their animal remains, and would give to its seas a condition quite peculiar.

On the opposite page an ideal landscape of the period is represented (PL. XXI.), in which the Iguanodon and Megalosaurus struggle for the mastery in the centre of a forest, which enables us also to convey some idea of the vegetation of the period.   Here we note a vegetation at once exotic and temperate—that of the tropics, and a flora resembling our own.   On the left we observe a group of trees, which resemble the dicotyledonous plants of our forests.   The elegant *Credneria* is there, whose botanical place is still doubtful, for its fruit has not been found, although it is believed to have belonged to plants with two seed-leaves, or dicotyledonous, and the arborescent Amentaceæ. An entire group of trees, composed of ferns and zamites, are in the background; in the extreme distance are some palms. We also recognize in the picture the alder, the wych-elm, the maple, and the walnut-tree, or at least species analogous to these.

1. The Neocomian beds in France are found in Champagne, in the department of the Aube, the Yonne, the Haute-Alps. It is largely developed in Switzerland at Neufchatel, and in Germany, where the lower beds consist of marls and greyish clay, alternating with small banks of calcareous limestone, which are very thick at Neufchatel and in the Drome. The fossils are *Spatangus retusus, Crioceras, Ammonites, Astierianus.*

2. *Urgonian.* The limestone of Orgon exists also at Aix-les-Bains, in Savoy, at Grenoble, and generally in the thick white calcareous beds which form the precipices of the Drome; the fossils *Chama ammonia, Pigaulus,* &c.

3. The *Aptien,* or greensand, consists generally of marls and argillaceous clay. In France it is found in the department of the Vaucluse, at Apt, in the department of the Yonne, whence its name; and in the Haute-Marne. Fossils, *Ancyloceras Matheronianus, Ostrea aquila,* and *Plicatula placunea.* These beds consist here of greyish clay, which is used for making tiles; there of bluish argillaceous limestone, in black or brownish flags. In the Isle of Wight it becomes a fine sandstone, greyish and a little argillaceous. At Havre, and in some parts of the country of Bray, it is a well-developed ferruginous sandstone.

We have already noted that the Neocomian formation, although a marine deposit, is in some respects the equivalent of the *Wealden clay,* a fresh-water formation of considerable importance on account of its fossils. We have seen that it was either formed at the mouth of a great river, or the river was sufficiently powerful for the fresh-water current to be carried out to sea, carrying with it some animals,

forming a fluvial or lacustrine fauna, on a small scale. These were
small crustaceans of the genera of *Cypris.* Some molluscous gasteropodes of the genera *Melania, Paludina,* and acephalous mollusks of
the five genera *Cyrena, Unio, Mytilus, Cyclas,* and *Ostrea.* Of
these, *Cypris spinigera* (Fig. 142) and *Cypris Waldensis* (Fig. 143)
may be considered as among the most characteristic fossils of this
local fauna.

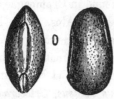

Fig. 142.—Cypris spinigera.   Fig. 143.—Cypris Waldensis.

The cretaceous rock is not alone interesting for its fossils; it presents also an interesting subject of study in a mineralogical point of
view. The white chalk, examined under the microscope by Ehrenberg,
shows a curious globiform structure. The green part of its sandstone
and limestone constitutes very singular compounds. According to
the results of Berthier's analysis, we must consider them as silicates
of iron. The iron shows itself here not in beds, as in the Jurassic
rocks, but in masses, in a species of pouch of the Urgonian beds.
They are usually hydrates in the state of hematites, accompanied by
quantities of ochre so abundant that they are frequently unworkable.
In the south of France these veins were worked to a great depth by
the ancient monks, who were the metallurgists of their age. But for
artists the powerful Urgonian beds possess a special interest: their
admirable vertical fractures, their erect, perpendicular peaks, each
surpassing the other in boldness, form his finest studies. In the Var,
the defiles of Vésubia, of the Esteron, of Tinea, are jammed up
between walls of peaks, for many hundreds of yards, between which
there is scarcely room for a narrow road at the side of the roaring
torrent. "In the Drome," says Fournet, "the entrance to the
beautiful valley of the Vercors is closed during a part of the year,

because, in order to enter, it is necessary to cross the two orifices, the *Great* and *Little Goulet*, through which the waters escape from the valley. Even during the dry season, he who would enter the gorge must take a foot-bath.

" This state of things could not last, and in 1848 it was curious to see miners suspended on the sides of one of these lateral precipices, some four hundred and fifty feet above the torrent, and about an equal distance below the summit of the chalk. There they began to excavate cavities, or niches in the face of the rock, all placed on the same level, and successively enlarged. These were united together in such a manner as to form a road practicable for carriages; now through a gallery, now covered by a corbeilling, to look over which affords a succession of surprises to the traveller.

" This is not all," adds M. Fournet: " he who traverses the high plateaux of the country finds at every step deep diggings in the soil, designated pits or *scialets*, the oldest of which have their concavities clothed in a curious vegetation, in which the *Aucolin* predominate: they find in these pits shelter from the cutting winds which rage so furiously in these culminating regions. Others form a kind of cavern, in which a temperature obtains sufficient to congeal water even in the middle of summer. These cavities form natural *glaciers*, which we again find upon some of the table-lands of the Jura.

" The cracks and crevasses of the limestone receive the waters produced by falling rain and melted snow: true to the laws of all fluid bodies, they infiltrate themselves through the rocks until they reach the lower and impervious marly beds, where they form sheets of water, which in course of time find some issue through which they discharge themselves. In this manner subterranean galleries, sometimes of great extent, are formed, in which are assembled all the marvels which ebullition, stalactites, stalagmites, placid lakes, and fugitive torrents can engender: finally, these waters, forcing themselves a passage through some external orifice, give birth to one of these fine cascades which pour out whole rivers at a single bound."

The *Albien* of Alc. D'Orbigny, which Lyell considers to be the

equivalent of the *gault*, French authors treat as the *glauconous* formation, the name being drawn from a rock composed of grey chalk beds,
grains of verditer, or silicate of iron, which is often mixed with the
limestone of this rock.   The fossils by which it is identified are very
varied.   Among the numerous types, we find crustaceans belonging
to the genera *Arcania* and *Corystes ;* many new mollusks of *Buccinum,
Solen, Pterodonta, Voluta, Chama ;* great numbers of molluscous
brachiopodes, forming well-developed submarine banks ; some echinoderms hitherto unknown, and especially a great number of zoophytes ;
some Foraminifera, and many Bryozoaires.   The glaucous beds are
formed of two classes : the *gault* clay and *glaucous* chalk.

The *gault* is the lowest member of the upper cretaceous group, and
consist of a bluish-black marl mixed with greensand, called " of the
gault," which occupies the lower beds.   It attains a thickness of about
100 feet on the south-east coast of England.   It extends into Devonshire, Mr. Sharpe considering the Black down beds of that county as its
equivalents.   It shows itself in the department of the Pas-de-Calais,
of the Ardennes, of the Meuse, of the Aube, of the Yonne, of the
Ain, of the Calvados, and of the Seine-Inférieure.   It presents very
distinct mineral forms, among which two predominate : the green
sandstone and the black or greyish clay.   It is important to know
this formation, for it is at this level that the Artesian waters flow in
the wells of Passy and Grenoble near Paris.

The *glaucous* chalk, or upper beds of the gault, represented typi-
cally in the departments of the Sarthe, of the Charente-Inférieure, of
the Yonne and the Var, is composed of quartzose sand, clay, sandstone,
and limestone.   In this formation, at the mouth of the Charente, we find
a remarkable bed, which has been described as a submarine forest.   It
consists of large trees with their branches embedded horizontally in
vegetable matter, kidney-shaped nodules of amber, and fossilized resin.

UPPER CRETACEOUS PERIOD.

During this phase of the terrestrial evolutions, the continents, to judge from the fossilized wood which we meet with in the rocks which now represent it, would be covered with a very rich vegetation; nearly identical, indeed, with that which we have described in the preceding sub-period: according to Adolphe Brongniart, the "age of angiosperms" had fairly set in: the cretaceous flora displays, he considers, a transitional character from the secondary to the tertiary vegetation: that the line between the gymnosperms, or naked-seeded plants, and the angiosperms, having their seeds enclosed in seed-vessels, runs between the upper and lower cretaceous formations. "We can now affirm," says Lyell, "that these Aix-la-Chapelle plants, called Credneria, flourished before the rich reptilian fauna of the secondary rocks had ceased to exist. The Ichthyosaurus, Pterodactyl, and Mosasaurus were of coeval date with the oak, the walnut and the fig."

The terrestrial fauna, consisting of some new reptiles haunting the banks of rivers, and birds of the genera of snipes, have certainly only reached us in small numbers. The remains of marine fauna occur, on the contrary, in great numbers, and so well preserved as to give a great idea of its riches: they even form a characteristic of its condition.

The sea of the upper cretaceous period bristled with numerous submarine reefs, occupying a vast extent of its beds, reefs formed by the singular family of mollusks called Rudistes by Lamarck, and by zoophytes in immense numbers, judging from the quantities of coral which remain. The Polypiers, in short, seem to attain here the principal epoch of their existence, and assume remarkable developments of form; the same occurs with the Bryozoaires and Amorphozoaires; while, on the contrary, the reign of the Cephalopodes seems to have been at an end. Beautiful types of these ancient reefs have been revealed to us, and we discover that they have been formed under the influence of submarine currents, which accumulated masses of these animals at certain points. Nothing is more curious than this assemblage of

rudiste productions, still perpendicular, isolated or in groups ; such, for
instance, as we find them at the summit of the mountains of the
Cornes, in the Corbieres, upon the banks of the pond of Berre, in Provence, and in the environs of Martigues, at La Cadiére, at Figuières, and
particularly above Beausset, near Toulon.

"It seems," says Alcide D'Orbigny, "as if the sea had retired
in order to show us, still intact, the submarine fauna of this period,
such as it was when in life. There are here enormous groups of *Hip-*
purites in their place, surrounded by polypi, echinoderms and
mollusks, which lived in union in animal colonies analogous to those
which still exist in the coral reefs of the Antilles and Oceania. In
order that these groups should be preserved it was necessary that they
should be covered at once and suddenly, by the sediment which
is now, after being destroyed by the action of the atmosphere,
revealing to us in their most secret details the nature of the ages
which have passed.

In the Jurassic period we have already met with these isles or reefs
formed by the accumulation of corals and other zoophytes : they even
constituted an entire formation in that period called the *coral-rag.*
The same phenomenon, reproduced in the cretaceous seas, gave birth to
similar calcareous formations ; we need not, therefore, insist farther on
them here. The coral isles or madrepores of the Jurassic epoch and the
reefs of Rudistes and Hippurites of the cretaceous period have the
same origin, and the *atols* of Oceania are reproductions in our days of
precisely the same phenomena.

The animals which characterize the cretaceous age are, among

Molluscous Cephalopodes,

Nautilus sublævigatus and *Danicus ; Ammonites rusticus ; Belemni-*
tella mucroneta.

Molluscous Gasteropodes.

Voluta elongata ; Phorus canaliculatus ; Nerinea bisulcata ; Pleu-
rotomaria Fleuriansa ; and *Santonensis, Natica supracretacea.*

ACEPHALOUS MOLLUSKS.

Trigonia scabra; Inoceramus problematicus and *Lamarkii; Clavigella cretacea; Pholadomya æquivalvis; Spondylus spinosus; Ostrea vesicularis; Ostrea larva; Janira quadricostata; Arca Gravesii.*

BRACHIOPODES RUDISTES.

Crania Ignabergensis; Terebratula obesa; Hippurites Toucasianus and *Organisans; Caprina Aguilloni; Radiolites radiosus,* and *acuticostus.*

BRYOZOAIRES AND ECHINODERMS.

Reticulipora obliqua; Ananchytes ovata; Micraster cor anguinum; Hemiaster bucardium and *Fourneli; Galerites albogalerus; Cidaris Forchammeri; Palæocoma Fustembergii.*

1. POLYPI; 2. FORAMINIFERA; 3. AMORPHOZOAIRES.

1. *Cycollites elliptica; Thecosmilia rudis; Enallocœnia ramosa; Meandrina Pyrenaica; Synhelia Sharpeana.* 2. *Orbitoides media; Lituola nautiloidea; Flabellina rugosa.* 3. *Coscinopora cupiliformis; Camerospongia fungiformis.*

Among the numerous beings which inhabited the upper cretaceous seas there is one which, by its organization, its proportions, and the despotic empire which it would exercise in the bosom of the waters, is certainly worthy of our attention. We speak of the *Mosasaurus,* which was long known as the great animal of *Maestricht,* because its remains were found near that city, in the most modern of the cretaceous deposits.

It was in 1780 that the head of some great Saurian was discovered in the quarries of Saint Peter's Rocks, near Maestricht, which may now be seen in the Museum of Natural History of Paris. This discovery baffled all the science of the naturalists of a period when the knowledge of these ancient beings was still in its infancy. One saw in it the head of a crocodile; another, that of a whale; memoirs and monographs rained down without throwing much light on the subject. It required all the efforts of Adrian Camper, joined to those of the immortal Cuvier, to assign its true zoological place to the Maestricht animal. The controversy over this fine fossil occupied the learned for the remainder of the last century and far into the present.

Maestricht is a city of Holland, built on the banks of the Meuse. At the gates of this city, in the hills which skirt the left or western bank of the Meuse, there rises a solid mass of chalky formation known as Saint Peter's Rocks. In their composition they correspond with the Meudon chalk beds, including the same fossils. The quarries are about a hundred feet thick: consisting in the upper part of twenty feet abounding in corals and Bryozoaires, succeeded by fifty feet of soft yellowish limestone, a fine building-stone, which extends up to the environs of Liege, which has been quarried from time immemorial; a few inches of greenish soil with encrinites, and then a very white chalk with layers of flints. This quarry is filled with marine fossils, often of great size.

These fossil débris naturally enough attracted the attention of the curious, and led many to visit the quarries: but of all the discoveries which attracted attention the greatest interest attached to the gigantic animal under consideration. Among those attracted by the discovery of these strange vestiges was an officer of the garrison of Maestricht, named Drouin. He purchased the bones of the workmen as the pick disengaged them from the rock, and concluded by forming a collection at Maestricht, which may be cited with admiration. In 1766, the trustees of the British Museum having purchased this collection, it was removed to London. Incited by the example of Drouin, Hoffman, the surgeon of the garrison, set about forming a similar collection, and his collection soon exceeded that of Drouin's Museum in riches. It was in 1780 he purchased of the workmen the magnificent fossil head, exceeding six feet in length, which has since so exercised the sagacity of naturalists.

Hoffman did not long enjoy the fruits of his precious prize however: the Chapter of the church of Maestricht claimed, with more or less foundation, certain rights of property, and in spite of all protest, the head of the *Crocodile of Maestricht*, as it was already called, passed into the hands of the Dean of the Chapter, named Godwin, who enjoyed his antediluvian trophy until an unforeseen incident changed the aspect of things. This incident was nothing less than the bombardment and surrender of Maestricht to the Army of the North under Kleber, in 1793.

The Army of the North did not enter on a campaign to conquer the cranii of crocodiles, but it had on its staff a savant who was devoted to such pacific conquests: Faujus de Saint-Fond, who was the predecessor of Cordier in the Zoological Chair of the Jardin des Plantes, was attached to the Army of the North as Commissaire des Sciences, and it is suspected that in soliciting this mission our naturalist had in his eye the already famous head of the crocodile of the Meuse. However that may be, Maestricht fell into the hands of the French, and Faujus eagerly claimed the famous fossil for the French nation, which was packed with the care due to a relic numbering so many thousands of ages, and despatched to the Museum of Natural History at Paris. On its arrival, Faujus undertook a labour which, as he thought, was to cover him with glory. He commenced the publication of a work entitled ' The Mountain of Saint Peter of Maestricht,' describing all the fossil objects found in the Dutch quarry there, especially of the *Great Animal* of Maestricht. He endeavoured to prove that this animal was a crocodile.

Unfortunately for the glory of Faujus, a Dutch savant had devoted himself to the same study. Adrian Camper was the son of the great anatomist of Leyden, Pierre Camper, who had purchased of the heirs of the surgeon Hoffman some parts of the skeleton of the animal found in the quarry of Saint Peter. He had even published in the 'Philosophical Transactions' of London, as early as 1786, a memoir, in which the animal is treated as a whale. At the death of his father, Adrian Camper re-examined the skeleton, and in a work which Cuvier quotes with admiration, he fixed the ideas which were until then floating about. He proved that the bones belonged neither to a fish, nor a whale, nor to a crocodile, but rather to a particular genera of Saurian reptiles, which approach the Iguano in one direction and

the Lacertian Monitors in another. So that long before Faujus had
finished his publication the work of Adrian Camper had appeared, and
totally changed the ideas of the world in respect to it. This, however,
did not hinder Faujus from continuing to call his animal the Crocodile
of Maestricht. He even announced some time after, that Adrian
Camper was also of that opinion. "Nevertheless," says Cuvier, "it
is as far from the Crocodile as it is from the Iguano; and these two
animals differ as much from each other in their teeth, bones, and
viscera, as the ape differs from the cat, or the elephant from the horse."

The masterly memoir of Cuvier, while confirming all the views of
Camper, has restored the individuality of this surprising being. And
it is now recognised that this ancient *Lizard of the Meuse* belonged

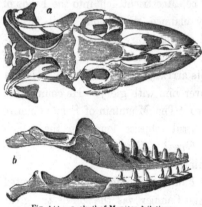

to a genera intermediate between the Saurians with forked
and extensible tongue, which
comprehend the Monitors
represented in the margin,
and the ordinary lizards; and
the short - tongued Saurians,
whose palate is armed with
teeth, a tribe which embraces
the Iguanidæ. In respect to
the crocodiles, it partakes of
them so far only, that all belong to the family of Saurians.

Fig. 144.—*a*, skull of Monitor Niloticus;
b, under-jaw of same.

The idea of a lizard, organised to live and move with rapidity in the
bosom of the waters, is not readily conceived; but studying the skeleton
of the Mosasaurus reveals to us this anatomical mechanism. The
vertebræ of the animal are concave in front and convex behind; they
were attached by means of orbicular or arched articulations, which
permitted it to execute easily movements of flexion in any direction.
From the middle of the back to the extremity of the tail, these
vertebræ are deficient in the articular processes which support and
strengthen the trunk of terrestrial vertebrated animals: they re-

sembled in this respect the vertebræ of the dolphins; an organisation necessary to render natation easy. The tail, compressed laterally at the same time that it was thick in a vertical direction, constituted a straight rudder, short, solid, and of great power. An arched bone was firmly attached to the body of each caudal vertebra in the same manner as in fishes, for the purpose of giving increased power to the tail: finally, the extremities of the animal could scarcely be called feet, but rather paddles, like those of the Ichthyosaurus, the Plesiosaurus, and the whale. We see in Fig. 145, in the margin, that the

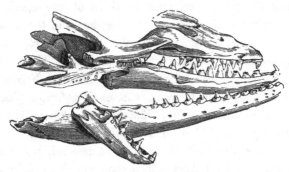

Fig. 145.—Head of Mosasaurus Camperi.

jaws are armed with numerous teeth, attached to their alveoles by an osseous base, both large and solid. Moreover, a dental system quite peculiar occupies the vault of the palate, as in the case of certain serpents and fishes, where the teeth are directed backwards, like the barb of a hook, thus opposing themselves to the escape of prey. This disposition of the teeth sufficiently proves the destructive character of this Saurian.

The dimensions of this aquatic lizard, estimated at twenty-four feet, are calculated to excite surprise. But we have already seen the Ichthyosaurus of the dimensions of a whale, and Teleosaurus thirty feet long; and the Iguanodon and Megalosaurus ten times the size of living Iguanidæ. In all these colossal forms we can only see a difference of dimensions, not any aggrandisement of a type: the laws which affect the organization of all these beings remain the same:

they were not errors of nature—*monstrosities*, as we are sometimes tempted to call them—but simply types, uniform in their structure, and only adapted by their dimensions to the physical conditions by which God has surrounded them.

On PL. XXII. is represented an ideal view of the earth during the *upper cretaceous* period. In the sea swims the Mosasaurus : mollusks, zoophytes, and other animals proper to the period are seen on the shore. The vegetation seems to approach that of our days ; it consists of ferns and pterophyllums, mingled with palms, willows and some dicotyledons of species analogous to those of our epoch. Some Algæ, then very abundant, compose the vegetation of the sea-shore.

We have said that the terrestrial flora of the upper cretaceous period was nearly identical with that of the lower chalk. The marine flora of these two epochs included some Algæ, Confervæ, Naïadæ, among which may be noted the following species : *Confervites fasciculata, Chondrites Mantelli, Sargassites Hynghianus.* Among the Naïadæ, *Zosterites Orbigniana* and *Z. lineata*, and several others.

The *Confervæ* are fossils which may be referred, but with some doubt, to the filamentous Algæ, which comprehend the great group of the Confervæ. These plants were formed of simple filaments, or branches, diversely crossing each other, or subdivided, and presenting traces of transversal partitions.

The *Chondrites* are fossil Algæ, with thick, smooth fronds, rameuse, pinnatifid, or divided into pairs, with divisions cylindrical, but with elliptical ends, or cylindric, and resembling *Chondruns Dumontia* and *Halymenia* among living genera.

The *Sargasses*, finally, have been vaguely referred to the genus *Sargassum*, so abundant in equatorial seas. These Algæ are distinguished by a filiform, branchy or rameuse stem, bearing foliaceous appendages, regular, often petiolate, like leaves, with vesicules, and supported by a small foot-stalk.

The rocks which actually represent the *upper cretaceous period* divide themselves naturally into three series ; but British and French

XXII.—Ideal landscape of the Cretaceous Period.

geologists make some distinction; the former dividing them into 1, *upper greensand*; 2, *chalky marl*; 3, *white chalk*, with *flints*; 4, *Maestricht* and *Faxoe* beds. The latter into 1, *Turonian*; 2, *Senonian*; 3, *Danian*.

The *Turonian* beds are so named because the province of Touraine, between Saumur and Montrichard, possesses the finest type of this strata. The mineralogical composition of the beds is a fine and grey marly chalk, as at Vitry-le-François; of chalk entirely white, of very fine grain, a little argillaceous, and poor in fossils, in Yonne, the Aube, and the Seine-Inférieure. Some tuffa chalks, granulated, and white or yellowish, mixed with spangles of mica, and containing ammonites, are found in Touraine and a part of the Sarthe, with white, grey, yellow, or bluish limestone, including Hippurites and Rudiolites. In England the lower chalk beds pass also into chalky marl, with ammonites, and then into beds known as the upper greensand, containing green particles of chloritic sand, mixed in Surrey with calcareous matter. In the Isle of Wight this formation attains the thickness of 100 feet. The *Senonian* beds take their name from the ancient *Senones*. The city of Sens is in the centre of the best-characterized portion of this formation; Epernay, Meudon, Sens, Vendôme, Royau, Cognac, Saintes, are the typical regions of the formation in France. In the basin of Paris, comprehending the Tournais beds, it attains a thickness of upwards of 1,500 feet, as was proved during the sinking of the Artesian well at Grenoble.

In its geographical distribution the chalk has an immense range; fine chalk of nearly similar aspect and composition being met with in all directions over hundreds of miles, alternating in its lower beds with layers of flints. In the higher beds in England it usually consists of a pure white calcareous mass, generally too soft for building-stone, but sometimes passing into a solid rock.

The *Danian* beds, which occupy the summit of the scale in the cretaceous formation, are finely developed at Maestricht, on the Meuse, and in the island of Zeeland, belonging to Denmark, where it is represented by a compact limestone, slightly yellowish, quarried for

the construction of the city of Taxoe.   It is slightly represented in
the basin of Paris, at Meudon, or at Lavirsines, in the department of
the Oise, by a white limestone often found in clods, known as *pisolithic*
limestone, from its pitchy fracture and colour.   In this formation
Ammonites Danica is found.   With respect to the *Danian* type of the
yellowish sandy limestone of Maestricht, besides some mollusks, polypi,
and bryozoaires, this limestone encloses remains of fishes, turtles, and
crocodiles.   But what rendered this rock so celebrated was that it
contained the remains of the celebrated *animal* of Maestricht, the
Mosasaurus.

At the close of the geological period, whose natural physiognomy
we have thus traced, Europe was still far from presenting the configuration which it now assumes.   A map of the period would represent the great basin of Paris, with the exception of a girdle of chalk,
and the whole of Switzerland, the greater part of Spain and Italy,
the whole of Belgium, Holland, Prussia, Hungary, Wallachia, and
North Russia as one sheet of water.   A band of Jurassic rocks still
united France and England at Cherbourg—a band which disappears
at a later period, and isolates the British islands from what is now
France.

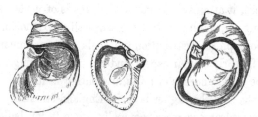

Fig. 146.   Exogyra conica.   Upper greensand and gault, from Blackdown Hill.

TERTIARY EPOCH.

A GREAT and new organic creation shows itself in the tertiary epoch; nearly all the animals are changed, and the most remarkable change in this new generation is the appearance of the great class of mammifera.

During the transition epoch, crustaceans and fishes predominate in the animal kingdom; in the secondary epoch the earth belonged to the reptiles; but during the tertiary epoch the mammifera were kings of the earth; nor do they appear in small numbers, or at distant intervals: great numbers of these beings appear to have lived on the earth, and at the same moment; many of them, so to speak, unknown and undescribed.

If we except the marsupials, which were imperfect mammifera, the first mammifers created would appear to have been the Pachyderms, to which the elephants belong. This order of animals long held the first rank; it represented the mammalia almost alone during the first of the three periods which constitute the tertiary epoch. In the second and third periods mammifers appear of species which have vanished, and which were alike curious from their enormous proportions, and by the singularity of their structure. Of those created during the last period of the epoch the greater part still exist. Among the new reptiles, some great salamanders, like crocodiles, are

T

added to the animal creation during the three periods of the epoch. During the same epoch birds appear, but in numbers much smaller than in the mammalia: here songsters, there rapacious birds, in other cases domestic; or, rather, some appear to wait the yoke of domestication from man, the supreme lord of the earth.

The seas were inhabited by a considerable number of beings of all classes, and nearly as varied as in our days; but we no longer find in the tertiary seas these ammonites, belemnites, and hippurites which filled the seas and multiplied with such astonishing profusion in the secondary epoch. Henceforth the shelled mollusks approximate in their forms to those of our days.

What occurs to us, however, as most remarkable in the tertiary epoch is the prodigious increase of animal life: it seems as if it had then attained its fullest extension. Swarms of shelled mollusks of microscopic proportions—the Foraminifera and Nummulites—must have encumbered the seas, crowding together in ranks so serried that the agglomerated débris of their shells form in some places beds hundreds of yards thick. It is the most extraordinary display which has appeared in the whole series of the creation.

Vegetation during the tertiary epoch presents well-defined characteristics. The tertiary flora approaches, and is sometimes identical with that of our days. The class of dicotyledons shows itself there in its fullest development; it is the epoch of flowers. The surface of the earth is embellished by the variegated colours of the flowers and of the fruits which succeed them. The white spikes of the Gramineæ display themselves upon the verdant meadows without limits; they seem provocative of the increase of insects, which now have singularly to multiply themselves. In the woods, crowded with trees, covered with flowers to the rounded summit, like our oak and birch trees, the birds increase in number. The atmosphere, purified and disembarrassed of the veil of vapour which has hitherto pervaded it, now permits animals with delicate pulmonary organs to live and breathe and increase their species.

During the tertiary epoch the influence of the central heat may

have ceased to make itself felt, in consequence of the constantly
increasing thickness of the terrestrial crust.   By the influence of the
solar heat the intended climate would be developed in the various
latitudes ; the temperature of the earth would still be nearly that of
the torrid zones, and at this epoch also cold would begin to make itself
felt at the poles.

Abundant rains would, however, continue to pour upon the earth
enormous quantities of water, which would give birth to important
rivers : new lacustrine deposits of fresh water were formed in great
numbers, and rivers, by their alluvial deposits, began to form new land.
It is, in short, during the tertiary epoch that we trace an alternate
succession of beds now containing organic beings of marine origin,
and now those belonging to fresh water. It is at the end of this epoch
that continents and seas take their respective places as we now see
them, and that the surface of the earth received its actual form.

The tertiary epoch embraces three very distinct periods, to which
the names of *Eocene, Miocene,* and *Pliocene* have been given.   The
etymology of these names are Eocene, from ἔος, dawn, χαινός, recent ;
Miocene, from μετον, less, χαινός, recent ; and Pliocene, from πλείον,
more, χαινός, recent : which is simply to say that each of these periods
is more or less remote from the dawn of life and from the present
time : the expressions are forced, and not very expressive, but usage
has consecrated them.

Fig. 147.—Trigonia margaritacea.

THE EOCENE PERIOD.

DURING this period *terra firma* has vastly extended itself over the domain of the sea : furrowed with rivers and rivulets, and here and there with great lakes and ponds, the landscape of this period presented the same curious mixture which we have noted in the preceding age; that is to say, a combination of the vegetation of the primitive ages with others analogous to those of our own epoch. Alongside the birch, the walnut, the oak, the elm, and the alder, rise lofty palm-trees, of species now extinct, such as the *Flabellaria* and the *Palmacites ;* with many evergreen trees, as the Conifers, for the most part belonging to genera still existing, as the *firs*, the *pines*, the *yews*, the *cypresses*, the *junipers* and the *thyas*, or tree of life.

The *Cupanioïdes* among the Sapindales ; the *Cucumites* among the Cucurbitales ; species analogous to our bryonias creep along the trunks of great trees and hang in festoons of aerial garlands from their branches.

The ferns were still represented by the genera *Pecopteris*, by the *Taucopteris, Asplenium, Polypodites*. Of the mosses, some *Hepaticas* formed a humble, but elegant and lively vegetation alongside the terrestrial and frequently ligneous plants which we have noted. The *Equisetii* and the *Charas* would still grow in the marshy places and on the banks of rivers and ponds.

It is not without some surprise that we observe here certain plants of our own epoch, which seem to have had the privilege of ornamenting the greater water-courses. Among these we may mention the water caltrop, *Trapa natans*, whose fine rosettes of green and dentated leaves float so gracefully in ornamental ponds, supported by their spindle-shaped petioles ; its fruit a hard coriaceous nut, with four horny spines, known in France as the water-chesnuts, which encloses a farinaceous grain not unpleasant to the taste; the pond-weed, *Potamogeton*, whose leaves form thick tufts of green, giving food and shelter to the fishes; the water-lilies, *Nymphœaceœ*, which blossom beside their large round and hollow leaves, so neatly adapted to float on the water, now supporting their flowers of deep yellow, like the *Nénuphar*, or of the pure white flowers of the *Nymphæa*.   Listen to Lecoq, as he describes the vegetation of the period : " The lower tertiary period," he says, " constantly reminds us of the tropical landscape of the actual epoch, in localities where water and heat together impress on vegetation a power and majesty unknown in our climate.   The Algæ, which have already been observed in the marine waters of the later cretaceous period, presented themselves under still more varied forms in the earlier tertiary deposits, where they have been formed in the bed of the sea.   Hepaticas and mosses grew in the more humid places; many pretty ferns, as *Pecopteris*, *Taucopteris*, and the *Equisetum stellare* vegetated in fresh and humid places.   The fresh waters are crowded with *Naiades*, with *Chara*, with *Potamogeton*, with *Caulinites*, with *Zosterites*, and with *Halochloris*.   Their leaves, floating or submerged, like those of our aquatic plants, concealed legions of mollusks whose remains have also reached us."

Great numbers of conifers lived during this period.   M. Brongniart enumerates forty-one different species, which for the most part remind us of the forms with which we are familiar—of pines, cypresses, thyas, and junipers, firs, yews and some Ephedra.   Some Palms mingled with these groups of evergreen trees; the *Flabellaria Parisiensis* of Brongniart, *F. rhapifolia* of Sternberg, *F. maxima* of Unger, and some *Palmacites*, would raise their broad-spreading crowns near the magnificent *Higtea*: Malvaceæ, without doubt arborescent, as many among them, natives of very hot climates, are in our days.

Of creeping plants, like the *Cucumites variabilis*, Brongniart, and the numerous species of *Cupanioïdes*, belonging, the one to the Cucurbitales, and the other to the Sapindales; they clasp their slender stems round the trunks, doubtless ligneous, of various Leguminaceæ.

The family of Betulaceæ of the Cupulifers show the form then new of Quercus, the oak; the Juglandeæ of the Ulmaceæ mingle with the Proteaceæ, now confined to the southern hemisphere. The *Derma-*

Fig. 148.—Branch of Eucalyptus restored.

tophyllites, preserved in the yellow amber, appears to have belonged to the family of the Ericineæ, and *Tropa Areturæ* of Unger, of the group of the Œnothereæ, floated on the same deep waters in which grew the *Charas* and the *Potamogetons*.

This numerous flora comprehended more than 200 known species, of which 143 belonged to the Dicotyledons, thirty-three to the Mono-cotyledons, and thirty-three to the Cryptogams.

Trees predominate here as in the·preceding period, but the great numbers of aquatic plants of the period are quite in accordance with

Fig. 149 —Fruit-branch of Banksia restored.

the geological facts, which intersect the continents and islands with extensive lakes and inland seas, and vast marine bays and arms of the sea, which penetrate deeply into the land.

It is moreover a peculiarity of this period that it comprehended a great number of those plants now confined to Australasia, which give a strange aspect to a country which seems in its vegetation, as in its animals, to have preserved in its warm latitudes much of the organic creation which belonged to the primitive world. As a type of dicotyledonous trees of the epoch, we present here a restored branch of *Eucalyptus* (Fig. 148) with its flowers. All the family of the Proteaceæ, which comprehends the *Banksias*, the *Hakeas*, the *Gerilea protea*, existed in Europe during the tertiary epoch. The family of Mimosas, which comprehends the *Acacia* and *Ingas*, which are only native in our age of the southern hemisphere, abounded in Europe during the same geological period. A branch of *Banksia* with its fructification, taken from the impressions discovered in the rocks of the period, is represented in Fig. 149 : it is different in species from any Banksia of our days.

Mammifera, birds, reptiles, fishes, insects, and mollusks form the terrestrial fauna of the eocene period. In the waters of the lakes, their surfaces deeply furrowed by the passage of large pelicans, lived mollusks of varied forms, as *Physæ*, *Limna*, *Planorbis*, and turtles floated about, as *Trionyx* and the *Emides*. Snipes made their retreat among the reeds which grew on the shore ; seagulls skimmed the surface of the waters or run upon the sands ; owls hid themselves in the cavernous trunks of old trees ; gigantic buzzards hovered in the air watching for their prey ; while heavy crocodiles slowly dragged their unwieldy bodies through the high marshy grasses. All these terrestrial animals have been discovered in England or in France, alongside the overthrown trunks of palm-trees. The temperature of these countries was, then, much more elevated than it is now. The mammifera which lived under the latitudes of Paris and London are only found now in the warm countries of the globe.

The Pachyderms, from the Greek παχύς, thick, δέρμα, skin, seem to have been the first mammifers which appeared in the eocene period, and they held the first rank from their importance

in number of species as well as size. Let us pause an instant over
these Pachyderms. Their predominance over other fossil mammifers, which exceeds considerably the number now living, is a fact
much insisted on by Cuvier. Among them were a great number of
intermediate forms, which we seek for in vain in existing genera. In
fact, the Pachydermata are separated in our days by intervals of
greater extent than we find in any other genera of mammifera; and
it is very curious to discover among the animals of the ancient world
the broken link which connects the chain of these beings, which have
for their great tomb the plaster quarries of Paris; Montmartre and
Panton being their latest refuge.

Each block which issues from these quarries encloses some fragment
of a bone of these mammifers; and how many millions of these bones
had been destroyed before attention was called to the subject! The
Palæotherium and the Anoplotherium were the first of these animals
which Cuvier reconstructed; and subsequent discoveries of other fragments of the same animal have only served to confirm what the
genius of the great naturalist divined. His studies in the quarries of
Montmartre gave the signal, as they became the model, for similar
researches and restorations all over Europe of the animals of the
ancient world—researches which, in our age, have withdrawn geology
from the state of infancy in which it languished, in spite of the magnificent and persevering labours of Steno, Werner, Hutton, and
Saussure.

The *Palæotherium*, the *Anoplotherium*, and *Xiphodon*, were herbivora, which must have lived in great herds. They appear to have
been intermediate, according to their organization, between the rhinoceros, the horse, and the tapir. There seems to have existed many
species of them, of very different sizes. After the labours of Cuvier,
nothing is easier than to represent the *Palæotherium* as it lived: the
nose terminating in a muscular fleshy trunk, or rather snout, somewhat
like the tapir; the eye small, and displaying little intelligence; the head
enormously large; the body squat, thick, and short; the legs short and
very stout; the feet supported by three toes, enclosed in a hoof; the size,

that of a large horse. Such was the great Palæotherium, peaceful flocks
of which must have inhabited the valleys of the plateau which surrounds

Fig. 150.—Palæotherium magnum restored.

the ancient basin of Paris; in the lacustrine formations of Orleans
and Argenton; in the tertiary formations of Issil and Puy-en-Velay, in
the department of the Gironde; in the tertiary formations near
Rome; and in the marly beds of the Bensted quarries in the Isle of
Wight.   Fig. 150 represents the great Palæotherium, after the
design in outline given by Cuvier in his work on *fossil bones*.

The discovery and rearrangement of these and other forms now
swept from the face of the globe are the noblest triumphs of the great
French zoologist, who gathered them, as we have seen, from heaps of
confused fragments, huddled together pell-mell, the portions of twenty
different species, all animals of a former world, and unknown within
the historic period.   The generic characters of Palæotherium give
them forty-four teeth, namely, twelve *molars*, two *canines*, and twenty-

eight others, three toes, a short proboscis, for the attachment of which
the bones of the nose were shortened, as represented in Fig. 151,

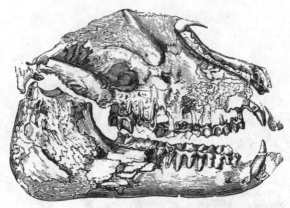

Fig. 151.—Skull of Palæotherium magnum.

leaving a deep notch below them. The molar teeth bear considerable resemblance to those of the rhinoceros : in the structure of that
part of the skull intended to support the short proboscis, and in the
feet, the animal seems to have resembled the tapir.

The geological place of the extinct Palæotherium seems to have
been the first great fresh-water formation of the eocene period, where
it is chiefly found with its congeners, of which nearly fifty species have
been found and identified by Cuvier ; and Dr. Buckland is not singular
in thinking that they lived and died on the margins of lakes and
rivers, as the rhinoceros and tapirs do now. He is also of opinion
that some retired into the water to die, and that the dead carcases of
others have been drifted into the lower parts, in seasons of flood.

The *Palæotherium* varied greatly in size, some being as large as
the rhinoceros, while others ranged between the size of the horse
and hog, or a roe. The smaller Palæotherium resembled the tapir
more than the horse, with its slim and light legs ; it must have been
very common in the north of France, where it would browse on the
grass of the wild prairies. Another species, the *P. minimum*, scarcely
exceeded the hare in size, and it seems to have had all the lightness

and activity of that animal. It seems to have lived among the bushy thickets of the environs of Paris, in Auvergne, and elsewhere.

All these animals lived upon seeds and fruits, on the green or subterranean stems, and the fleshy roots of the plants of the period, and in the neighbourhood of fresh water.

The *Anoplotherium*, so called from ἄνοπλος, defenceless, and θηρίον, animal, had posterior molars analogous to those of the rhinoceros, the feet terminating in two great toes, an equally-divided hoof, like the ox and other ruminants, and with tarsus with toes nearly like those of the camel. It would be about the size of the ass, its head light; but what would distinguish it most must have been an enormous tail of at least three feet in length, and very thick at its junction with the body. This tail evidently served it as a rudder and propeller when swimming in lakes or rivers, which it frequented, not to seize fish, for it was strictly

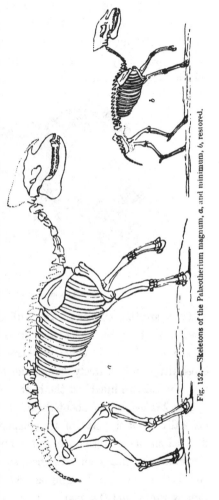

Fig. 152.—Skeletons of the Palæotherium magnum, *a*, and minimum, *b*, restored.

herbivorous, but in search of roots and stems of succulent aquatic plants. "Judging from its habits of swimming and diving," says Cuvier, "the Anoplotherium would have the hair smooth, like the otter; perhaps its skin was even half naked. It is not likely either that it had long ears, which would be inconvenient in its aquatic kind of life; and I am inclined to think that, in this respect, it resembled the hippopotamus and other quadrupeds which frequent the waters much." To this description Cuvier had nothing more to add. His memoir upon the *Pachydermous fossils* of Montmartre is accompanied by a design in outline of *Anoplotherium communis*, of which a copy is given in Fig. 153.

Fig. 153.—Common Anoplotherium. Half natural size.

There are the remains of Anoplotherium of very small size. *A. leporinum*, or the hare, whose feet are evidently arranged for running: *A. minimum* and *A. obliquum* were of still smaller size; the last, especially, would scarcely exceed the size of a rat. Like the water-rats, this species inhabited the banks of brooks and small rivers.

The *Xiphodon* was about three feet in height at the withers, and generally about the size of the chamois, but lighter in its form, and its head smaller. Inasmuch as the appearance of the *Anoplotherium communis* was heavy and sluggish, so was *Xiphodon gracilis* graceful and active: light and agile as the gazelle or the goat, it would rapidly run round the marshy meadows and ponds, pasturing on the aromatic herbs of the dry lands, or browsing on the sprouts of the young shrubs. "Its course," says Cuvier, in the memoir already quoted, "was not embarrassed by a long tail; but, like all active herbivorous animals, it was probably timid, and with great and very mobile ears, like those of the stag, announcing the slightest approach of danger. Neither is there any doubt that its body was covered by short smooth hair; and consequently we only require to know its colour in order to paint it as it already animated this country, where it has been dug up after so many ages." Fig. 154 is the reproduc-

Fig. 154.—Xiphodon gracilis.

tion from the design in outline with which Cuvier accompanies the description of this animal, which he classes among the Anoplotheriums, and which has received in our days the name of *X. gracilis*.

The gypsum of the environs of Paris includes, moreover, the remains of other Pachyderms: the *Chæropotomus*, or water-hog, from χοῖρος, ποταμός, which has some analogy with the pecari, though much larger; the *Adapis*, which reminds us of the hedgehog, of which, however, it was three times the size. It seems to have been a link between the Pachydermata and the insectivorous carnivora. The *Lophiodon*, the size of which varied with the species, from that of the rabbit to that of the rhinoceros, was still nearer to the tapirs than to the Anoplotherium; it is found in the lower beds of the gypsum, that is to say, in the "Calcaire Grossier," or coarse limestone. The remains of these animals are, however, so imperfect that little can be advanced with certainty upon their organization and habits.

A Parisian geologist, M. Desnoyers, librarian of the Museum of

Natural History there, has discovered in the gypsum of the Valley of
Montmorency, and elsewhere in the neighbourhood of Paris, as at
Dautin, at Clichy, and at Dammartin, the imprints of the footsteps of
some mammifera, of which there seems to be some question whether
they were Anoplotherium and Palæotherium. Imprints of steps of
turtles, birds, and even carnivora sometimes accompany these curious
traces, which have the form of some kind of kernel more or less lobed,
according to the divisions of the hoof of the animal, and which recall
completely, in their mode of production and preservation, those imprints of the steps of the turtle and of Labyrinthodon which have
been mentioned in the triasic period. This discovery is interesting,
as furnishing a means of comparison between the imprints and the
animals which have produced them. It brings into view, as it were,
the material traces left in their walks upon the soil by animals whose
species has been annihilated, but who occupied the mysterious sites
of the ancient world.

It is interesting to present to ourselves the vast pasturages of the
eocene period swarming with herbivora of all sizes. The country
now surrounding the city of Paris belongs to the period, and
not far from its gates, the woods and plains were crowded with
"game" of which the Parisian sportsman little dreams, but which
would nevertheless singularly animate the earth at this distant epoch.
The absence of great carnivora explains the rapid increase of the
agile and graceful denizens of the wood, whose race seems to have
been so multiplied then, but which were speedily annihilated under
the teeth of the more ferocious beasts of prey which soon appear on
the scene.

The same novelty, riches, and variety which distinguished the
mammifera of the tertiary period extended to other classes of animals.
The class of birds, of which we can only name the most remarkable,
was represented by the curious fossil known as the bird of Montmartre.
The bones of other birds have been obtained from Hordwell, as well as
the remains of quadrupeds. Among the latter the *Hyænodon*, supposed to be the oldest known example of a true carnivorous animal in

the series of British fossils, and the fossil bat known as the *Vespertilio Parisiensis*. Among the reptiles the crocodile, which bears the name of Isle of Wight Alligator, *Toliapicus*. Among the turtles the *Trionyx*, of which there is a fine example in the Museum of Natural History at Paris.

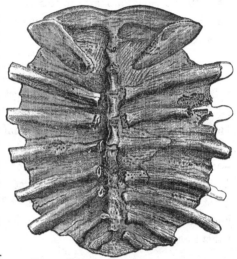

In the class fishes we now see the first *Pleuranectes*, or flat-fish, of which *Platax altissimus* and *Rhombus minimus* are well-known examples. Among the crustaceans we see the first crabs. At the same time multitudes of new mollusks

Fig. 155.—Trionyx, or turtle, of the tertiary period.

make their appearance : *Olivas, Tritons, Cassis, Harpa,* and *Crepidulas.*

The forms hitherto unknown of *Schizaster* are remarkable among the echinoderms ; the zoophytes are also abundant, especially the *Foraminifera,* which seem to make up in their numbers for deficiency in their dimensions. It was in this period, in the bosom of its seas, and far from shore, that the *Nummulites* had their birth, whose calcareous envelopes play such a considerable part as the elements of some of the tertiary formations. The shelly agglomerates of these mollusks constitute very important rocks of the period. The Nummulitic limestone, forms in the chain of the Pyrenees entire mountains of great height ; in Egypt, it constitutes banks of great extent, and it is from these rocks that the ancient pyramids were formed. What time must have been necessary to convert the spoil of these small shells into beds many hundreds of yards thick ? The species of *Milliolites*

U

were also so abundant in the eocene seas, that we owe to their agglo-meration the greater part of the calcareous rocks out of which Paris has been built—agglutinated in this manner these shells form round Paris the continuous beds which are quarried at Gentilly, at Vaugirard, and at Chatillon.

On the opposite page we present, in PL. XXIII., an ideal landscape of the eocene period. We remark amongst its vegetation a mixture of the fossil species with others belonging to the present age. The alders, the wych-elms, and the cypresses, mingle with the *Flabellaria*: some palms of an extinct species. A great bird—a wader, the *Tantalus* —occupies the projecting point of a rock on the right: the turtle, *Trionyx*, floats on the river, in the midst of Nymphæas and Nenuphars, and other aquatic plants. Whilst a herd of Palæotheriums, Anoplotheriums and Xiphodons peacefully browse the grass of the wild meadows of this tranquil oasis.

With a good general resemblance in their fossils, nothing can be more dissimilar than the mineral characters of the eocene deposits of France and England; "those of our own island," says Lyell, "being almost exclusively of mechanical origin—accumulations of mud, sand, and pebbles; while in the neighbourhood of Paris we find a great suc-cession of strata composed of limestones, some of them siliceous, and of crystalline gypsum and siliceous sandstone, and sometimes of pure flint used for millstones. Hence it is by no means an easy task to institute an exact comparison between the various members of the English and French series. It is clear that on the sites both of Paris and London a continual change was going on in the fauna and flora by the coming in of new species and dying out of others, while proportionate changes in the geographical conditions were in progress in consequence of the rising and sinking of the land and of the bottom of the sea. A particular subdivision of time was occasionally repre-sented in one area by land, in another by an estuary, in a third by sea; and even where the conditions were in both areas marine, there was

XXIII.—Ideal landscape of the Eocene Period.

often shallow water in one, and deep sea in another, producing a want of agreement." The eocene rocks, as developed in France and England, may be tabulated as follows :—

ENGLISH.			FRENCH.	
Lower Eocene	1. Thanet sands. 2. Plastic clay and sand. 3. London clay.	Gypseuse formation . .	1. Bracheux sand. 2. Plastic clay and lignite. 3. London clay.	
Middle Eocene	1. No equivalent. 2. Bagshot beds.	Calcaire Grossier	1. Milliolitic limestone. 2. Nummulitic limestone. 3. Cerithitic limestone.	
Upper Eocene	1. Barton clay and white sand. 2. Osborn and Headon beds. 3. Bembridge beds.	Argile Plastic	1. Beauchamp sandstones. 2. Siliceous limestone. 3. Gypsum of Montmartre.	

The *Argile Plastic,* or plastic clay, are extensive beds of sand with occasional beds of potter's clay, which lie at the base of the tertiary formation in France. Habitually variegated, sometimes grey or white, it is employed as a potter's earth in the manufacture of delf ware. This deposit seems to have been formed chiefly in fresh water. At its base it is a conglomerate of chalk and of divers calcareous breccia cemented by siliceous sand, in which was found at Bas-Meudon some remains of reptiles, turtles and crocodiles, of mammifera, and, more, remains of a large bird, equalling the ostrich in size, and named the *Gastornis,* which Professor Owen classes among the wading rather than among aquatic birds. In the Soissonais there is found at the same level a great mass of lignite, inclosing some shells and bones of the most ancient Pachyderm yet discovered, the *Coryphodon,* which partakes at once of the Anoplotherium and of the Cochon. The *Sables Inférieure,* or Bracheux sands, form a marine bed of great thickness near Beauvais ; they are principally sand, but include beds of calcareous clay and banks of shelly sandstone, and are considered to be older than the plastic clay and lignite, and to correspond with the Thanet sands in England. They are rich in shells, including many Nummulites. At La Fere, in the department of the Aisne, a fossil skull of *Arctocyon primæus*, supposed to be related to the bear, and to be the oldest tertiary mammifer, was found in a deposit of this age.

The *Calcaire grossier*, consisting generally of limestone of various kinds, and of a coarse—sometimes compact grain, is suitable for mason-work.   These beds, which form the most characteristic member of the Paris basin, naturally divide themselves into three groups of beds, characterized, the first, by *Nummulites*, in which the London clay is slightly represented ; the second by *Milliolites ;* and the upper beds by *Cerithites ;* after which the beds are sometimes named Num-mulitic limestone, Milliolitic limestone, and Cerithitic limestone. Above these a great mass, generally sandy, is developed.   It is marine at the base, and there are indications of brackish water in its upper parts ; it is called Beauchamp sandstone, or Sable Moyen, and divides the *Grossier calcaire* from the gypseous beds.   The *siliceous limestone,* or lower travertin, is a compact siliceous limestone ex-tending over a wide area, and resembles a precipitate from mineral waters.   The *gypseous* formation consists of a long series of marly and argillaceous beds, of a greyish, green, or white colour ; in the intervals of which a powerful bed of gypsum, or sulphate of lime, is intercalated. This gypsum bed is found in its greatest thickness at Montmartre and at Pantin, near Paris. Its existence is probably due to the action of the free sulphuric acid upon the carbonate of lime of the formation ; the sulphuric acid itself being produced by the transformation of the gaseous masses of sulphureted hydrogen emanating from volcanic mouths into that element, by the action of air and water.   It was, as we have already said, in the gypsum quarries of Montmartre that the numerous bones of Palæotherium and Anoplotherium were found. It is exclusively at this level that we find the remains of these animals, which seem to have been preceded by the *Coryphodon,* and after-wards by the *Lophodon ;* an order of successive appearance which is now pretty well established.   It may be added that round Paris the eocene formation, from its lowest beds to the summit, is composed of beds of plastic clay of the Calcaire grossier, with *Nummulites* and *Mil-liolites,* followed by the gypseous beds ; the series terminating in the

Fontainebleau sandstone beds, remarkable for their thickness and also for their fine landscapes, without reckoning their usefulness in furnishing paving-stone for the capital. In Provence the same series of rocks are continued, and attain an enormous thickness. This upper part of the eocene deposit is entirely of lacustrine formation. Grignon has procured from a single spot, where they were embedded in a calcareous sand, no less than four hundred fossils, chiefly formed of comminuted shells, in which, however, were well-preserved species both of marine, terrestrial, and fresh-water shells. Of the Paris basin, Sir Charles Lyell says, "Nothing is more striking in the assemblage of fossil testacea than the great proportion of species referable to the genus *Cerithium*. There occur no less than a hundred and thirty-seven such species in the Paris basin, and almost all of them in the Calcaire grossier. Most of the living *Cerithia* inhabit the sea near the mouths of the rivers where the waters are brackish; so that their abundance in the marine strata now under consideration is in harmony with the hypothesis that the Paris basin formed a gulf into which several great rivers flowed."

To give the reader some idea of the formation, first come the limestones and lower marls, which contain fine lignite or wood coal produced from vegetable matter buried in moist earth, and excluded from all air; a material which is worked in some parts of the south of France as actively as a coal-mine. In these lignites *Anodontas* and other fresh-water shells are found.

From the base of Saint-Victoire to the other side of Aix, we trace a conglomerate characterized by its red colour, but which loses its unity in its prolongation towards the west. This conglomerate contains terrestrial *Helix* of various sizes, mixed with fresh-water shells. Upon this conglomerate, comprising therein the marls, rests a thick bank of limestone with gypsum of Aix and Manosque, which is believed to correspond with that of Paris, some beds being remarkably rich in sulphur. The calcareous, marly flagstones which accompany the gypsum of Aix contain some insects of various kinds, and some fishes resembling *Lebias cephalotes*. Finally, the whole terminates at

Manosque in new beds of marl and of sandstones, intersected with banks of limestone with *Limna* and *Planorbis*. At the foot of this series are found three or four beds of lignite more fusible than coal, which also gives out a very sulphureous oil. We may form some estimate of the thickness of this bed, if we add that, above the beds of fusible lignite, we may reckon sixty others of dry lignite, some of them capable of being profitably worked if this part of Provence were provided with more convenient roads.

" The Nummulitic formation, with its characteristic fossils," says Lyell, " plays a far more conspicuous part than any other tertiary group in the solid framework of the earth's crust, whether in Europe, Asia, or Africa. It often attains a thickness of many thousand feet, and extends from the Alps to the Carpathians, and is in full force in the North of Africa. It has been traced from Egypt, where it was largely quarried of old for building the Pyramids, into Asia Minor, and across Persia, by Bagdad, to the mouths of the Indus. It occurs in Cutch, in the mountain ranges which separate Scinde from Persia, and which form the passes leading to Cabul, and eastward as far as Bengal and the frontiers of China."

" When we have once arrived at the conclusion," he adds, " that the Nummulitic formation occupies a middle place in the eocene series, we are struck with the comparatively modern date to which some of the greatest revolutions in physical geography must be referred. All the mountain chains, such as the Alps, Pyrenees, Carpathians, and Himalayas, into the composition of whose central and loftiest parts the Nummulitic strata enter bodily, could have had no existence till after the middle eocene period."

In England the mottled clay of Hampshire and the Isle of Wight is often seen in contact with the chalk, but at other points, Mr. Prestwich shows that the Thanet sands intervene between the chalk and the Woolwich beds, which lie below the London clay.

This formation consists of tenacious brown and bluish-grey clay, with layers of the concrete called Septaria, well known on the Essex coast, where it is collected for making Roman cement. Above this great bed, which extends from Norfolk to the Isle of Wight, lie the Bracklesham and Bagshot beds, which consist of beds of light-yellow sand with an intermediate layer of dark-green sand and brown clay, over which lie the Barton clay and white sands, the Headon beds, and Bembridge series, the latter consisting of limestone, clay, and marl, of marine, brackish, and fresh-water origin.

The eocene strata, Professor Ramsay thinks, extended in their day *much further* west, " because," he says, " here, at the extreme edge of the chalk escarpments, you find outlying fragments of them," from which he argues that they were originally deposited all over the chalk as far as these points, but being formed of soft strata they were "denuded " backwards.

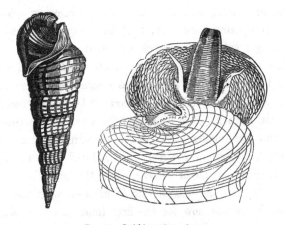

Fig. 156.—Cerithium telescopium.

MIOCENE PERIOD.

It is on the European continent that we find the most striking characteristics of the miocene period. In our own islands traces of it are few
and far between. In the Island of Mull certain beds of shale, interstratified with basalt and volcanic ash, are described by the Duke of Argyll as
of miocene date ; and again, miocene clay is found interstratified with
bands of imperfect lignite. The vegetation which distinguished the
period is a mixture of the vegetable forms proper to the burning
climate of Equatorial Africa with those which grow to-day in temperate Europe, such as palms, bamboos, laurels, Combretaceæ, with
the grand leguminales of warm countries, as *Phaseolites, Erythrina*,
Bauhinia, Mimosites, and Acacias ; some Apocyneæ analogous to
those of our equatorial regions ; a Rubiaceæ quite tropical in *Stein-
hauera,* mingle with some maples, walnut-trees, beeches, elms, oaks,
and wych-elms, genera suitable to temperate and even cold countries.

Besides these, there were still mosses, mushrooms, some charas, evergreens, fig-trees, plane-trees, and poplars. " During the second period
of the tertiary epoch," says Le Coq, " the algæ and marine monocotyledons were less abundant than in the preceding age ; the ferns also
diminished, the mass of conifers were reduced, and the species of
palms multiplied. Some of those cited in the preceding period still
belong to this, and the magnificent *Flabellaria,* with the fine *Phœ-
nicites,* which we see now for the first time, gave animation to the
landscape. Among the conifers some new genera appear: among
them we distinguish *Podocarpens,* a southern form of vegetation of the
present age. Almost all the arborescent families have their representatives in the forests of this period, where for the first time types so
different are united. The waters are covered with *Nymphæa, Arithnæ,*
Brongniart, and with *Myriophyllites capillifolius,* Unger, *Culmites
animalis,* Brongniart, and *C. Gœpperti,* Munster, spring up in profusion upon their banks, and the grand *Bambusinites sepultana* threw
the shadow of its long articulated stem across them. Some analogous
species occupy the banks of the great rivers of the New World,

even one umbellifera, the *Pompenellites zizioides*, is indicated by Unger.

Of this period date some powerful beds of lignite resulting from the accumulation for ages of all these different trees. It seems that arborescent vegetation had then attained its apogee. Some *Smilacites* interlaced like the wild vines with these grand vegetables, which fell from decay : some parts of the earth even now present these grand scenes of vegetation. They are still described by travellers who traverse the tropical regions, where nature often displays the utmost luxury under the screen of clouds which permits no rays of the sun to reach the earth. M. D'Orbigny relates an interesting incident which is much to the point. "I have reached a zone," he says, speaking of Rio Chapura, "where it rains regularly all the year. We can scarcely perceive the rays of the sun in the intervals through the screen of clouds which almost constantly veils it. This circumstance, added to the heat, gives an extraordinary development to its vegetation. The wild vines fall on all sides in garlands from the loftiest branches of trees whose summits are lost in the clouds."

The fossil species of this period, to the number of 133, resemble those which enrich our landscapes. Already the trees of the tropics mingle with the vegetables of temperate climates, but not yet with the same species. The oaks grow beside the palms, the birch with the bamboos, the elms with the laurel; the maples are united to the Combretaceæ, to the leguminales, and to the tropical Rubiaceæ. The forms of the species belonging to temperate climates are rather American than European.

The luxuriance and diversity of the miocene flora has been employed, as none but a German savant could, in identifying and classifying the middle tertiary, or miocene strata. We are indebted to Professor Heer, of Zurich, for the restoration of more than 9 0 plants, which he classified and illustrated in his ' Flora Tertiara Helvetiæ.' In order to appreciate the value of the learned Professor's undertaking, it is only necessary to remark that, where Cuvier had to study the position and character of a bone, the botanist had to study the outline, nervature, and microscopic structure of a leaf. Like the great French naturalist, he had to construct a science at the threshold of his great work.

The miocene formations of Switzerland are called *molasse*, or soft, which is applied to a soft, incoherent, greenish sandstone, occupying the country between the Alps and the Jura, and they may be divided into lower, middle, and upper miocene; the two first being fresh-water

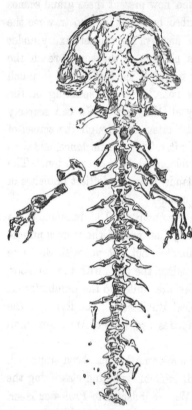

formations, the upper marine. The upper or fresh-water is best seen at Œningen, in the Rhine valley, where, according to Sir R. Murchison, it ranges ten miles east and west from Berlingen, on the right bank, and Œningen, near Steen, on the left bank. In this formation the Professor enumerates twenty-one beds. 1, a bluish-grey marl seven feet thick, without organic remains. 2, resting on limestone, with fossil plants, including poplar, cinnamon, and pond-weed. 3, is a bituminous rock, with *Mastodon angustidens*. 5, two or three inches thick of fossil fish. 9, the stone in which a skeleton of the great salamander *Andrias Scheuchzeri*, represented in the margin, was found. Below this, other strata with fish, tortoises, the great salamander, as before, with fresh-water mussels

Fig. 157.—Andrias Scheuchzeri.

and plants. In 16, Sir R. Murchison obtained the fossil fox of Œningen. In these beds Professor Heer had, as early as 1859, determined 475 species of fossil plants, and 900 insects.

The Swiss plants of the miocene period have been obtained from a country not one-fifth the size of Switzerland, yet such an abundance of species, which Heer places at 3000, does not exist in any area of equal extent in Europe. It exceeds in variety, he considers, after making every allowance for all not living at the same time, and other considerations :—the Southern American forests, and rivals such tropical countries as Jamaica and Brazil. European plants occupy an insignificant place, while the evergreens, oaks, maples, poplar, and plane-trees, robinias, and taxodiums of America and the Atlantic Islands, occupy such an important place in the fossil flora, that Unger was induced to suggest the hypothesis that in the miocene period the present basin of the Atlantic was dry land, and this hypothesis is ably advocated by Heer.

The terrestrial animals which lived in the miocene period were mammifers, birds, and some reptiles. Many new mammifers had been created during the preceding period, among others, apes, cheiropteras, bats, carnivora, marsupials, gnawers, dogs. Among the first we find *Pithecus antiquus* and *Mesopithecus ;* the bats, dogs, and coati inhabited the Brazils and Guiana; the rats inhabited North America ; the genettes, the marmottes, the squirrels, and opossums having some resemblance to the American opossums. Thrushes, sparrows, storks, flamingoes, and crows represent the birds. Among the reptiles appear some snakes, frogs, and salamanders. The lakes and rivers were inhabited by perches and shad. But it is among the mammifers we must seek for the most interesting species of animals. They are at once numerous, and remarkable for their dimensions and peculiarities of form ; but the species which appeared in the miocene period, as in those which preceded it, are only known by their fossils and bones.

The *Dinotherium,* which is one of the most remarkable of these mammifers, is the largest which has ever lived. For a long time we possessed only very imperfect portions of the skeleton of this animal, and Cuvier was induced erroneously to place it among the tapirs. The discovery of a lower jaw nearly perfect, armed with defensive tusks descending from its lower jaw, demonstrated that this hitherto mysterious animal was the type of a genus altogether new and singular. Nevertheless, as it was known that there were some animals of the ancient world in which both jaws were armed, it was thought for some time that such was the case with the Dinotherium. But in 1836 a head, nearly entire, was found in the beds already celebrated at Eppelsheim, in the Grand Duchy of Hesse Darmstadt. In 1837 this fine fragment was carried to Paris, and exposed to public view. It was nearly a yard and a half long. The defences, it was found, were enormous, and were carried at the extremity of the lower maxillary bone, and much curved inwards, as in the morse. The molar teeth were in many respects analogous to those of the tapir, and the great holes under the orbits, joined to the form of the nasal bone, rendered the

Fig. 158.—Dinotherium.

existence of a proboscis or trunk very probable. But the most remarkable bone which has yet been found is an omoplate or scapula, which reminds us of the form of the mole.

This colossus of the ancient world, respecting which so much has been said, somewhat approaches the Mastodon : it seems to announce the elephant, but its dimensions were infinitely greater than the living elephants, superior even to that of the Mastodon and the Mammoth, both fossil elephants, of which we shall have to describe the remains.

From its kind of life, and its frugal regimen, this pachyderm scarcely merited the formidable name imposed on it by naturalists, of δεινὸς, terrible, θηρίον, animal.   Its size was, no doubt, frightful enough, but its habits seem to have been harmless.   It is supposed to have inhabited the fresh-water lakes, or the mouths of great rivers and lagunes, by preference.   Herbivorous like the elephant, it employed its proboscis probably in seizing the herbage suspended over the waters, or floating on their surface.   We know that the elephants are very partial to the roots of herbaceous vegetables growing in flooded plains. The Dinotherium appears to have been similarly organized, and probably sought to satisfy the same tastes.   With the powerful mattock which Nature had supplied him for penetrating the soil, he would be able to tear from the bed of the river or lake feculent roots like the nymphæas, or even much harder roots, for which the mode of articulation in the jaws, and the powerful muscles intended to move them, as well as the large surface of the teeth, so well calculated for grinding, were evidently intended.

The *Mastodon* was, to all appearance, very nearly of the size of

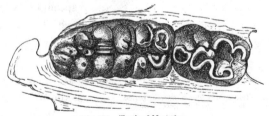

Fig. 159.—Teeth of Mastodon.

our elephant ; his body, however, being somewhat longer, and his members a little thicker. He had tusks, and very probably a trunk, and is chiefly distinguished from the elephant by the form of his molar teeth, which form the most distinctive character in his organiza-

tion. These teeth are nearly rectangular, and present on the surface of their crown great conical tuberosities, with rounded points disposed in pairs to the number of four or five, according to the species. Their form is very distinct, and may be easily recognized. They have nothing of the carnivorous in them, but are strictly herbivorous, and resemble those of the

Fig. 160.—Molar teeth of Mastodon worn.

hippopotamus. The molar teeth are at first sharp and pointed, but when the conical points are ground down by mastication, they assume the appearance represented in Fig. 160. When, from con-

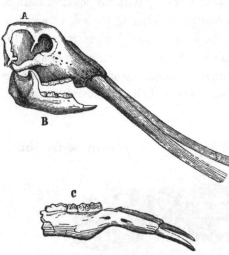

tinued grinding, the conical teat-like points are more deeply worn, they begin to assume the appearance of Fig. 159. In Fig. 161 we represent the lower jaw of the animal, in which are two projecting tusks or defences, corresponding with two of much larger dimensions projecting from the upper jaw.

It was towards the middle of last century that the Mastodon first attracted attention in Europe. In the year 1705, it is true, some traces of the animal had been found at Albany, in America, now the capital

Fig. 161.—Head of Mastodon of miocene period.
A, B, the whole head; C, lower jaw.

of New York, but the discovery attracted little attention. In 1739, a French officer, M. de Longueil, was traversing the virgin forests of the Ohio, in order to reach the great river Mississippi, and the savages who escorted him discovered quite accidentally, on the bank of a marsh, the bones of some unknown animal. In this turfy marsh, which the natives designated the Great Salt Lake, in consequence of the many streams charged with that mineral which lose themselves in it, herds of wild ruminants still seek its banks, attracted by the salt, of which they have a great fondness — the same cause probably attracted these great animals of more remote ages to its banks. M. de Longueil carried some of these bones with him; and, on his return to France, he presented them to Daubenton and Buffon: they consisted of a femur, one extremity of the defence, and three molar teeth. Daubenton, after mature examination, declared them to be the teeth of a hippopotamus; the defence and the femur bone, according to his report, belonged to an elephant; so that they were not even considered as belonging to the same animal. Buffon did not share this opinion, and he soon converted Daubenton, as well as other French naturalists, to his views. Buffon declared that the bones belonged to an elephant whose race had lived in the primitive ages of the globe. It was then that the fundamental notion of extinct species of animals exclusively belonging to the ancient ages of the world began to be entertained for the first time by naturalists—a notion of which they dreamed during nearly a century before it carried the admirable fruits which have since so enriched the natural sciences.

Buffon gave the fossil the name of the *Elephant of the Ohio*, but he deceived himself as to its size; he believed it to be eight times the size of our elephant, an estimate to which he was led by an erroneous appreciation of the number of the elephant's teeth. The *animal of the Ohio* had only four molars, while Buffon was led to believe that it must have had sixteen, confounding the germs, or supplementary teeth, which exist in the young animal, with the teeth of the adult animal. In reality, however, the Mastodon was much larger than the African elephant.

The discovery of this animal had produced a great impression in Europe. Being masters of Canada by the peace of 1763, the English sought eagerly for more of these precious remains. The geographer

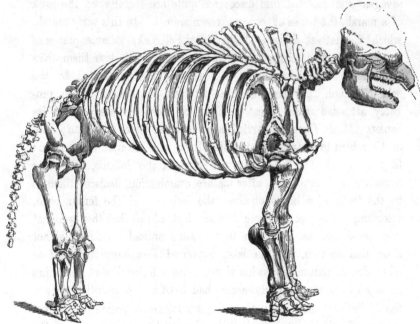

Fig. 162.—Skeleton of Mastodon giganteus.

Croghan traversed anew the region of the Great Salt Lake pointed out by De Longueil, and found there some bones of the same nature. In 1767 he forwarded many cases to London, addressing them to divers naturalists. Collinson, among others, the friend and correspondent of Franklin, who had his share in this consignment, took the opportunity of sending a molar tooth to Buffon.

It was not, however, till 1801 that the perfect remains of the skeleton were discovered. An American naturalist named Peale was fortunate enough to get together two skeletons of this important animal nearly complete. Having been apprised that many large bones had been found in the marly clay on the banks of the Hudson, near Newburg, in the State of New York, Mr. Peale proceeded to that locality. In the spring of 1801 a considerable part of one skeleton was found by the farmer who had been employed to extract the soil, but unfortunately it was much mutilated by his awkwardness, and by the precipitancy of the workmen. Having purchased these fragments, Mr. Peale sent them on to Philadelphia.

In a marsh situated five leagues west of the Hudson the same gentleman discovered, six months after, a second skeleton of the Mastodon, consisting of a perfect jaw and a great number of bones. With the bones thus collected, the naturalist managed to construct two skeletons nearly complete. One of these still remain in the Museum of Philadelphia; the other was sent to London, where it was exhibited publicly.

Fig. 163.—Mastodon restored.

Discoveries nearly analogous to these followed, the most curious of which was made in this manner by Mr. Barton, a Professor of the University of Pennsylvania. At six feet depth in the soil, and under a great bank of chalk, bones of the Mastodon were found sufficient to

x

form another skeleton.   One of the teeth found weighed about seven-
teen pounds (Fig. 164); but the circumstance which rendered this
discovery remarkable was, that in the middle of the bones, and enveloped in a sort of sac which was probably the stomach of the animal,

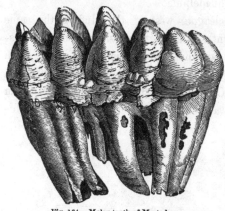

a mass of vegetable mat-
ter was discovered, partly
bruised, composed of small
leaves and branches, among
which a species of rush
has been recognized which
is yet common in Vir-
ginia.   We cannot doubt
that these were the remains,
undigested, of the food
which the animal had
browsed on just before his
death.

Fig. 164.—Molar tooth of Mastodon.

The indigenous natives of North America called the animal the *father
of the ox.* A French officer named Fabri wrote thus to Buffon in 1748.
The natives of Canada and Louisiana, where these remains are abundant,
speak of the Mastodon as a fantastic creature which mingles in all their
traditions and in their ancient national songs.   Here is one of these
songs, which Fabri heard in Canada.   " When the great *Manitou*
descended to the earth in order to satisfy himself that the creatures
he had created were happy, he interrogated all the animals.   The
bison replied that he would be quite contented with his fate in the
grassy meadows, where the grass reached his belly, if he were not also
compelled to keep his eyes constantly turned towards the mountains to
catch the first sight of the *father of oxen,* as he descended with fury
to devour him and his companions."
    The Chavanais Indians have a tradition that these great animals
lived in former times, conjointly with a race of men whose size was
proportionate to their own, but that the *Great Being* destroyed both by
repeated strokes of his terrible thunders.

The native Indians of Virginia had another legend. As these gigantic elephants destroyed all other animals specially created to supply their wants, God, the thunderer, destroyed them; a single one only succeeded in escaping. It was " a great male, which presented its head to the thunderbolts and shook them off as they fell; but being at length wounded in the side, he took to flight towards the great lakes, where it remains hidden till this day." All these simple fictions prove at least that the Mastodon has lived upon the earth at some not very distant epoch. We shall see, in fact, that it is contemporaneous with the Mammoth, which only preceded a little the appearance of man.

Buffon, as we have said, gave to this great fossil animal the name of the Elephant of the Ohio; it has also been called the Mammoth of the Ohio. In England it was received with astonishment. Dr. Hunter showed clearly enough, from the femur bone and the teeth, that it was no elephant; but having heard of the existence of the Siberian Mammoth, he at once came to the conclusion that they were bones of that animal. He even declared the teeth to be carnivorous, and the idea of a *carnivorous elephant* became one of the wonders of the day. Cuvier at once dissipated the clouds which had gathered round the subject, pointing out the osteological differences between the several species, and giving to the American animal the appropriate name of Mastodon, from μαστὸς, a teat, and ὀδοὺς, a tooth, or teat-like tooth.

Many bones of the Mastodon have been found in America since those days, but the remains are rarely met with in Europe, except as fragments—as the portion of a jaw-bone discovered in the Red Crag near Norwich, which Professor Owen has named *M. angustidens*. It was even thought for a long time, with Cuvier, that the Mastodon belonged exclusively to the New World; but the discovery of many of the bones mixed with those of the mammoth *Elephas primogenius* has dispelled that opinion in our days. The bones have been found in great numbers in the Val D'Arno. In 1858 a magnificent skeleton was discovered at Turin.

The form of the teeth of the Mastodon show that it feeds, like the

x 2

elephant, on the roots and fleshy parts of vegetables; and this is confirmed by the curious discovery made in America by Barton. It lived,
no doubt, on the banks of rivers and on moist and marshy lands.
Besides the Mastodon of which we have spoken, there existed in
Europe a Mastodon one-third smaller than the elephant.

There are some curious historical facts in connection with the
remains of the Mastodon which ought not to be passed over in silence.
On the 11th of January, 1613, the workmen in a sand-quarry situated
near the Castle of Chaumont, in Dauphiny, between the cities of
Montricourt and Saint-Antoine, upon the left bank of the Rhone,
found some bones, many of which were broken up by them. These
bones belonged to some great mammiferous fossil; but the existence
of such animals was at this time wholly unsuspected. Informed of
the discovery, however, a surgeon of the country, named Mazuyer,
purchased the bones, and gave out that he had himself discovered
them in a tomb, thirty feet long by fifteen broad, built of bricks, upon
which he found the inscription *Teutobocchus Rex*. He added, that in
the same tomb he found half-a-hundred medals of Marius. This
Teutobocchus was a barbarian king, who invaded Gaul at the head of
the Cimbri, and who was vanquished at *Aquæ Sextiæ* (Aix in Provence)
by Marius, who led him to Rome, where he took part in his triumph.
To support his invention, Mazuyer reminded the public that, according
to the testimony of Roman authors, the head of the Teuton king surpassed in height all the trophies borne upon the lances in the triumph.
The skeleton which he exhibited was five-and-twenty feet in length
and ten broad.

Mazuyer exhibited the skeleton of the pretended Teutobocchus in all
the cities of France and Germany, and also before Louis XIII., who
took great interest in contemplating this marvel. It gave birth to a
long controversy, or rather dispute, in which the anatomist Riolan distinguished himself—arguing against Habicot, a physician, whose name
is buried in forgetfulness. Riolan attempted to prove that the bones
of the pretended king were those of an elephant. Numerous *brochures*
were exchanged, as well as great personalities, when the opponents

met. We learn, also, from Gassendi, that a Jesuit of Tournon, named Jacques Tissot, was the author of the notice published by Mazuyer. Gassendi also proves that the pretended medals of Marius were con-

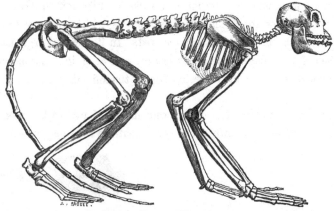

Fig. 165.—Skeleton of Mesopithecus.

troverted on the ground that they partook of Gothic characteristics. It seems strange that these bones should have been taken for a moment for human remains. The skeleton of Teutobocchus remained at Bourdeaux till 1832, when it was sent to the Museum of Natural

Fig. 166 —Mesopithecus restored. One-fifth natural size.

History of Paris, where M. de Blainville declared that they belonged to the Mastodon.

The apes make their appearance at this period.   In the ossiferous beds of Sansan M. Lartet discovered the *Dryopithecus*, as well as *Pithecus antiquus*, but only in imperfect fragments. Al. Albert Gaudry was more fortunate : in the miocene rocks of Pikermi, in Greece, he discovered the entire skeleton of *Mesopithecus*, which we present here, together with the same animal restored.   In its general organization it resembles the dog-faced baboon or ape, a piece of information which has guided the artist in the restoration of the animal.

The seas of the miocene period were inhabited by great numbers of beings altogether unknown in earlier formations : we may mention no

Fig. 167.—Cerithium plicatum.

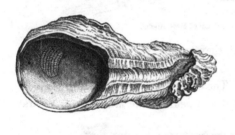

Fig. 168.—Astrea longirostris. One quarter natural size.

Fig. 169.—Murix Turonensis.

less than seventy marine genera which appear here for the first time, some of them species which have come down to our epoch.   Among these, the molluscous gasteropodes, such as *Conus, Turbinella, Ranilla, Dolium*, predominate.

The foraminifera are also represented by new genera, among which are the *Bolivina, Polystomella*, and *Dendritina*.

Finally, the crustaceans include the genera *Pagurus*, or the crabs ; *Astacus*, the lobsters, and *Portunus*, or paddling crabs.   Of the first, it is doubtful if any fossil species have been found : of the last, species have been discovered bearing some resemblance to *Podophthalmus vigil*,

XXIV.—Ideal landscape of the Miocene Period,

as *P. Defrancii*, which only differs from it in the absence of the sharp
spines which terminate the lateral angles of the carapace in the
former: while *Portunus leucodon*, Desmarest, bears some analogy to
Lupea.

An ideal landscape of the miocene period, which is given (Pl. xxiv.)

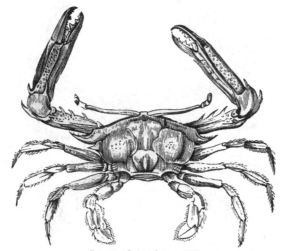

Fig. 170.—Podophthalmus vigil.

on the opposite page, represents the Dinotherium lying in the marshy
grass, the rhinoceros, the Mastodon, and an ape of great size, the
Dryopithecus, suspended from the branches of a tree.		The products
of the vegetable kingdom are, for the greater part, analogous to those
of our days.	It is remarkable for its abundance, and for its graceful
and serried vegetation: it still reminds us in some respects of the
vegetation of the carboniferous period.	It is, in fact, a continuation
of the characteristics of that period, and from the same cause, namely,
the submersion of the land under marshy waters, which has given
birth to a sort of coal which is often found in the miocene formation,
and which we call *lignite*.	This imperfect coal does not quite re-
semble that of the transition epoch, because it is of too recent date,
and because it has not been subjected to the same internal heat accom-

panied with the same pressure of superincumbent beds which produced
the older coal-beds of the transition epoch.

The *lignites*, which we find in the miocene, as in the eocene period,
constitute, however, a combustible which is worked and utilized in

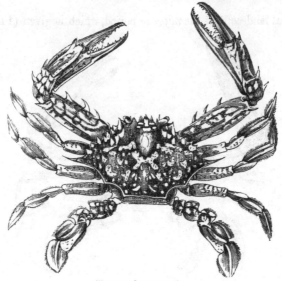

Fig. 171.—Lupea pelagica.

many countries, especially in Germany, where it is made in many
places to serve in place of coal. These beds sometimes attain the
thickness of twenty yards, but in the environs of Paris they form beds
of a few inches only, which alternate with clay and sand. We cannot
doubt that the lignites, like the coal, are the remains of the buried
forests of the ancient world ; in fact, the substance of the woods of our
forests, often in a state perfectly recognizable, are frequently found in the
lignite beds ; and the studies of modern botanists have demonstrated
that the species of which the lignites are formed belong to a vegetation
closely resembling that of Europe in the present day.

Another very curious substance is found in the lignite — yellow
amber, or *succin*. It is a resin, altered by time, which seems to have
flowed from the trees of the tertiary epoch ; the waves of the Baltic

Sea, corroding the lignite which lies under its bed, and detaching the amber in fragments, the substance, being lighter than water, rises to the surface, floats upon the waves, and is thrown upon the shore. For ages the Baltic has supplied commerce with this substance. The Phœnicians ascended its banks to collect this much-sought-for substance, which is now chiefly found between Dantzic and Memel, where it is a government monopoly collected by contractors, who are protected by a law making it theft to gather or conceal it.

Amber, which has lost some of its value in commerce, retains all its palæontological value; fossil insects are often found imprisoned in the nodules, where they have been preserved in all their brilliant colouring and integrity of form : as the poet says—

> "The things themselves are neither rich nor rare,
> The wonder's how the devil they got there."

The anti-putrescent and aromatic qualities have embalmed them, in order to transmit to our age the smaller beings and the most delicate organisms of the primitive world.

The miocene rocks are very imperfectly represented in the Paris basin, and they change their composition according to the localities. They are divided into two series of beds: 1. *Molasse*, or soft clay ; 2. *Faluns*, or shelly marl.

The *molasse* presents at its base some quartzose sand of great thickness, sometimes pure, sometimes a little argillaceous or micaceous. The beds include banks of sandstone, of which the quarries of Fontainebleau, of Orsay, and Montmorency, are worked for paving-stone for the streets of Paris and the neighbouring cities. This last formation is altogether marine. To these sands and sandstones succeeds a freshwater deposit, formed of a white limestone, partly siliceous, which forms the soil of the plateau of La Beauce, between the valley of the Seine and that of the Loire, where it is called the *Calcaire de la Beauce*. It is there mixed with a reddish earth more or less sandy, containing small blocks of millstone silex, easily recognized by its yellow-ochreous colour and numerous winding, mazy cavities, which form their tissues.

The *faluns* in the Paris basin consist of divers beds formed of shells and polypi, almost entirely broken up. In many parts of the country, and especially in the environs of Tours and Bourdeaux, they are dug out for manuring the land. To the falun beds near Auch belonged the soft lime-water which composed the celebrated bed of Lausan, in which M. Lartet found a considerable number of bones of turtles, birds, and mammifers, such as Mastodons and Pithecus antiquus. Isolated masses of these faluns occur near the mouth of the Loire and to the south of Tours, and in Brittany.

Caryophylla cyathus.

PLIOCENE PERIOD.

This last period of the tertiary epoch was marked in some parts of Europe by great movements of the terrestrial crust, always due to the same cause, namely, the continual and gradual cooling of the globe. This leads us to recall what we have repeatedly stated, that this cooling, during which the outer zone of the fluid mass passes to the solid state, leads to furrows and wrinkles in the external surface, sometimes accompanied by fractures through which the viscous semi-fluid matter pours itself; leading afterwards to the upheaval of mountain ranges through these gaping interstices: thus during the pliocene period many mountains and mountainous chains were formed in Europe by basaltic and volcanic eruptions. These upheavals would be preceded by unexpected and irregular movements of the elastic mass of the soil—earthquakes, in short; phenomena which have been already sufficiently explained.

In order to appreciate the vegetation of the period as compared with that with which we are familiar, let us listen to M. Lecoq: "Arrived, finally," says that author, "at the last period which pre-ceded our own epoch—the epoch in which the temperate zones were still embellished by equatorial forms of vegetation, which were how-ever slowly declining, driven out as it were by a cooling climate and by the invasion of more vigorous species—great terrestrial com-motions took place: mountains are covered with eternal snow; conti-nents now take their actual forms; but many great lakes now dried up still existed; great rivers flowed majestically through smiling countries, whose surface man had not yet come to modify."

"Two hundred and twelve species compose this rich flora, in which the ferns of the earlier ages of the world are scarcely indicated, where the palms seem to have quite disappeared, and we see forms much resembling those which are constantly under our observation. The *Culmites avrundinaceus*, Unger, abounds near the water, where also grows the *Cyperites tertiarius*, Unger; where floats *Dotamogeton geniculatus*, Braun.; and where we see submerged *Isoctites Brunnii*, Unger. Great conifers still form the forests. This fine family has, as we have seen, passed through every epoch and still remains with us; its elegant form and persistent evergreen foliage, as seen in the *Taxodites*, *Thuyoxylons*, *Abietites*, *Pinites*, *Eleoxylons* and *Taxites*, being still the forms most abundant in our natural forests.

" The predominating character of this period is the abundance of the group of the Amentaceæ : whilst the conifers are thirty-two in number, of the other we reckon fifty-two species, among which are many European genera, such as *Alnus; Quercus*, the oak ; *Salix*, the willow ; *Fagus*, the bean ; *Betula*, the birch ; &c.

The following families constitute the arborescent flora of the period besides those already indicated : Balsaminaceæ, Laurinaceæ, Thymelaceæ, Santalaceæ, Corneæ, Myrtaceæ, Calycantheæ, Pomaceæ, Rosaceæ, Amygdaleæ, Leguminaceæ, Anacardaceæ, Juglandaceæ, Rhamnaceæ, Celastrinaceæ, Sapindaceæ, Melacineæ, Acerineæ, Tiliaceæ, Magnoliaceæ, Capparidaceæ, Sopoteaceæ, Styraceæ, Oleaceæ, Junceaceæ, Ericineæ.

" In all these families great numbers of European genera are found, even more abundant in species than now. Thus, as Brongniart observes, in this flora we reckon fourteen species of maple ; three species of oaks ; and these species proceed from two or three very circumscribed localities, which would not probably at the present time represent in a radius of several leagues more than three or four species of these genera."

An important difference is observable in the pliocene as compared with the preceding period : it is the absence of the family of palms in the European flora, as noted by Lecoq, which forms such an essential botanical feature in the miocene period. We remark on this, because, in spite of the general analogy which exists between the vegetation of the Pliocene period and that of temperate regions in the present, it does not appear that there is any one species of the period identical with any one now growing in Europe. Thus the European vegetation even at the most recent geological epoch differs in its species from the vegetation of our age, although a general resemblance is observable between them.

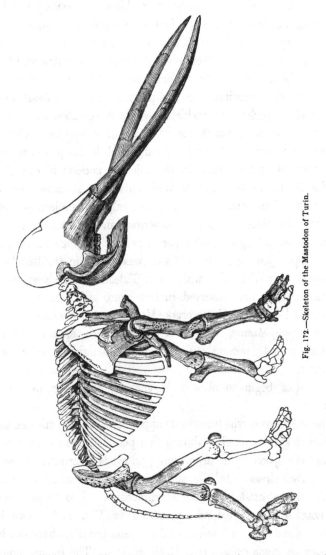

Fig. 172.—Skeleton of the Mastodon of Turin.

The terrestrial animals of the pliocene period present us with a great number of creatures alike remarkable from their proportions and from their structure. The mammifera and the batrachian reptiles are alike deserving of our attention in this age. Among the former the Mastodon, which makes its first appearance in the miocene formation, continues to be found, but becomes extinct apparently before we reach the upper beds. Others present themselves, unknown till now: some of them, such as the *hippopotamus*, the *camel*, the *horse*, the *ox* and the *deer*, continuing to the present age. The fossil horse, of all animals, is perhaps that which presents the greatest resemblance to existing individuals ; but it was small, not exceeding the ass in size.

The *Mastodon*, which we have considered in the preceding period, still lived, as we have said, in the pliocene period: in Fig. 172 the species living in this age is represented; it is called the Turin Mastodon. This species, as we see, has only two projecting tusks or defences, in place of four as described in the American species. Other species belonging to this period are not uncommon: the portion of an upper jaw-bone with a tooth was found in the Red Crag at Postwick, near Norwich, which Dr. Falconer has shown to be a pliocene species, first observed in Auvergne, and named by Messrs. Croizet and Jobert, its discoverers, *Mastodon Avernensis*.

The *hippopotamus*, tapir and camel, which appear during the pliocene period, present no peculiar characteristics to arrest our attention.

The apes begin to abound in species: the stags were already numerous.

The *rhinoceros*, which made its appearance in the miocene period, appears in greater numbers during this period. The species proper to the tertiary epoch is *R. tichorhynus*, which is descriptive of the bony partition which separated its two nostrils, an anatomical arrangement which does not exist in the actual rhinoceros. Two horns surmount the nose of this animal, as represented in Fig. 173. Two living species, namely, the rhinoceros of Africa and Sumatra, have two horns, but they are much smaller than *R. tichorhynus*. The Indian rhinoceros has only one horn.

The body of *R. tichorhynus* was covered very thickly with hairs, and its skin was without the rough and callous scales which we remark on the skin of the African rhinoceros.

At the same time with this gigantic species there existed a dwarf species about the size of our hog, and along with it several intermediate species, whose bones exist in sufficient numbers to reconstruct the skeleton. The curvature of the nasal bone of the fossil rhinoceros and its gigantic horn have given birth to many tales and popular legends. The *Roc*, which played so great a part in the fabulous myths of the people of Asia, originated in the discovery in the bosom of the earth of the cranium and horns of a fossil rhinoceros. The famous

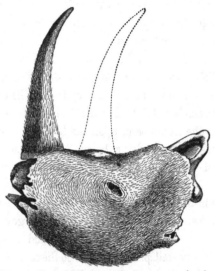

Fig. 173.—Head of Rhinoceros tichorhynus, restored under the direction of Eugene Deslongchamps.

dragon of western tradition has the same origin.

In the city of Klagenfurth, in Carinthia, is a fountain on which is sculptured the head of a monstrous dragon with six feet, and a head surmounted by a stout horn. According to the popular tradition still in full force at Klagenfurth, this dragon maintained itself in a cave, whence it issued from time to time to frighten and ravage the country. An undaunted chevalier kills the dragon, paying with his life this trait of his courage. It is the legend which is familiar to every country, including that of the valiant St. George and the Dragon and the legend of St. Martha. Nearly about the same age appeared the fabulous Tarasque of the Languedocean city, which bears the name of Tarascon.

But at Klagenfurth the popular legend has happily found a mouthpiece—the head of the pretended dragon, killed by the valorous knight, is preserved in the Hôtel-de-ville, and this head the sculptor has moulded for his fountain, and in it Herr Unger, of Vienna, recognized at a glance the cranium of the fossil rhinoceros ; its discovery in some cave had probably originated the fable of the knight and the dragon. And all legends are capable of similar explanation when we can ascend to their sources, and reason upon the testimony on which they are founded.

The traveller Pallas give a very interesting account of a *Rhinoceros tichorhynus* which he saw with his own eyes withdrawn from the ice in which its skin, hair, and flesh had been preserved. It was in December, 1771, that the body of the rhinoceros was observed buried in the frozen sand upon the banks of the Viloui, a river which throws itself into the Lena below Yakoutsk, in Siberia, in 64° north latitude. " I ought to speak," the learned naturalist says, " of an interesting discovery which I owe to the Chevalier de Bril. Some Yakouts hunting this winter near the Viloui found the body of a large, un-known animal. The Sieur Ivan Argounof, inspector of the Zimovic, had sent on to Irkutsk the head and a fore and hind foot of the animal, all very well preserved." The Sieur Argounof, in his report, states that the animal was half-buried in the sand ; it measured as it lay three ells and three-quarters Russian in length, and he estimated its height at three and a half ; the animal, still retaining its flesh, was covered with skin which resembled tanned leather ; but it was so corrupted that he could only remove the fore and hind foot and the head, which he sent to Irkutsk, where Pallas saw them. " They appeared to me at first glance," he says, " to belong to a rhinoceros ; the head especially was quite recognizable, since it was covered with its leathery skin, and the skin had preserved all its exterior characters and many short hairs. The eyelids had even escaped total corruption, and in the cranium here and there, under the skin, I perceived some matter which was evi-dently the remains of putrified flesh. I also remarked in the feet the remains of the tendons and cartilages where the skin had been removed. The head was without its horn, and the feet without hoofs. The place of the horn, and the raised skin which had surrounded it, and the division which existed in both the hind and fore feet, were evident proofs of its being a rhinoceros. In a dissertation addressed to the Academy of St. Petersburg, I have given a full account of this singular discovery. I give there reasons which prove that a rhinoceros had penetrated nearly to the Lena, in the most northern regions, and which have led to the discovery of the remains of other strange animals in Siberia. I confine myself here to describing the country where these curious remains were found, and to the cause of their preservation."

" The country watered by the Viloui is mountainous; all the stratification of these mountains is horizontal. They consist of selenitic and calcareous schists and clay, mixed with numerous beds of pyrites. On the banks of the Viloui we meet with coal much broken; probably coal beds exist higher up near to the river. The brook Kemtendoï skirts a mountain entirely formed of selenite or crystallized sulphate of lime and of rock salt, and this mountain of alabaster is more than three hundred versts (about two hundred miles), in ascending the Viloui, from the place where the rhinoceros was found. Opposite to the place we see near the river a low hill, about a hundred feet high, which, though of sand, contains some beds of millstone. The body of the rhinoceros had been buried in coarse gravelly sand near to this hill, and the nature of the soil, which is always frozen, had preserved it. The frozen soil near the Viloui descends to a great depth, for, although the rays of the sun will soften the soil to the depth of two yards in more elevated sandy places, in the valleys, where the soil is one half sand and the other clay, it is still frozen at the end of summer at half an ell below the surface. Without this intense cold the skin of the animal and many of its parts would long since have perished. The animal could only have been transported from some southern country to the frozen north at the epoch of the deluge, for the most ancient chronicles speak of no changes of the globe more recent to which we could attribute the deposit of these remains and of the bones of elephants which are found dispersed all over Siberia."

In this extract from the voyages of Pallas, the author, not to repeat

Y

himself, refers to a memoir inserted in the 'Commentarii' of the Academy of St. Petersburg. This memoir, written in Latin, bears the title 'Upon some Animals of Siberia.' After some general considerations, he relates the circumstances attending the discovery of the fossil rhinoceros, with some official documents affirming their correctness: the manner in which the facts were brought under his notice by the Governor of Irkutsk, General Bril. "The skin and tendons of the head and feet still preserved considerable flexibility, imbued as it were with humidity from the earth, but the flesh exhaled a fetid odour, resembling the ammonia of the latrine. Compelled to cross the Backal Lake before the ice broke up, I could neither make a sufficiently careful description or design of the parts of the animal, but I made them place the remains, without leaving Irkutsk, upon a furnace, with orders that after my departure they should be dried by slow degrees and with the greatest care, continuing the process for some time, for the viscous matter which continued to ooze out could only be dissipated by great heat. It happened, unfortunately, that during the operation the posterior part of the upper thigh and the foot were burnt in the overheated furnace, and they were thrown away; the head and the extremity of the hind foot only remaining intact and undamaged in the process of drying. The odour of the softer parts, which still contained viscous matter in their interior, was changed by the desiccation into one resembling that of flesh decomposed in the sun.

"The rhinoceros to which the members belonged was neither large for its species nor advanced in age, as the bones of the head attested, yet it was evidently an adult from the comparison made of the size of the cranium as compared with that of others of the same species more aged, which were afterwards found in the fossil state in divers parts of Siberia. The entire length of the head from the upper part of the nape of the neck to the extremity of the denuded bone of the jaw was thirty inches; the horns were not with the head, but we could still see evident vestiges of two horns, the nasal and frontal. The front, unequal and a little protuberant between the orbits, and of a rhomtoidal egg-shape, is deficient in the skin, and only covered by a light, horny membrane, bristling with straight hairs hard as a horn.

"The skin which covers the greater part of the head is, in the dried state, a tenacious, fibrous substance, like curried leather, of a brownish-black on the outside and white in the inside; when burnt, it had the odour of common leather; the mouth, in the place where the lips should have been soft and fleshy, was corrupted and much lacerated; the extremity of the maxillary bone was naked. Upon the left side, which had probably been longest exposed to the air, the skin was here and there decomposed and rubbed on the surface, while the right side was so well preserved that the pores, or little holes from which the hairs had fallen, were still visible all over that side as well as the front and round the orbits. In the right side of the jaw there were in certain places numerous hairs in groups or fascicules, for the most part rubbed down to the roots, but still retaining their full length, here and there standing erect, rigid, and of an ashy colour, but with one or two black, and stiffer than the others, in each fascicule.

"What was most astonishing, however, was the fact that the skin which covered the orbits and formed the eyelids was so well preserved and so healthy that the openings of the eyelids could be seen, though deformed and scarcely penetrable to the finger; the skin which surrounded the orbits, though desiccated, formed circular furrows. The cavities were filled with matter, either argillaceous or animal, such as still occupied a part of the cavity of the cranium. Under the skin the fibres and tendons still subsisted, and above all the remains of the temporal muscles: finally, in the throat hung some great bundles of muscular fibre. The denuded bones were young and less solid than in other fossil cranii of the same species. The bone which gave support to the nasal horn was not yet attached to the *vomer*; it was deficient in the articulations of the processes of the young bones. The extremities of the jaws preserved no vestige either of teeth or alveola, but they were covered here and there with integumentary remains. The first molar was distant about four inches from the extreme edge of the jaw.

"The foot which remains to me, and which, if I am not mistaken, belongs to the left hind limb, has not only preserved its skin quite intact and furnished with hair, or their roots, but also the tendons and ligaments of the heel in all their strength and flexibility up to the knee. The place of the muscles was filled with black mud. The extremities of the foot are defended by three angles, the bony parts of which, with the periosteum, still remain here and there: the horny hoof had been detached. The hairs adhering in many places to the skin are short, stiff, and ashy-coloured, which proves that the feet were entirely covered with hair, in bundles and hanging down.

" We have never, so far as I know, observed so much hair on any rhinoceros which has been brought to Europe in our times, as appears to be indicated on the head and feet of that described. I leave you then to decide if our rhinoceros of the Lena was born, or not, in the temperate climate of Central Asia. In short, the rhinoceros, as I gather from the relations of travellers, belongs to the forests of Northern India; and it is likely enough that these animals in the more hairy skin differ from those which live in the burning zones of Africa, just as other animals of warm climates are less warmly clothed than those of the same genera in temperate countries."

Of all the fossil ruminants perhaps the largest, and certainly not the least curious, is the *Sivatherium*, whose remains have been found in India, in the Sivalik mountains, one of the spurs of the Himalayas. Its name is taken from that of Siva, the Indian deity worshipped in this part of India.

The *Sivatherium* was about the size of the elephant; it belongs to the genera of stags, *Cervus*, and is probably the most gigantic which ever existed. It somewhat resembles the actual elan, but is much larger, and greatly more massive. The head presents an arrangement which has not been observed in any other animal known: it carried four sets of horns; two rising above the region of the ears in broad tines, and two projecting forward from above the eyes. These four horns were very divergent, and calculated to give the colossal stag a very strange aspect. Fig. 174 is a representation of the *Sivatherium* restored. At the same time it is to be noted that hitherto only the head has been discovered.

Fig. 174.—Sivatherium restored.

As if to rival these gigantic mammifera, great numbers of reptiles seem to have lived in the pliocene period, although they are no longer of the same importance as in the secondary epoch. Only one of these, however, need occupy our attention: it is the *Salamander*. The living salamanders are amphibious batrachians, with smooth skins, which rarely attain the length of twenty inches. The Salamander of the tertiary epoch had the dimensions of a crocodile; its discovery opens a pregnant page in the history of geology: the skeleton of this reptile was long considered to be that of a human victim of the deluge, and was spoken of as *homo diluvii testis*. It required all the efforts of Camper and Cuvier to eradicate this error from the minds of the learned, and probably in the minds of the vulgar it survived them both.

Upon the left bank of the Rhine, not far from Constance, a little above Stein, and near the village of Œningen, there are some fine quarries of schistose limestone: in consequence of their varied products these quarries have often been described by naturalists; they are of the tertiary epoch, and were visited, among others, by Horace de Saussure, and are described in the third volume of his 'Voyage dans les Alps.'

In 1725 a large block of stone was laid open, incrusted on which a skeleton was discovered, remarkably well preserved; and Scheuchzer, a Swiss naturalist of some celebrity, who added to his scientific pursuits the study of theology, was called upon to pronounce as to the nature of this relic of ancient times. He thought he recognized in the skeleton that of a man. In 1726 he gave a description of these fossil remains in the 'Philosophical Transactions' of London; and in 1731 he made it the subject of a special dissertation, which had for its title 'Homo diluvii testis'—Man's testimony to the deluge. This dissertation was accompanied by an engraving of the skeleton. Scheuchzer returned to the subject in another of his works, 'Physica Sacra,' saying, "It is certain that this schist is the half, or nearly so, of the skeleton of a man: that the substance even of the bones, and, what is more, of the flesh and of parts still softer than the flesh, are there incorporated in the stone: in a word, it is one of the rarest relics which we have of that cursed race which was buried under the waters. The figure shows us the contour of the frontal bone, the orbits with the openings which give passage to the great nerves of the fifth pair. We see there the remains of the brain, of the sphenoidal bone, of the roots of the nose, a notable fragment of the os maxilla, and some vestiges of the liver."

And our pious author exclaims, this time taking the lyric form—

"Betrübtes Beingerüst von einem altem Sunder
Erweiche, Stein, das Herzder neuen Bosheitskinder!"

"Oh cursed, deplorable, ancient skeleton,
With so much in thy aspect of the repentant fisherman."

The reader has before him the fossil of the schistose rocks of Œningen (Fig. 175). It is obviously impossible to find in this skeleton what the enthusiastic savant wished to see. And we can appreciate from this instance the errors to which a preconceived idea blindly followed may lead. How a naturalist who really was so capable as Scheuchzer could have found in this enormous head, and in these upper members, the least resemblance with the same osseous parts in man is incomprehensible !

The preadamite "witness to the deluge" made a great noise in Germany, and no one there dared to dispute the opinion of the Swiss naturalist, under the double aspect of theologian and savant, which probably is the reason why Gesner, in his 'Traité des Petrifactions,' published in 1758, describes with admiration the fossil of Œningen, which he attributes, with Scheuchzer, to the *antediluvian man.*

Pierre Camper alone dared to oppose himself to this, the public opinion then universally professed throughout Germany. He went to Œningen in 1787 to study the celebrated fossil ; he had no difficulty in detecting the error into which Scheuchzer had fallen. He recog-

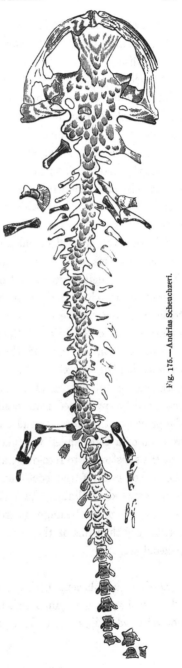

Fig. 175.—Andrias Scheuchzeri.

nized at once that it was a reptile, but he deceived himself nevertheless as to the family to which it belonged; he took it for a Saurian. "A petrified lizard," wrote Camper; " is it possible it could pass for a man ?" It was accordingly left to Cuvier to place in its true family the fossil of Œningen; in a memoir on the subject he demonstrates that this skeleton belonged to one of the amphibious batrachians which bear the name of salamanders. "Take," he says, in this memoir, "a skeleton of a salamander and place it alongside the fossil without allowing yourself to be misled by the difference of size, just as you could easily do in comparing a drawing of the salamander of natural size with one of the fossil reduced to a sixteenth part of its size, and everything is explained in the clearest manner.

"I am even persuaded," adds the great naturalist, in a subsequent edition of the memoir. " that, if we could rearrange the fossil and look closer into the details, we should find numerous proofs in the articular faces of the vertebræ, in those of the jaws, in the vestiges of the small teeth, and even in the labyrinth of the ear." And he invited the proprietors or depositaries of the precious fossil to proceed to such an examination, and he had the satisfaction of making himself the examination he suggested. Finding himself at Haarlem, he asked permission of the director of the museum to examine the stone which contained the pretended man fossil. The operation was carried on in the presence of the director and another naturalist. A drawing of the skeleton of a salamander was placed near the fossil by Cuvier, who had the satisfaction to recognize, as the stone was chipped away under the chisel, each of the bones announced by the drawing as they were exposed to light. In natural sciences there are few instances of more triumphant results — few demonstrations more satisfactory than this of the advantages of method and induction in palæontology.

During the pliocene period some birds of species very numerous, which still exist, gave animation to the vast solitudes which man had not yet filled. Vultures and eagles, among the rapacious birds; and among other genera, the gulls, swallows, pies, parroquets, pheasants, jungle-fowl, and ducks.

In the marine fauna of the period we see, for the first time, marine mammifera or cetacea—the *Dolphin* and *Balænodon* belonging to the period, although science knows very little of the fossil species belonging to the two genera. Some bones of dolphins, found in different parts of France, apprise us, however, that the ancient species differed from those of our days. The same remark may be made respecting the narval. This cetacean, so remarkable for its long tusk in form of a horn, has at all times been an object of curiosity.

These whales, whose remains are found in the rocks of eocene formation, differ little from the actual whales. But the observations geologists have been able to make upon these gigantic remains of the ancient world are too few to permit of any very precise conclusions. It is certain, however, that the fossil whale does differ from the actual whale in certain characteristics drawn from the bones of the cranium. The discovery of an enormous fragment of a fossil whale, made at Paris in 1779, in the catacombs of a wine-merchant in the Rue de Dauphiné, made a great sensation. Science pronounced without much hesitation that these were genuine remains; but the public failed to comprehend the existence of a whale in the Rue de Dauphiné. It was in making further excavations in his cellars that the wine-merchant made this interesting discovery. His workmen met under the pick, buried in a yellow clay, an enormous piece of bone. Its complete extraction caused him a great deal of labour, and presented many difficulties. Little interested in making further discoveries, our wine-merchant contented himself with raising, with the help of a chisel, a portion of the monstrous bone. The piece thus detached weighed 227 pounds. It was exposed by the wine-merchant, and large numbers of the curious went to see it. Among others, Lamanon, a naturalist of that day, who, examining it, conjectured that the bone had belonged to the head of a whale. As to the bone itself, it was purchased for the Museum of Teyler, at Haarlem, where it still remains.

There exists in the Museum of Natural History of Paris only a copy of the bone of the cetacea of the Rue de Dauphiné, which retains the name of *Balænodon Lamanona*: its examination by Cuvier led him to recognize it as a bone belonging to one of the antediluvian Balæna, which differ not only from the living species, but from all others known at the time.

Since the days of Lamanon, other bones of Balænæ have been discovered in the soil in different countries, but the study of these fossils has always left something to be desired. In 1806 a Balæna fossil was disinterred at Monte Pulgnasco by M. Cortesi; another was found in Scotland. In 1816 many bones of this animal were discovered in a little valley formed by a brook running into the Chiavana, one of the affluents of the Po; but no very rigorous study of either of these bones was undertaken.

Cuvier has established, among the cetacean fossils, a particular genera which he designates under the name of *Ziphius*. The animals to which he gave the name, however, are not identical either with the whales, *Balænæ*, the Cachelots or sperm whales, the Hyperoodons. They hold, in the order of cetaceans, the place that the Palæotheriums and Anoplotheriums occupy among the pachyderms, or that which the Megatherium and Megalonyx occupy in the order of the Edentatæ. The *Ziphius* still live in the Mediterranean.

The species of mollusks which distinguish this period from all others are very numerous. They include the Cardiums, Panopæa, Pectens, Fusus, Murex, Cypræa, Voluta, Chenopus, Buccinum, and many others.

The *pliocene rocks* are divided into lower and upper, or *older* pliocene, including the red crag of Suffolk, containing marine shells, of which sixty per cent. are extinct species, resting on a white or coralline crag; and *newer* pliocene, which includes, 1. the Bridlington beds, with arctic marine fauna ; 2. Chillesford beds, with marine shells, chiefly arctic ; 3. Norwich crag, with marine shells, and bones of *Mastodon arvenensis*.

The foreign equivalents of the older pliocene are found in the *sub-*

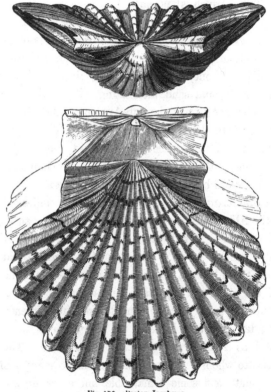

Fig. 176.—Pecten Jacobæus.

Apennine strata. These rocks are sufficiently remarkable in the county of Suffolk, where they consist of a series of marine beds of quartzose sand, coloured red by ferruginous matter.

At the foot of the Apennine chain, which forms the backbone, as it were, of Italy, throwing out many spurs, the formations on either side and on both sides of the Adriatic are tertiary strata; they form, in many cases, low hills lying between the Apennines of secondary

formation and the sea, the strata generally being a light-brown or
bluish marl covered with yellow calcareous sand and gravel, with some
fossil shells, which, according to Broche, are found all over Italy. But
this wide range includes some older tertiary formations, as in the
Superga of Turin, which is miocene.

The *Antwerp* crag, which is of the same age with the red crag of
Suffolk, forms great accumulations upon divers points of Europe: at
Antwerp, in Belgium, at Carentan and Perpignan, and, we believe, in
the basin of the Rhone, in France. The thickest deposits of this
rock consist of clay and sand, alternating with marl and arenaceous
limestone. This constitutes the sub-Apennine hills, alluded to above
as extending on both slopes of the Apennines. This deposit occupies
the Upper Arno. Its presence is recognized over a great part of
Australia. Finally, the seven hills of Rome are composed in part of
marine tertiary rocks belonging to the pliocene period.

The *coralline crag* ranges over about twenty miles between the
Stour and the Alde, with a breadth of four—a mass of shells, bryozoa,
and small corallines, mixed with marl, passing occasionally into a
soft building-stone. It is so thick in Oxfordshire, that in some
quarries the bottom has not been reached at the depth of fifty feet.

Many of the corals and bryozoa belong to extinct genera: they are
supposed to indicate an equable climate free from intense cold; an
inference rendered more probable by the prevalence of northern forms
of shells, such as *Glycimeris* or *Cyprini*, and *Astarte*. The late Professor Forbes, to whom science is indebted for so many philosophical
deductions, points out some remarkable inferences drawn from the
fauna of the Phocian seas. It appears that in the glacial period, which
we shall shortly have under consideration, many shells, previously
established in the temperate zone, retreated southwards, to avoid an
uncongenial climate. The Professor gives a list of fifty which inhabited the British seas while the red crag was forming, but which are
all wanting in the glacial deposits; from which he infers that they
migrated at the approach of the glacial period, and returned when the
temperate climate was restored.

XXV.—Ideal landscape of the Pliocene Period.

The *newer pliocene* prevails over Norfolk, Suffolk, and Essex, where it is popularly known as the Crag. In Essex it rests directly on the London clay. Between Weyborne and Cromer it rests on the chalk. Near Orford, in Suffolk, it rests on the *Chillesford* beds conformably, the *Bridlington* beds being nearly contemporaneous with them.

In Pl. xxv. an ideal landscape of the pliocene period is given under European latitudes. At the bottom of the picture, a mountain recently thrown up reminds us that the period was one of frequent convulsions, in which the soil was disturbed and overthrown, and mountains and mountain ranges made their appearance. The vegetation is nearly identical with that of our days. We see assembled in the foreground the more important creations of the period—the fossil species have been restored from the remains which have been found, others from their descendants where they still exist.

At the close of the pliocene period, and in consequence of the deposits left by the seas of the tertiary epoch, the map of Europe is nearly what it is now: few permanent changes have occurred to disturb its general outline. Although the point does not admit of demonstration, there is strong presumptive proof that in this period, or those immediately subsequent to it, the entire European area, with some trifling exceptions, including the Alps and Apennines, emerged from the deep. In Sicily, lower pliocene rocks, covering half the surface of the island, have been raised from two to three thousand feet above the level of the sea. Fossil shells have been observed at the height of eight thousand feet in the Pyrenees; and, as if to fix the date of upheaval, there are great masses of granite which have penetrated the lias and the chalk. Fossil shells of the period are also found at ten thousand feet in the Alps, at thirteen thousand in the Andes, and at eighteen thousand in the Himalayas.

In the mountainous regions of the Alps it is always difficult to determine the age of beds, in consequence of the contorted state of the strata: for instance, the lofty chain of the Swiss Jura consists of many parallel ridges, with intervening longitudinal valleys;

the ridges formed of curved fossiliferous strata, which are extensive in proportion to the number and thickness of the formations
which have been exposed on upheaval.   The proofs which these
regions offer of comparatively recent elevation are numerous : in the
central Alps, cretaceous, oolitic, liasic, and eocene strata, are found at
the loftiest summits, graduating insensibly into metamorphic rocks of
granular limestone, and into talcose and mica schists.   In the eastern
parts of the chain the older fossiliferous rocks are recognized in similar
positions, presenting proofs of tremendous Plutonic action; oolitic and
cretaceous strata having been raised twelve thousand feet, eocene ten
thousand, and miocene four and five thousand feet above the level of
the sea.   Equally striking proofs of recent elevation exist in the
Apennines : the celebrated Carrara marble, once supposed, from its
mineral texture and the absence of fossils, and from its resting, 1. on
talcose schists ; 2. on quartz and gneiss, to be very ancient, now turns
out to be an altered limestone of the oolitic series, and the underlying
crystalline rocks metamorphosed secondary sandstone and shale.   Had
all these rocks undergone complete metamorphose, another page in the
earth's history would have been obliterated.   As it is, the proofs of
what we state are found in the gradual approach of the rocks to their
unaltered condition as the distance from the intrusive rock increases.
This intrusive rock, however, does not always reach the surface, but
it exists below at no great distance, and is observed piercing through
the talcose gneiss, and passing up into secondary strata.

At the close of this epoch, therefore, there is every probability that
Europe and Asia had pretty nearly attained their present general
outline.

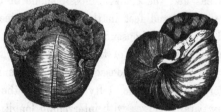

Fig. 177.—Bellerophon heuleus.

QUATERNARY EPOCH.

THE quaternary epoch of the history of our globe follows the tertiary epoch, and brings the narrative of its revolutions down to our own times.

The tranquillity of the globe is only troubled during this epoch by certain cataclysms whose sphere is limited and local, and by a passing trouble which most unexpectedly affects its temperature : the *deluges* and the *glacial* period—these are the two remarkable peculiarities which distinguish this epoch. But the fact which predominates in the quaternary epoch, and distinguishes it from all other phases of the earth's history, is the appearance of man, the culminating and supreme work of the Creator.

In this last phase of the history of the earth geology recognizes three chronological divisions :

1. The European deluges.
2. The glacial period.
3. The creation of man and subsequent Asiatic deluge.

Before describing the three orders of events which occurred in the quaternary epoch, we shall present a brief sketch of the organic kingdoms of nature, namely, the animals and vegetables which flourished in this epoch, and the new formations which arose. Lyell and some other geologists designate this the POST TERTIARY EPOCH, which they divide into—1. *The Post Pliocene Period ;* 2. *Recent,* or Upper Post Pliocene.

Post Pliocene Period.

In the days of Cuvier, the tertiary formation was considered a chaos of superficial deposits, having no distinct relations with other epochs. It was reserved for the English geologists, with Sir Charles Lyell at their head, to throw light upon this dark page of the earth's history: from the study of fossils, science has not only reanimated the animals; it has reconstructed the theatre of their existence. We see the British Islands now a straggling archipelago, and then the mouth of a vast river, of which the continent is lost; for, says Professor Ramsay, "We are not of necessity to consider Great Britain as having always been an island: it is an accident that it is an island now; and it has been an island many times before." In the tertiary epoch, we see it surrounded, then, by shallow seas swarming with humble forms of animal life; islands covered with bushy palms; banks on which turtles basked in the sun; vast basins of fresh or brackish water, in which the tide made itself felt, and which abounded with various species of sharks; rivers in which crocodiles increased and multiplied: woods which sheltered numerous mammifers and some serpents of large size; fresh-water lakes which received the spoils of numerous shells. Dry land has increased immensely. Groups of ancient isles we have seen united, and become continents, with lakes, bays, and perhaps inland seas. Gigantic elephants, vastly larger than any now existing, close the epoch, and probably usher in the succeeding one; for we are not to suppose any distinct feature distinguishes one period from another in nature, although it is convenient to arrange them so for the purposes of description. If we may judge from their remains, these animals must have existed in great numbers, for it is stated that on the coast of Norfolk alone the fishermen in trawling for oysters fished up between 1820 and 1833 no less than two thousand molar teeth of elephants. If we consider how slowly these animals multiply, these quarries of ivory, as we may call them, supposes many centuries for their production.

The same lakes and rivers were at the same time occupied by the hippopotamus, as large and as formidably armed as that now inhabiting the African solitudes; beside it, the two-horned rhinoceros; three species of Bos, one of which was hairy and bore a mane. Some stags of gigantic size, as compared with living species, bounded over the plains. In the same savannahs lived the reindeer, the goat, a horse of small size, the ass, the bear and the roe; for the mammalia had succeeded to the ichthyosaurii of a former age. Nevertheless, the epoch had its tyrants also. A *tiger*, as large as the largest of the Bengal tigers, hunted its prey in the British jungles. Another animal of the feline race, the *Machairodus*, was probably the most ferocious and destructive of carnivora; bands of hyenas and a terrible bear, surpassing in size that of the Rocky Mountains, had established itself in the caverns; two beavers and an ape were ushered into creation.

The discovery of the remains of most of these animals in caverns was perhaps among the most interesting discoveries of geology. The first discovery was made in the celebrated Kirkdale Caves in Yorkshire, which have been described by Dr. Buckland, and afterwards at Kents-hole, near Torquay. This pleasant Devonshire town is built in a creek, shut out from exposure on all sides except the south. In this creek, hollowed out of the rocks, is the great fissure or cavern known as Kents-hole: like that of Kirkdale, it has been under water, from whence, after a time more or less in extent, it has emerged, but had remained entirely closed till the moment when chance led to its discovery. The principal cavern is six hundred feet in length, with many crevices or fissures of smaller extent ranging in various directions in the rocks. A bed of hard stalagmite of very ancient formation, which has been again covered with a thin bed of soil, forms the floor of the cavern, which is a red sandy clay. In this bed was disinterred a mass of fossil bones belonging to extinct species of bears, tigers, lions and hyenas.

Such an assemblage gave rise to all sorts of conjectures. It was generally thought that the dwelling of some beasts of prey had been discovered, and that these bones of horses, stags and others were the remains of their prey. Others asked if, in some cases, instinct did not

z

impel sick animals, or animals broken down by old age, to seek such places as the family sepulchre. While others, again, suggested that these bones had been engulfed pell-mell in the hole in some ancient inundation. However that may be, the remains discovered in these caves show that all these mammalia existed at the close of the tertiary epoch, and that they all lived in England. What were the causes which led to their extinction?

It was the opinion of Cuvier and the early geologists that the ancient species were destroyed in some great and sudden catastrophe, from which none made their escape. But recent geologists trace their extinction to slow, successive, and determinative action due to local causes; the chief one being the gradual lowering of the temperature. We have seen that at the beginning of the tertiary epoch, in the old eocene age, palms, cocoa-nuts and acacias, resembling those now met with in countries more favoured by the sun, grew in our island. The miocene flora still has indications of a warm, but less tropical climate; and the pliocene period, which follows, contains remains which announce an approach to the existing climate. In following the vegetable productions of the tertiary epoch, the botanist meets with the flora of Africa, South America and Australia, and finally settles in the flora of temperate Europe. Many circumstances demonstrate this decreasing temperature until we arrive at what geologists call the *glacial period* —the winter of the ancient world.

But before entering on the evidences which exist of the glacial period we shall glance at the picture presented by the animals of the period; the vegetable products we need not dwell on; it is, in fact, that of our own era—the flora of temperate regions in our own epoch. The same remark would apply to the animals, but for some signal exceptions: in this epoch man appears, and some of the mammalia of the last epoch, but of larger dimensions, have long disappeared. The more remarkable of these extinct animals we shall describe, as we have done those belonging to anterior ages. They are not numerous; those of our hemisphere being the Mammoth, *Elephas primogenius;* the bear, *Ursus spelæus;* gigantic tiger, *Felis spelæa;* hyena, *Hyena spelæa;*

the ox, *Bos priscus* and *primogenius*; the gigantic stag, *Cervus megaceros*; to which we may add the *Dinornis* and *Epeornis*, among birds. In America there existed in the quaternary epoch some

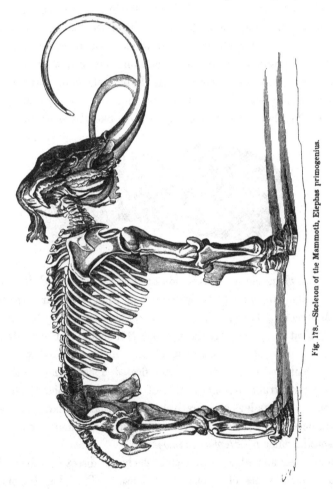

Fig. 178.—Skeleton of the Mammoth, Elephas primogenius.

Edentata of colossal dimensions and of very peculiar structure: these were *Megatherium*, the *Megylonyx* and the *Mylodon*: we shall pass these animals in review, beginning with those of our own hemisphere.

z 2

The Mammoth, the skeleton of which is represented in Fig. 178, surpassed the largest elephants of the tropics in size, for it was sixteen to eighteen feet in height. The monstrous tusks with which it was armed were twelve to thirteen feet in length, curving into a half circle. The form of its teeth permits of their being distinguished from its congener the Mastodon; for while the teeth of the latter have pointed tuberosities in the form of a mammelon on their surface, those of the Mammoth, like those of the Indian elephant, have a broad united surface, with regular furrowed lines of large curvature. The teeth of the Mammoth are four in number, like the elephants, two in

Fig. 179.—Tooth of the Mammoth.

each jaw, when the animal is adult: its head elongated, its frontal concave, its jaws curved and truncated in front, are characteristics of the Mammoth. It has been an easy task, as we shall see, to recognize the general form and structure of the Mammoth, and even its skin. We know beyond a doubt that it was thickly covered with long shaggy hair, and that a copious mane floated upon its neck and along its back; its trunk resembled that of the Indian elephant; its body was heavy, and its legs comparatively shorter than the latter animal, of which, nevertheless, it had many of the habits. Blumenbach gave it the specific name of *Elephas primogenius*.

In all ages and almost all countries chance discoveries have been made of fossil bones of elephants in the soil. Pliny has transmitted to us a tradition collected by the historian Theophrastus, who wrote three hundred and twenty years before Jesus Christ, of the existence of bones or fossil ivory in the soil of Greece, that the bones were sometimes transformed into stones: "These bones," the historian gravely tells us, "were both black and white, and born of the earth." Some of the elephants' bones having a slight resemblance to those of man, they have often been taken for human bones. In the earlier historic times these great bones, accidentally disinterred, have passed as having belonged to some hero or demigod; at a later period they were taken for the bones of giants. We have already spoken of the mistake made by the Greeks in taking the patula of a fossil elephant for the knee-bone of Ajax; in the same manner the bones revealed by an earthquake, and attributed by Pliny to a giant, belonged, no doubt, to a fossil elephant. To a similar origin we may assign the pretended body of Orestes, thirteen feet in length, which was discovered at Tegea by the Spartans; those of Asterius, the son of Ajax, discovered in the Isle of Ladea, of ten cubits (about eighteen feet), according to Pausanius; finally, such were the great bones found in the Isle of Rhodes, of which Phlegon, of Tralles, speaks in his 'Mundus Subterraneus.'

We might fill volumes with the history of the remains of pretended giants found in ancient tombs. The books, in fact, which exist formed a voluminous literature in the middle ages—its title *gigantology*. All the facts, more or less real, true or imaginative, may be explained by the accidental discovery of the bones of some of these gigantic animals. We find in this mediæval work the history of a pretended giant, discovered in the fourth century, at Trapani, in Sicily, of which Boccaccio speaks, which may be taken for Polyphemus; of another, found in the sixteenth century, according to Fasillus, near Palermo; others, according to the same author, at Melilli, between Leontium and Syracuse, at Calastri and Petralia, at each of which places the bones of pretended giants were disinterred. P. Kercher speaks of three other giants found in Sicily, of which only the teeth remained perfect.

In 1577 a storm having uprooted an oak near the cloisters of Reyden, in the Canton of Lucerne, some large bones were exposed to view. Seven years after a physician and Professor of Basle, Felix Plater, being at Lucerne, examined these bones, and declared they could only proceed from a giant. The Council of Lucerne consented

to send the bones to Basle for more minute examination, and Plater
thought himself justified in attributing to the giant a height of nine-
teen feet. He designed a human skeleton on this scale, and returned
the bones with the drawing to Lucerne. In 1706 there only remained
of these bones a portion of the *scapula* and a fragment of the wrist-
bone : the anatomist Blumenbach, who saw them at the beginning of
the century, easily recognized them for the bones of an elephant.
Let us not omit to add, as a compliment to this bit of history, that
the inhabitants of Lucerne adopted the image of this pretended giant
as the supporters of the city arms.

Spanish history preserves many stories of giants. The tooth of
St. Christopher, shown at Valence, in the church dedicated to the
saint, was certainly the molar tooth of a fossil elephant; and in 1789
the canons of St. Vincent carried through the streets in public pro-
cession, to procure rain, the pretended arm of a saint, which was
nothing more than the *femur* of an elephant.

In France, under the reign of Charles VII., some of these bones of
pretended giants appear in the bed of the Rhone. A repetition of the
phenomenon occurred near Saint-Peirat, opposite to Valence, when the
Dauphin, afterwards Louis XI., then residing at the latter place, caused
the bones to be gathered together, and sent to Bourges, where they long
remained objects of public curiosity in the interior of Saint-Chapelle.
In 1564, a similar discovery took place in the same neighbourhood.
Two peasants observed on the banks of the Rhone, along a *talus*, some
great bones projecting from the earth. They carried them to the
neighbouring village, where they were examined by Cassanion, who
dwelt at Valence. It was no doubt apropos to this that Cassanion
wrote his treatise ' De Gigantibus.' The description given by the
author of a tooth, sufficed, according to Cuvier, to prove that it
belonged to an elephant ; it was a foot in length, and weighed eight
pounds. It was also on the banks of the Rhone, but in Dauphiny,
as we have seen, that the skeleton of the famous Teutobocchus, of
which we have spoken in a previous chapter, was found.

In 1663, Otto de Guericke, the illustrious inventor of the pneumatic machine, was witness to the discovery of the bones of an elephant, buried in the shelly limestone, or musselkalk. Along with it were found its enormous tusks, which should have sufficed to establish its zoological origin. Nevertheless they were taken for horns, and the illustrious Leibnitz composed out of the remains a strange animal, carrying a horn in the middle of its forehead, and in each jaw a dozen molar teeth, a foot long. Having fabricated this fantastic animal, Leibnitz named it also; he called it the *fossil unicorn*. In his ' Protogn,' a work remarkable besides as the first attempt at a theory of the earth, Leibnitz gave the description and a drawing of this imaginary being. During more than thirty years the unicorn of Leibnitz was universally accepted throughout Germany; and nothing less than the discovery of the entire skeleton of the Mammoth in the valley of Unstrutt was required to make them renounce the idea. This skeleton was at once recognized by Tinzel, librarian to the Duke of Saxe-Gotha, as an elephant, and was established as such; not, however, without a keen controversy with adversaries of all kinds.

In 1700, a soldier of Wurtemberg accidentally observed some bones showing themselves projecting out of the earth, in an argillaceous soil, near the city of Canstadt, not far from the banks of the Necker. Having addressed a report to the reigning Duke, he caused the place to be excavated, which occupied nearly six months. A veritable cemetery of elephants was discovered, in which were not less than sixty tusks. Those which were entire were preserved; the fragments were abandoned to the court physician, and they became a mere vulgar remedy. In the last century the fossil bones of bears, which were abundant in the country, were administered in Germany as a medicament. It was then called the *dicorn fossil*. The magnificent tusks of the Mammoth found at Canstadt helped to combat fever and colic. What an intelligent being this court physician of Wurtemberg must have been !

Numerous discoveries like those we have quoted distinguished the eighteenth century, but the progress of science has rendered such mistakes as we have had to relate impossible. These bones were now univer-

sally recognized as belonging to an elephant, but erudition now intervened, and helped to obscure a subject which was otherwise perfectly clear. Some learned pedant declared that the bones found in
Italy and France proceeded from the elephants which Hannibal led from
Carthage with the army in his expedition against the Romans. The
part of France where the most ancient bones of these elephants were
found is in the environs of the Rhone, and consequently on the route
of the Carthaginian general, and this consideration appeared a
particularly triumphant answer to the naturalist's reasoning, to these
terrible savants. Again, at a later period, Domitius Ænobarbus conducted their armies, which followed with a number of elephants, armed
for war. Cuvier scarcely took the trouble to refute this insignificant
objection. It is merely necessary to read in his learned dissertation of
the number of elephants which would remain to Hannibal when he
had passed through Gaul.

But the best reply that can be made to this strange objection is to
show how extensively these fossil bones of elephants are scattered, not
in Europe only, but over the world : there is no region of the globe in
which the remains are not found. In the north of Europe, in Scandinavia, in Ireland, in Germany, in Central Europe, in Poland, in middle
Russia, in South Russia, in Greece, in Spain, in Italy, in Africa, in
Asia, and, as we have seen, in England. In the New World, in a word,
we have found, and continue to find still, tusks, molar teeth and bones
of the Mammoth. What is most singular is, that these remains exist
more especially in great numbers in the north of Europe, in the frozen
regions of Siberia; regions altogether uninhabitable for the elephant
in our days. "There is not," says Pallas, "in the whole of Asiatic
Russia, from the Don to the extremity of the Tchutchian promontory,
any brook or river, especially of those which flow in the plains, on the
banks of which some bones of elephants and other animals foreign to
the climate have not been found. But in the more elevated regions,
the primitive and schistose chains, they are wanting, as are the marine
petrifactions. But in the lower slopes and in the great muddy and
sandy plains, above all, in places which are swept by rivers and brooks,

they are always found, which proves that we should not the less find them throughout the whole extent of the country if we had the same means of searching for them."

Every year in the season of thawing, the vast rivers which descend to the Frozen Ocean in the north of Siberia sweep down with their waters numerous portions of the banks, and expose to view the bones buried in the soil and in the excavations left by the rushing waters. Cuvier gives a long list of places in Russia in which interesting discoveries have been made of elephants' bones, and it is certainly curious that the more we advance towards the north in Russia the more numerous and extensive do the bone repositories become. In spite of the undoubted testimony, often repeated, of numerous travellers, we can scarcely credit the statements made respecting some of the islands of the glacial sea near the poles, situated opposite the mouth of the Lena and of the Indigirska. Here, for example, is an extract from 'Belling's Voyage' concerning these isles :—"All the island nearest to the main land, which is about thirty-six leagues in length, except three or four small rocky mountains, is a mixture of sand and ice, so that when the thaw sets in and its banks begin to fall many Mammoth bones are found. All the isle," Belling adds, according to his translator, "is formed of the bones of this extraordinary animal, of the horns and cranii of buffaloes, or of an animal which resembles them, and of some rhinoceros horns."

New Siberia and the Isle of Lachon are for the most part only an agglomeration of sand, of ice, and of elephants' teeth. At every tempest the sea casts ashore new quantities of Mammoths' tusks, and the inhabitants of Siberia carry on a profitable commerce in this fossil ivory. Every year during the summer innumerable fishermen's barks direct their course towards this isle of bones, and during winter immense caravans take the same route, all the convoys drawn by dogs, returning charged with the tusks of the Mammoth, weighing each from 150 to 200 pounds. The fossil ivory thus withdrawn from the frozen north is imported into China and Europe, where it is employed for the same purposes as ordinary ivory, which is furnished, as we know, by the elephant and hippopotamus of Africa and Asia.

The *isle of bones* has served as a quarry of this valuable materia. for export to China for five hundred years, and it has been exported tc Europe for upwards of a hundred. But the supply from these strange mines remains undiminished. What a number of accumulated genera- tions of these bones and tusks does not this profusion imply !

The abundance of the remains of fossil elephants in the Russian steppes has given birth to a legend of a very ancient origin. The Russians of the north believe that these bones proceed from an enormous animal which lived, like the mole, in holes which it dug in the earth ; it could not support the light, says the legend, but died when exposed to it.

It was in Russia that the fossil elephant received the name of the *Mammoth*, and its tusks mammoth horns. Pallas asserts that the name originates in the word mamma, which in Tartar idiom signifies earth. According to other authors, the name proceeds from the Arabic word behemoth, which, in the Book of Job, designates an unknown animal, or of the epithet mehemot, which the Arabs have been accustomed to add to the name of the elephant when of unusual size. A circumstance curious enough is that this same legend of an animal living exclusively underground has spread to China. The Chinese call it *tien-schu*, and we read in the great Chinese work on *Natural History*, composed in the sixteenth century, " The animal named *tien-schu*, of which we have already spoken in the ancient work upon ceremonies, entitled ' Ly-ki,' a work of the fifth century before Jesus Christ, is called also *tyn-schu* or *yn-schu*, that is to say, the mouse which hides itself. It constantly confines itself to subterranean caverns ; it resembles a mouse, but is of the size of a buffalo or ox. It has no tail ; its colour is dark ; it is very strong and excavates caverns in places full of roots, and covered with forests." Another writer, quoting the same passage, thus expresses himself: " The *tyn-schu* haunts obscure and unfrequented places. It dies as soon as it is exposed to the rays of the sun or moon ; its feet are short in proportion to its size. Its tail is as long as that of a Chinese. Its eyes are small, its neck short. It is very stupid and sluggish. When the inundations of the river *Tam-schuanu-my* took place, in 1571, it often showed itself in the plain; it is nourished by the roots of the plant *fu-kia.*"

The existence in China of the bones and tusks of the Mammoth is sufficiently confirmed by the following extract from an old Russian traveller, Isbrant Ides, who, in 1692, traversed the Chinese empire. In the extract which follows we remark the very surprising fact of the discovery of a head and foot of the Mammoth which had been preserved in ice, with all the flesh. "In the mountains which lie to the north-east of the river Kala we found," says the traveller, "the teeth and bones of the Mammoth; they were found on the banks of the rivers Senizea, Trugau, Mungazea, and the Lena, in the environs of the city of Takutskoi and up to the Frozen Ocean. All these rivers pass across mountains, of which we shall have occasion to speak, and in the season of thawing their frozen courses are so impetuous that they tear up the mountains and sweep with their waters masses of earth of prodigious size. The inundation over these masses of earth remain on the banks, and becoming dry, we find in the middle of them the teeth of the Mammoth, and sometimes even the Mammoth entire. A traveller who lived with me in China, and who employed a whole year in seeking for their teeth, assured me that he once found in a piece of frozen earth the head of one of these animals, with the flesh decomposed, with the tusks attached to the muzzle like those of the elephants, and that he and his companions had great trouble in tearing asunder the bones of the head, and among others that of the neck, which was still stained with blood; that having finally searched more forward in the same mass of earth, he found there a frozen foot of monstrous size, which he carried to the city of Tragau. This foot was, from what the traveller said to me, of the circumference of a large man about the middle of the body.

"The people of the country have various opinions about these animals. The idolaters, like the Lakoutas, the Tunguses, and the Ostiakas, say that the Mammoth lived in spacious caverns which they never left; that they could wander here and there in these caverns;

but that while they have lived in these places, the caverns have been raised and afterwards sunk into a deep abyss, forming now a profound precipice; they are also convinced that a Mammoth dies the instant he sees the light, and they maintain that it is thus those have perished which are found on the banks of rivers near their dens, from which those individuals have inconsiderately advanced.

" The old Russians of Siberia believe that the Mammoths are only elephants, though the teeth found be a little more curved and thicker in the jaw than in that animal. ' Before the deluge,' they say, ' the country was warmer and the elephants which basked in the waters, and were afterwards interred in the mud, more numerous. The climate became very cold after this catastrophe. The mud much frozen : and with it the bodies of these elephants, which the frozen earth preserved uncorrupted till the time when the thaw revealed them.' "

This recital may seem suspicious to some readers. We have ourselves some difficulty in believing that this head and foot were taken from the ice, with the flesh and skin, when we consider that this animal has been extinct probably more than ten thousand years. But the assertion of Isbrant Ides, who lived in 1692, is confirmed by respectable testimony of more recent date. In 1800 a Russian naturalist, Gabriel Sarytschen, travelled in Northern Siberia. Having arrived in the neighbourhood of the Frozen Ocean, he found upon the banks of the Alasæia, which throws itself into this sea, the entire body of a Mammoth wrapped up in a shroud of ice. The body was in a complete state of preservation, for the permanent contact of the ice had kept out the air and prevented decomposition. It is well known that at zero and below it, animal substances will not putrefy, so that in our households we can preserve all kinds of animal food as long as we can surround them with a bed of ice, and this is what happened to the Mammoth found by Gabriel Sarytschen in the ice of the Alasæia. The rolling waters had disengaged the mass which had imprisoned the monstrous pachyderm for thousands of years. The body in a complete state of preservation and enclosed in its flesh and its skin, to which long hairs adhered in certain places, found itself again nearly erect on its four feet.

The Russian naturalist Adams again, in 1806, made a discovery quite as extraordinary as the preceding. We borrow his recital from the "Commentarii" of the Academy of St. Petersburg, as translated by Cuvier. In 1799, a Tungusian fisherman observed among the icebergs, upon the banks of the Frozen Sea, near the mouth of the Lena, an odd-shaped block which he did not examine. The following year he perceived that this mass was a little more disengaged, but he had not yet divined what it could be. Towards the end of the year following, one entire side was exposed, and he now discovered the whole side and tusks of the Mammoth distinctly issuing from the ice. It was not till the fifth year, when the ice having melted quicker than usual, this enormous mass was stranded on a bed of sand on the coast. In the month of March, 1804, the fisherman removed the tusks. It was not till two years after this that Mr. Adams, of the St. Petersburg Academy, who was travelling with Count Golovkni, sent by the Russian Czar on an embassy to China, having been informed of the discovery at Yakoutsk, betook himself to the place. He found the animal already much mutilated. The Yakoutskes of the neighbourhood had cut off the flesh to feed their dogs; wild beasts had also mangled it. Nevertheless, the skeleton remained nearly entire, with the exception of one fore-foot. The spine of the back, a scapula, the pelvis, and the remains of the three limbs were still connected by the ligaments and a portion of the skin; the missing scapula was found a little distance off. The head was covered with a dry skin; one of the ears, well preserved, was furnished with a tuft or mane; the balls of the eyes were still distinguishable; the brain still occupied the cranium, but was dried up; the under lip had been rubbed and the upper lip had been destroyed to expose the jaws; the top of the neck was furnished with a flowing mane; the skin was covered with tufts of black hairs and reddish wool. What remained of the animal was so heavy that ten persons could scarcely carry it. There was recovered, according to Mr. Adams, more than thirty pounds of hair and wool which the white bears had taken from it, and buried in the moist soil, while devouring the flesh. The animal was a male; its tusks were more than nine feet in length, and its head and tusks weighed more than four hundred pounds.

Mr. Adams took every care to collect all that remained of this unique specimen of an ancient creation. He succeeded in repurchasing the tusks at Yakoutsk, and the Emperor of Russia, who became the purchaser of this precious monument, paid him eight thousand roubles; it is deposited in the Academy of St. Petersburg. "We have yet to find," says Cuvier, "any individual equal to it."

The Mammoth (Fig. 180) is, without doubt, the most important of

Fig. 180.—Mammoth restored.

all the animals of the ancient world yet discovered, and reconstituted, so to speak, by modern science. It is on these grounds we have dwelt at unusual length upon the principal traits of its history, in the course of which we have been able, besides, to give many interesting

E. MEUNIER

XXVI.—Skeleton of the Mammoth in the St. Petersburg Museum.

facts drawn from the annals of different countries in which their remains have been found.

Dogs and other voracious animals had devoured the flesh of the Mammoth found on the shore of the Frozen Ocean. Adams transported its bones to St. Petersburg, and they now form the finest skeleton of the *Elephas primogenius* which exists. Beside the skeleton of this famous Mammoth, there is placed that of an Indian elephant, and another elephant with skin and hair, in order that the visitor may have a proper appreciation of the vast proportions of the Mammoth as compared to them. PL. XXVI., on the opposite page, represents the saloon of the Museum of St. Petersburg, which contains these three interesting remains.

We cannot doubt, after such testimony, of the existence in the frozen north of remains of the Mammoth almost entire. The animals seem to have perished suddenly; seized by the ice at the moment of their death, their bodies have been preserved from decomposition by the persistent action of the cold. If we suppose that one of those animals had fallen accidentally into the crevasse of some glacier, we could explain to ourselves that its body, buried immediately under eternal ice, had been maintained intact during these thousands of years.

In Cuvier's great work on *fossil bones*, he gives a long and minute enumeration of the diverse regions of Germany, France, Italy, and other countries which have furnished in our days bones or tusks of the Mammoth. We venture to quote two of these descriptions. "In October, 1815," he says, "there was discovered at Canstadt, in Wurtemberg, near which some remarkable discoveries were made in 1700, a very remarkable deposit, which the King, Frederick I. caused to be excavated and its contents collected with the greatest care. We are even assured that the visit which the prince, in his ardour for all that was great, paid to this spot contributed to the malady of which he died a few days after. An officer, Herr Natter, had commenced some excavations, and in four-and-twenty hours had discovered twenty-one teeth, or parts of teeth, and a great many bones. The king having ordered him to continue the excavations, on the second day they came

upon a group of thirteen tusks placed near each other, and along with
them some jaw-teeth, lying as if they had been piled up there. It was
on this discovery that the king caused himself to be transported thither,
and ordered all the surrounding soil to be dug up and every object to be
carefully preserved in its original position. The largest of the tusks,
though it had lost its points and its root, was still eight feet long and
one foot in diameter. Many isolated tusks were also found, with a
quantity of jaw-teeth, from two inches to a foot in length, some still
adhering to the jaw. All these fragments were better preserved than
those of 1700, which was attributed to the depth of the bed, and, perhaps, to the nature of the soil. The tusks were generally much curved.
In the same deposit some bones of horses and stags were found, together
with a quantity of teeth of the rhinoceros, and others which we thought
to belong to a bear, and one specimen which was attributed to the
tapir. The place where this discovery was made is named Seelberg;
it is about six hundred paces from the city of Canstadt, but on the
opposite side of the Necker."

"All the great river basins of Germany have, like those of the
Necker, yielded fossil bones of the elephant; those especially abutting
on the Rhine are too numerous to be mentioned, nor is Canstadt the
only place in the valley of the Necker where they are found."

But of all parts of Europe, that in which they are found in
greatest numbers is the valley of the Upper Arno. We find there
a perfect cemetery of elephants. These bones were at one time
so common in this valley that the peasantry employed them in-
discriminately with stones in constructing walls and houses. Since
they have learned their value, however, they reserve them for sale
to travellers.

The bones and tusks of the Mammoth are met with in America as
well as in the Old World. Cuvier enumerates several places on that
continent where their remains are met with mingled with those of the
Mastodon. The Russian Captain Kotzebue found them also on the
north coast of America, where they were so common that the
sailors burnt many pieces in their fires. Adalbert de Chamisso, who

accompanied Kotzebue's expedition as naturalist, brought to Europe a tusk four feet long and five inches in diameter.

It is very strange that the East Indies, that is, one of the two regions which is now the home of the elephant, should be the only country in which the fossil bones of these animals have not been discovered. In short, from the preceding enumeration, it appears that, during the geological period whose history we are recording, the gigantic Mammoth inhabited all the regions of the globe where the climate agreed with that in which elephants are now found, namely, tropical countries: from which we must draw the conclusion to which so many other inferences lead, that, at the epoch in which these animals lived, the temperature of the earth was infinitely higher than in our days.

Among the antediluvian carnivora one of the most formidable seems to have been the *Ursus spelæus*, or cavern bear. (Fig. 181.) This species must have been a fifth, if not a fourth, larger than the brown bear of our days. It was also more squat: some of the skeletons we possess are from nine to ten feet long, and only about six feet high. The *U. spelæus* abounded in France, in Belgium and Germany; and so extensively in the latter country that the teeth of the antediluvian bear, as we have already stated, formed for a long time part

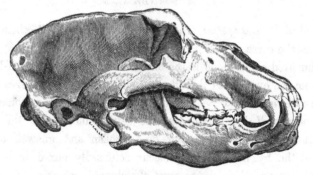

Fig. 181.—Head of Ursus spelæus.

of its materia medica, under the name of the *bicorn fossil*. Fig. 181 represents the head of the cavern bear.

2 A

At the same time with the *Ursus spelæus* another carnivora, the *Felis spælus,* or gigantic tiger, lived in Europe. Twice the size of the present race of tigers, this ánimal combined many of the characteristics of both lion and tiger. Its remains, which are rarely met with, would assign to the animal a length of more than four yards, with a size exceeding that of the largest bull.

The hyænas of our age consist of two species, the striped and the spotted hyænas. The last presents considerable conformity in its structure with that of the post pliocene period, which Cuvier designates under the name of the spotted fossil hyæna. It seems to have only been a little larger than the existing one. Fig. 182 represents

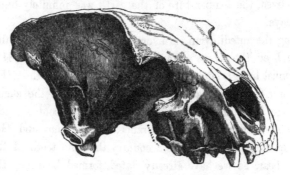

Fig. 182.—Head of Hyæna spelæa.

the head of *Hyæna spelæa,* whose remains, with those of others, were found in the caves of Kirkdale and Kents-hole; the remains of about three hundred being found in the former. Dr. Buckland satisfied himself, from the quantity of their dung, that the hyænas had lived there. In the cave were found remains of the ox, young elephants, rhinoceros, horse, bear, wolf, hare, water-rat, and several birds. All the bones have the appearance of having been broken and gnawed by the teeth of the hyænas, and they occur confusedly mixed in loam or mud, or dispersed through the crust of stalagmite which covered them.

The horse, then, ascends to the quaternary epoch, if not to the last period of the tertiary epoch. Its remains are found in the same rocks with those of the Mammoth and of the rhinoceros. It is distinguished from our horse only by its size, which was smaller: its remains abound in the post pliocene rocks not only in Europe, but in America; so that an aboriginal horse existed in the New World before it was carried thither by the Spaniards, although we know that it was unknown at the date of their arrival. This species then was extinct, and its disappearance can in no manner be attributed to the action of man.

The oxen of the period were, if not identical, at least very near to our living species: there were three species—the *Bos priscus*, *primigenius* and *Pallasii*: the first with slender legs, with convex frontal, broader than high, and differing slightly from the *Aurochs*, but taller and with immense horns. The remains of *B. priscus* are found in France, Italy, Germany, Russia and America. *B. primigenius* was, according to Cuvier, the source of our domestic cattle. The Bos Pallasii is found in America and in Siberia, and resembles in many respects the musk-ox of Canada.

Where these great mammifera are found we generally discover the fossil remains of several species of stags. The palæontological question as regards these stags is very obscure, and it is often difficult to determine if the remains belong to an extinct or existing species. This doubt does not extend, however, to the gigantic forest-stag, *Cervus megaceros*, one of the most magnificent of the antediluvian animals, whose remains are still frequently found in Ireland in the neighbourhood of Dublin; more rarely in France, in Germany, in Poland and in Italy. Intermediate between the stag and the elk, the *Cervus megaceros* partakes of the elk in its general proportions and form of its cranium, but it approaches the stag in its size and in the disposition of its horns. These magnificent appendages, however, while they decorated the head of the animal and gave a most imposing appearance to it, must have sadly impeded its progress through the thick and tangled forests of the ancient world. The length of these horns was between nine and ten feet, and they were so divergent that, measured from one extremity to the other, they occupied a space of between three and four yards.

The skeleton of the *Cervus megaceros* is found in the calcareous

2 A 2

tufa deposits, which underlie the peat-moss in Ireland ; sometimes in the turf itself, near the Curragh in Kildare ; in which position they sometimes present themselves in little heaps piled up in a small space, and nearly always in the same attitude : the head aloft, the neck stretched out, the horns reversed and thrown downwards towards the back, as if the animal, suddenly immersed in marshy ground, were under the necessity of throwing up its head in search of respirable air. In the Geological Cabinet of the Sorbonne, at Paris, there is a magnificent skeleton of *Cervus megaceros ;* another belongs to the College of Surgeons in London, and a third at Vienna.

The most remarkable creatures of the period, however, were the great Edentata : the Glyptodon, the gigantic Megatherium, the Mylodon and the Megalonyx. The order of ancient Edentata is characterized by the absence of teeth in the fore part of the mouth. The masticating apparatus of the Edentata consists only of molars and canine teeth ; sometimes, indeed, they are absent altogether, as the order feed chiefly on insects or tender leaves of plants. The armadillo, ant-eater and pangolins, are the living examples of the order. We may add, yet farther to characterize them, largly developed claws at the extremities of the toes. The order seems thus to establish itself as a zoological link in the chain between the hoofed mammifera and the unguicalate animals, or those armed with claws. All these animals belonged to the American continent.

The *Glyptodon,* which appears during this period, belonged to the family of armadillos, and their most remarkable feature was the presence of a hard scaly shell composed of numerous scales, which cover the entire upper surface of the animal from the head to the tail ; in short, a mammiferous animal, which appears to have been enclosed in a shell like the turtles : it resembles in many respects the *Dasypus,* or ant-eater, and had sixteen teeth in each jaw. These teeth were channeled laterally with two broad and deep lines, which divided the surface of the molars into three parts, whence it was named the Glyptodon. The hind feet were broad and massive, and evidently designed to support a vast incumbent mass; it presented phalanges armed with nails or claws, short, thick and depressed. The animal was, as we have said, enveloped in, and protected by, a cuirass, or solid carapace, composed of plates which, seen from beneath, appeared to be hexagonal and united by dentated sutures. Above they represented double rosettes. The habitat of *Glyptodon clavipes* was in the pampas of Buenos Ayres, and on the banks of an affluent of the Rio Santo, near Monte Video: specimens have been found not less than six feet in length.

The tesselated carapace of the Glyptodon was long thought to belong to the Megatherium, but Professor Owen shows, from the anatomical structure of the two animals, that the cuirass belonged to one of them only, and that the Glyptodon.

The *Schistopleuron* does not differ essentially from the Glyptodon,

Fig. 183.—Schistopleuron typus. One-twentieth natural size.

but is supposed to have been a different species of the same genera; the chief difference between the two animals being in the structure of

the tail, which is massive in the first and in the other is composed of
half-a-score of rings. In other respects the organization and habits
are the same, both being herbivorous, and feeding on roots and vegetable débris. Fig. 183 is the *Schistopleuron typus* restored, and as it
is supposed to have lived.

The *Megatherium*, or Animal of Paraguay, as it was called, is, at
first view, the oddest and most extraordinary being we have yet had
under consideration, where all have been strange, fantastic and formidable. The animal creation still goes on as if—

"Nature made them and then broke the die."

If we throw a glance at the skeleton on the opposite page (PL. XXVII.)
representing the first, which was found at Buenos Ayres in 1788, and
which is preserved in the Museum of Madrid, it is impossible to avoid
being struck with its unusually heavy form, at once awkward and fantastic in all its parts. It is allied to the sloths, which Buffon tells us is
"of all the animal creation that which has received the most vicious
organization—a being to which nature has forbid all enjoyment:
which has only been created for hardships and misery." This notion
of the romantic Buffon is, however, hardly just. An attentive examination of the *Animal' of Paraguay* shows that its organization cannot
be considered either odd or awkward when viewed in connection
with its kind of life and individual habits. The special organization
which renders the movements of the sloths so heavy, and apparently
so painful on the level ground, gives them, on the other hand, marvellous assistance when they live in trees, whose leaves form their
exclusive food. In the same manner, if we consider that the
Megatherium was created to burrow in the earth and feed upon the
roots of trees and shrubs, every organ of its heavy frame would
appear to be perfectly appropriate to its kind of life, and well adapted
to the special purpose which has been assigned to it by the Creator.
We ought to place the Megatherium between the sloths and ant-eaters.
Like the first, it fed exclusively on the leaves of trees; like the

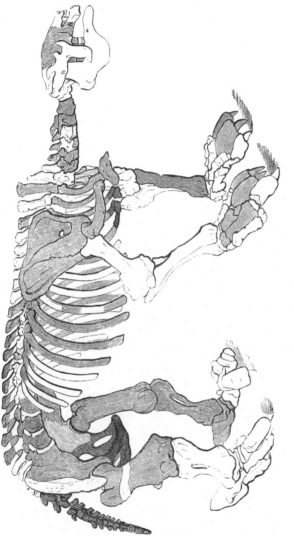

XXVII.—Skeleton of the Megatherium (Clift).

second, it burrowed deep in the soil, finding there at once nourishment and shelter. It was large as an elephant or rhinoceros of the
largest species. Its body measured twelve or thirteen feet in length,
and it was between six and seven feet high. The engraving on the
opposite page (PL. XXVII.) will convey more accurately than any words
can the form and proportions of the animal.

To the zeal and energy of Sir Woodbine Parish, the English reader
is chiefly indebted for the materials on which our naturalists have been
enabled to reconstruct the history of the Megatherium. The remains
collected by him were found in the river Salada, which runs through
the pampas to the south of Buenos Ayres. A succession of three
unusually dry seasons had left the waters so low as to expose the
pelvis to view as it stood upright in the mud in the bed of the river.
Further inquiries led to the discovery of two other complete skeletons,
not far from the spot where the first had been found; and not far
from them an immense shell or carapace, the bones connected with
which · crumbled to pieces after exposure to the air. The osseous
structure of this enormous animal, as furnished by Mr. Clift, an
eminent anatomist of the day, under whose superintendence the
skeleton was drawn, must have exceeded fourteen feet in length, and
upwards of eight feet in height. The deeply-shaded parts show the
portions which are deficient in the Madrid skeleton.

Cuvier pointed out that the skull very much resembled the sloths,
but that the rest of the skeleton bore relationship partly to the sloths
and partly to the ant-eaters.

The large bones which descend from the zygomatic arch along the
cheek-bones would furnish a powerful means of attaching the motor
muscles of the jaws. The anterior part of the muzzle is so fully developed, so riddled with holes for the passage of the nerves and vessels
which must have been there, not for a trunk, which would have been
useless to an animal furnished with a very long neck, but in the
shape of a snout analogous to that of the tapir.

The jaw and dental apparatus cannot be exactly stated, because the
number of teeth in the lower jaw is not known. The upper jaw,

Professor Owen has shown, contains five on each side ; and from comparison and analogy with the *Scelidotherium* it may be conjectured that the *Megatherium* had four on each side of the lower jaw. Being without incisors or canines, the structure of its eighteen molars proves that it was not carnivorous : they each resemble the composite molars of the elephant.

The vertebræ of the neck, as exhibited in the foreshortened figure (Fig. 184), taken from Pander and D'Alton's work, showing nearly a front view of the head, as well as the anterior and posterior extremities

Fig. 184.—Skeleton of Megatherium foreshortened.

of the Madrid skeleton, although powerful, are nothing compared to the volume of those of the other extremity of the body ; for the head seems to have been relatively light and defenceless. The lumbar vertebræ increase in a degree corresponding to the enormous enlargement of the pelvis and the inferior members. The vertebræ of the tail are enormous, as is seen in the margin, which represents the bones of the pelvis and hind foot, discovered by Sir Woodbine Parish, now in the Museum of the College of Surgeons. If we add to these

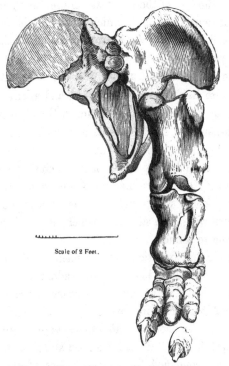

Scale of 2 Feet.

Fig. 185.—Bones of the pelvis of the Megatherium.

osseous organs the muscles, tendons and teguments which covered them, we must admit that the tail of the *Megatherium* could not be less than two feet in diameter. It is probable that, like the armadillo, it employed the tail to support the enormous weight of its body: it would also be a formidable defensive arm when employed as it is by the pangolins and crocodiles. The hind feet would be about three feet long and one foot broad. They would form a powerful implement for excavating the earth at great depths where the roots of vegetables penetrated. The fore feet would pose themselves on the soil at its full length. Thus solidly supported by the two hind feet and the tail, and in advance by one of the fore feet, the animal would employ the fore foot at liberty in hollowing out the earth or tearing up the roots of trees ; the toes of the fore feet were for this purpose furnished with large and powerful talons, which lie at an oblique angle in relation to the soil, much like the burrowing talons of the mole.

The solidity and size of the pelvis must have been enormous ; its iliac bones are nearly at right angles with the vertebral column ; their external edges are distant more than a yard and a half from each other when the animal is standing. The femur is three times the thickness of the thigh-bone in the elephant, and the many peculiarities of structure in this bone appear to have been intended to furnish to the whole frame a solidity by means of short and massive proportions. The two bones of the leg are, like the femur, short, thick and solid ; presenting proportions which we only meet with in the armadillos and ant-eaters, burrowing animals with which, as we have said, its two extremities seem to connect it.

The anatomical organization of these members denotes heavy, slow, and powerful locomotion, but solid and admirable combinations for supporting the weight of an enormous sedentary creature ; a sort of excavating machine, nearly immoveable, and of incalculable power for its own purposes. In short, the *Megatherium* exceeded in dimensions all the existing Edentata. It had the head and shoulders of the sloth ; the feet and legs combined the characteristics of the ant-eaters, the sloths, and the chlamyphores, of enormous size, since it was at least eight feet when full grown, its feet armed with gigantic claws, and its tail at once a means of supporting this vast body and an instrument of defence. An animal built in such massive proportions could evidently neither creep nor run : its walk would be excessively slow. But what necessity was there for rapid movement in a being only occupied in burrowing under the earth, seeking for roots, and which would consequently rarely change its place ? What occasion to fly from its enemies, when it could overthrow the most formidable of animals, the crocodile, with a sweep of its tail ? Sheltered by its habits from the attempts of all other animals, this robust herbivora, of which the following is a restoration, ought to have lived peacefully and respected in the solitary pampas of America.

The immediate cause of the extinction of the Megatheriæ is probably

Fig. 186.— Megatherium restored.

to be found in causes still in operation in South America. The period between the years 1829 and 1830 is called the "gran seco," or the great drought, in South America, and, according to Darwin, the loss of cattle in Buenos Ayres alone exceeded a million. One proprietor at San Pedro, in the middle of the finest pasture country, lost 20,000 in these years. "I was informed by an eye-witness," he adds, "that the cattle in thousands rushed into the Parana, and being exhausted by hunger they were unable to crawl up the muddy banks, and were thus drowned. The arm of the river which runs by San Pedro was so

full of putrid carcases, the master of a vessel told me, that the smell
rendered it quite impassable. All the small rivers became highly
saline, and this caused the death of vast numbers in particular spots;
for when an animal drinks of such water it does not recover. Azara
describes the fury of the wild horses, on a similar occasion, rushing
into the marshes; those which arrived first being overwhelmed and
crushed by those which followed." The upright position in which
the various specimens of Megatheriæ were found indicates some
such cause of death, as if the ponderous animal approaching the banks
of the river when shrunk within its banks had been bogged in soft
mud, sufficiently adhesive to hold it there till it perished.

Like the Megatherium, the *Mylodon* closely resembles the sloth,

Fig. 187.—Mylodon robustus.

and it belonged exclusively to the New World. Smaller than the
Megatherium, it differs from it chiefly in the form of the teeth. These

organs presented only molars with smooth surface, indicating that the animal fed on vegetables, probably the leaves and tender buds of trees. As the Mylodon presents at once hoofs and claws on each foot, it has been thought that it formed the link between the hoofed and unguilic animals. Three species are known which lived in the pampas of Buenos Ayres.

In consequence of some hints given by the illustrious Washington, Mr. Jefferson, one of his successors as President of the United States, discovered in a cavern of Western Virginia some bones which he declared to be the remains of some carnivorous animal. They consisted of a femur, a humerus, an ulna, and three claws, and half a dozen other bones of the foot. These bones Mr. Jefferson believed to be analogous to those of the lion. Cuvier saw at once the true analogies of the animal. The bones were the remains of a species of gigantic sloth, the complete skeleton of which was subsequently discovered in the Mississippi, in a state of preservation so complete that the cartilages still adhering to the bones were not decomposed. Jefferson called this species the *Megalonyx*. It partook in many points of the sloth : its size was that of the largest ox ; the muzzle was pointed ; the jaws armed with cylindrical teeth ; the anterior members much longer than the posterior ; the articulation of the feet oblique to the leg ; two great toes, short, armed with long and very powerful claws ; the index finger more slender, armed with a claw less powerful also ; the tail strong and solid : such were the salient points of the organization of the *Megalonyx*, whose form was a little lighter than the *Megatherium*.

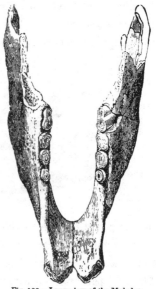

Fig. 188.—Lower jaw of the Mylodon.

The country in which the Megatheriæ have been found is described by Mr. Darwin as belonging to the great pampean formation, consisting partly of a reddish clay and partly of calcareous marly rock.   Near the coast there are some plains formed from the wreck of the upper one and from mud and gravel thrown up by the sea during the slow elevation of the land, as shown in the raised beds of recent shells.   At Punta Alta a highly-interesting section of the lately-formed plain is seen, in which many of these gigantic animal remains have been found.   " These con-sisted," says Mr. Darwin, " first, of parts of three heads and other bones of Megatherium, the huge dimensions of which are expressed by its name; secondly, of the *Megalonyx*, a great allied animal; thirdly, the *Sceiidotherium*, also an allied animal, of which I obtained a nearly perfect skeleton; it must have been as large as a rhinoceros; fourthly, the *Mylodon*, a closely-allied genus, of a size little inferior."

The remains on which Scelidotherium is founded included the cranium, which is nearly entire, with the teeth and part of the os hyoides, seven cervical, eight dorsal, and five sacral vertebræ, both the scapulæ, and some other bones.   The remains of the cranium

Fig. 189.—Skull of Scelidotherium.

indicated that its general form was an elongated, slender, compressed cone, beginning behind by a flattened vertical base, expanding slightly to the cheek-bone, and thence contracting to the anterior extremity. All these parts were discovered in their natural relative positions, indi-cating, as Mr. Darwin observes, that the gravelly formation in which they were discovered had not been disturbed.

The lower jaw-bone of *Mylodon*, which Mr. Darwin discovered at the base of the cliff called Punta Alta, in Northern Patagonia, had the teeth entire on both sides: they are implanted in deep sockets, and only about one-sixth of the last molar projects above the alveolus, but the proportion of the exposed part increases gradually in the inner teeth.

"The habits of these megatheroïd animals," says Mr. Darwin, "were a complete puzzle to naturalists, until Professor Owen, with remarkable ingenuity, solved the problem. The teeth indicate that they lived on vegetable food, and probably on the leaves and small twigs of trees. Their ponderous forms and great curved claws seem so little formed for locomotion, that some naturalists have actually believed that, like the sloths, to which they are intimately related, they subsisted by climbing, back downwards, on trees, and feeding on the leaves. It was a bold, not to say preposterous idea, to conceive even antediluvian trees with branches strong enough to bear animals as large as elephants. Professor Owen, with far more probability, believes that, instead of climbing on the trees, they pulled the branches down to them, and tore the smaller ones up by the roots, and so fed on their leaves. The colossal breadth and weight of their hinder quarters, which can hardly be imagined without being seen, become, on this view, of obvious service instead of being an incumbrance; their apparent clumsiness disappears. With their great tails and huge heels firmly fixed like a tripod in the ground, they could freely exert the full force of their powerful arms and great claws. The *Mylodon* was furnished with a long extensile tongue, like that of the giraffe, which by one of those beautiful provisions of Nature thus reaches its leafy food."

Two gigantic birds seem to have lived in New Zealand during this epoch. The *Dinornis*, which, if we may judge from the *tibia*, which is upwards of three feet long, and from its eggs, which are much larger than those of the ostrich, must have been of most extraordinary size for a bird. In Fig. 190 an attempt is made to restore the Dinornis. As to the *Epiornis*, the egg only has been found.

On the opposite page (PL. XXVIII.) an attempt is made to represent the appearance of Europe during the epoch we have under consideration. The bear is seated at the mouth of its den—the cavern, thus reminding us of the origin of its name of *Ursus spelæus*, where it gnaws the bones of the elephant. Above the cavern the *Hyæna*

Fig. 190.—Dinornis.

spelæa looks out with savage eye for the moment when it will be prudent to dispute possession of these remains with its formidable rival. The great wood-stag, with other great animals of the epoch, occupies the farthest shore of a small lake, where some small hills rise out of a valley crowned with the trees and shrubs of the period. Mountains recently upheaved rise on the distant horizon, covered with

XXVIII.—Ideal landscape in the Quaternary Epoch.—Europe.

a mantle of frozen snow, reminding us that the glacial period is approaching, and already manifests itself.

All these fossil bones belonging to the great mammalia which we have been describing are found in the quaternary formation; but the most abundant of all are those of the elephant and the horse. The extreme profusion of the bones of the Mammoth crowded into the upper beds of the globe is only surpassed by the prodigious quantity of the bones of the horse buried in the same beds. The singular abundance of the remains of these two animals prove that during the quaternary epoch the earth gave nourishment to immense herds of the horse and the elephant. It is probable that from one pole to the other, from the equator to the two extremities of the axis of the globe, the earth must have formed a vast prairie without limits, while an immense carpet of verdure covered its whole surface; and such abundant pastures would be absolutely necessary to sustain this prodigious quantity of herbivorous animals of great size.

The mind can scarcely realize the immense and verdant plains of the primitive world animated by the presence of an infinity of such inhabitants. In its burning temperature, pachyderms of monstrous forms, but of peaceful habits, traversed the tall vegetation composed of grasses of all sorts. Stags of gigantic size, the head ornamented with enormous horns, consorted with the heavy phalanges of the Mammoth; while the horse, small in size and compact of form, galloped and frisked round this magnificent horizon of verdure which no human eye had yet contemplated.

Nevertheless, all was not quiet and tranquil in the landscapes of the ancient world. Voracious and formidable carnassiers made a bloody war on the inoffensive herds. The tiger, the lion, and the ferocious hyæna, the bear, and the jackal, there selected their prey. On the opposite page an attempt is made to represent the great animals among the Edentatæ which inhabited the American plains during the quaternary epoch (PL. XXIX.). We observe there the Glyptodon, the Megatherium, the Mylodon, and along with them the Mastodon. A

2 B

small ape, the orcopithecus, which first appeared in the miocene
period, occupies the branch of a tree in the landscape. The vegetation is that of tropical America at the present time.

The deposits of this age consist, in the British Islands, of unstratified red till, with large boulders of granitic gneiss, quartzose of
variable thickness, contorted beds of drift, sometimes stratified, with
coarse gravel and large boulders, with superficial sand and gravel,
extending over great part of the lowlands of Scotland. In Norfolk,
along the coast, sections ranging from thirty to three hundred feet are
exhibited for fifty miles near Cromer, consisting of—1, chalk with
flints in horizontal strata; 2, Norwich crag, a marine formation of
new pliocene; 3, forest beds, chiefly of vegetable matter, with
scattered cones of Scotch and spruce firs, with bones of elephants
and living mammalia; in this formation stumps of trees stand erect
with their roots in the ancient soil; 4, alternate fresh-water and marine
strata, with lignite beds and shells of recent periods; 5, laminated
blue clay without fossils, on which rest boulder clay of the glacial
period, with transported blocks, some of them polished and scratched;
6, contorted drift.

In all directions, however, proofs are being gradually obtained that
about this period movements of submersion under the sea were in
progress, all north of the Thames. Near Blackpool, about fifty miles
from the sea, and at a height of 560 feet above its level, "till" is found
containing rounded stones and marine shells of *Turrilites communis.*
Ramsay points out indications, first of an intensely cold period, when
land was much more elevated than it is now; then of submergence beneath the sea; and, lastly, re-elevation attended by glacial
action. "When we speak of the vegetation and quadrupeds of
Cromer Forest being pre-glacial," says Lyell, "we merely mean that
their formation preceded the era of the general submergence of the
British Isles beneath the waters of the glacial sea." "The successive
deposits seen in direct superposition on the Norfolk coast," adds Sir
Charles, "imply, at first, the prevalence over a wide area of the

newer pliocene sea. Afterwards, the bed of this sea was converted into dry land, and underwent several oscillations of level, so as to be, first, dry land supporting a forest; then an estuary; then again land; and, finally, a sea near the mouth of a river, till the downward movement became so great as to convert the whole area into a sea of considerable depth, in which much floating ice, carrying mud, sand, and boulders melted, letting its burthen fall to the bottom. Finally, over the till with boulders stratified drift was formed; after which, but not until the total subsidence amounted to more than 400 feet, an upward movement began, which re-elevated the whole country, so that the lowest of the terrestrial formations, or the forest bed, was brought up to nearly its pristine level, in such a manner as to be exposed at low tide. Both the descending and ascending movement seem to have been very gradual."

Fig. 191.—Palæophognos Gesneri. Fossil toad.

THE EUROPEAN DELUGES.

The tertiary formations in many parts of Europe are covered by beds of heterogeneous débris, which fill up valleys more or less extensive. These beds are composed of divers elements, but proceed invariably from fragments detached from neighbouring rocks : the erosions which we remark at the bottom of the hills have greatly enlarged the valleys already existing ; the mounds of earth accumulated at the same point which is formed of rounded materials, that is to say, rocks ground smooth and round by continual friction during a long period in which they have been transported from one point to another—all these signs indicate that the denudations of the soil, the displacement and conveyance of heavy bodies to great distances, are due to the violent and sudden action of large currents of water. An immense wave has been thrown suddenly on the surface of the earth, making great ravages in its passage, furrowing the earth and driving before it débris of all sorts in its disorderly course. Geologists give the name of *diluvium* to the formation thus removed and overturned, which from its heterogeneous nature brings under our eyes, as it were, the rapid passage of an impetuous torrent, a phenomenon which is commonly designated as a *deluge*.

To what cause are we to attribute these sudden and apparently temporary invasions of dry land by passing currents of water ? In all probability to the upheaval of some vast extent of dry land, to the formation of some mountain or mountain range in the neighbourhood of the sea or in the basin of the sea itself. The land suddenly elevated by an upward movement of the terrestrial crust, or by forming ridges and furrows on the surface, has by its reaction violently agitated the waters of the globe, and by this new impulse they have been thrown with great violence over the earth, inundating the plains and valleys, and for the moment covering the soil with their furious waves, mixed with the earth, sand and mud, 'of which the surface has been denuded in the abrupt invasion. The phenomena has been sudden, like the upheaval of a mountain or ridge of mountains, which is presumed to have been the cause, but of short duration; but probably often repeated : witness the valleys which occur in every country, especially those in the neighbourhood of Lyons and of the Durance, where the strata occur in successive beds. Besides this, the displacement of blocks of minerals from their normal position are proofs easily recognizable of the phenomena.

We have had, doubtless, during the epochs anterior to that of which we write many deluges such as we are considering. Mountains and chains of mountains through all the ages we have been describing were formed by upheaval of the crust into ridges where it was too elastic or too thick to be fractured. Each of these subterranean commotions would be provocative of momentary irruptions of the waves. This fact is demonstrated in the coal formations. We sometimes find in these rocks *conglomerates ;* that is, masses of rock or stones ground into shape by the action of water, and fixed in a cement so as to form immense blocks, altogether foreign to other parts of the strata.

But the visible testimony to this phenomena—the living proofs of this denudation, of this ravishment of the soil, are found nowhere so strikingly as in the beds superposed far and near upon the tertiary formation which bears the geological name of *diluvium.* This term was long employed to designate what is now better known as the "boulder" formation, an alluvial deposit which is abundant in Europe, north of the 50th, and in America, north of the 40th parallel, reappearing again in the southern hemisphere; but is altogether absent in equatorial regions.   It consists of sand and clay, sometimes stratified, mixed with rounded and angular fragments of rock, generally of the same district; and their origin has generally been ascribed to a series of diluvial waves raised by hurricanes, earthquakes, or the sudden upheaval of land from the bed of the sea, which had swept over continents, carrying with them vast masses of mud and heavy stones, and forcing these stones over rocky surfaces so as to polish and impress them with furrows and striæ.   Other circumstances occurred, however, to establish a connection between this formation and the glacial drift. The size and number of erratic blocks increases as we travel towards the Arctic regions; some intimate association exists therefore between it and the accumulation of ice and snow which characterize the approaching glacial period.

There is very distinct evidence of two deluges succeeding each other in our hemisphere during the quaternary epoch.   One we shall distinguish as the two *European deluges;* the other as the *Asiatic.*   The two European deluges were anterior to the appearance of man; the Asiatic deluge is posterior to that event: the human race, then, in the early days of its existence certainly suffered from a great deluge.   In the present chapter we confine ourselves to the two cataclysms which overwhelmed Europe in the quaternary epoch.

The first occurred in the north of Europe, and it was produced by the upheaval of the mountains of Norway.   Commencing in Scandinavia, the wave spread and carried its ravages into those regions which now constitute Sweden, Norway, European Russia, and the north of Germany, sweeping before it all the loose soil on the surface, and covering the whole of Scandinavia—all the plains and valleys of Northern Europe—with a mantle of shifting soil.   As the regions in the midst of which this great mountainous upheaval occurred—as the seas surrounding these vast spaces were partly frozen and covered with ice, from their elevation and neighbourhood to the pole—the wave which swept these countries carried along with it enormous masses of ice: the collision of these several solid blocks of congealed water would only contribute to increase the extent and intensity of the ravages occasioned by this violent cataclysm, which is represented in PL. xxx.

The physical proofs of this *deluge of the north* exist in the vast mantle of unstratified earth which covers all the plains and depressions of Northern Europe, with the attendant phenomena of striated and dome-shaped rocks and far-transported erratic blocks, which become more characteristic as we ascend to higher latitudes, as in Norway, Sweden and Denmark, the southern borders of the Baltic, and in the British Islands generally, in all of which deposits of marine fossil shells occur which demonstrate the submergence of large areas of Scandinavia, of the British Isles and other regions during parts of the glacial period. Some of these rocks characterized as *erratic* are of very considerable volume : such, for instance, is the granite block which has been made the pedestal of the statue of Peter the Great at St. Petersburg. This block was found in the interior of Russia, where the whole formation is *Permian,* and its transport thither can only be explained by supposing it carried in some vast iceberg by the diluvian current. This hypothesis alone enables us to account for another block of granite, weighing about three hundred and forty tons, which was found on the sandy plains in the north of Prussia, an immense model of which was made for the Berlin Museum. The last of these erratic blocks deposited in Germany covers the grave of the King Gustavus Adolphus, of Sweden, killed in the battle of Lutzen, in 1632. He was interred beneath the rock. Another similar block has been raised in Germany into a monument to the geologist Leopold Von Buch.

These erratic blocks which are met with in the plains of Russia, Poland and Prussia, and ·in the eastern parts of England, are composed of rocks entirely foreign to the region where they are found. They belong to the primitive rocks of Norway; they have been transported to their present sites, protected by a covering of ice, by the waves of the northern deluge. How vast must have been the impulsive force which could carry such enormous masses across the Baltic,

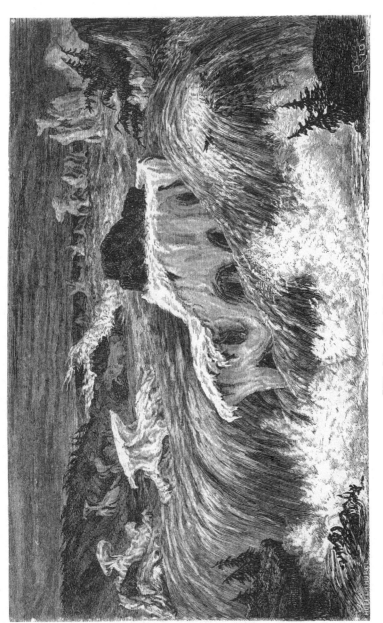

XXX.—Deluge of the North of Europe.

and so far inland as the places where they are deposited for the contemplation of the thoughtful !

The second European deluge is supposed to have been the result of the formation and upheaval of the Alps, which has filled with débris and moveable soil all the valleys of France, of Germany and Italy in a circumference which has the Alps for its centre. The proofs of a great convulsion at a comparatively recent geological date are numerous. The Alps may be from eighty to a hundred miles across, and the probabilities are that their existence is due, as Sir Charles Lyell supposes, to a succession of unequal movements of upheaval and subsidence : that the Alpine region had been exposed for countless ages to the action of rain and rivers, and that the larger valleys were of pre-glacial times, is highly probable. In the eastern part of the chain some of the primary fossiliferous rocks, as well as oolitic and cretaceous rocks, and even tertiary deposits are observable ; but in the central Alps these disappear, and more recent rocks, in some places even eocene strata, graduate into metamorphic rocks, in which oolitic, cretaceous and eocene strata have been altered into granular marble, gneiss and other metamorphic schists ; showing that eruptions continued after the deposit of the middle eocene formations. Again, in the Swiss and Savoy Alps oolitic and cretaceous formations have been elevated to the height of twelve thousand feet, and eocene strata ten thousand feet above the level of the sea ; while in the Rothal, in the Bernese Alps, a mass of gneiss occurs a thousand feet thick, which rests on strata containing oolitic fossils, which have been again covered with the same strata.

Besides these proofs of recent upheaval, we can trace effects of two different kinds, resulting from the powerful action of masses of water violently displaced by this gigantic upheaval. At first broad tracks have been hollowed out by the diluvial waves, which have at these points formed deep valleys. Afterwards these valleys have been filled up by materials borrowed from the mountain and transported into the valley ; these materials consisting of rolled stones, argillaceous and sandy mud, generally calcariferous and ferriferous. This double effect shows itself with more or less distinctness in all the great valleys of the centre and south of France, where the valley of the Garonne is in respect to these phenomena classic ground.

As we leave the little city of Muret, three successive levels will be observed on the left bank of the Garonne. The lowest of the three is that of the valley, properly so called, while the loftiest corresponds to the plateau of Saint-Gaudens. These three levels are distinctly marked in the Toulousan country, which illustrates the diluvial phenomena in a remarkable fashion. The city of Toulouse reposes upon a slight eminence of diluvian formation. The flat diluvian plateau contrasts strongly with the mammalous hills of Gascony and Languedoc. They are essentially constituted of a bed of gravel, round or oval stones, mixed and again covered with deposits of sand and earth. The pebbles are principally quartzose, brown or black exteriorly, with portions of hard " old red " and new red sandstone. The soft earth which accompanies the pebbles and gravel is a mixture of argillaceous sand of a red or yellow colour, caused by oxide of iron which enters into its composition. In the valley, properly so called, we find the pebbles again associated with other minerals which are rare in the upper level. Some teeth of the Mammoth, or *Rhinoceros tichorhynus*, have been found in several parts of the river banks.

The small valleys tributary to the principal valley would appear to have been excavated secondarily, partly out of diluvian deposits, and their alluvium, essentially earthy, have been formed at the expense of the tertiary formation, and even of the diluvium itself. Among other celebrated sites, the diluvial formation is largely developed in Sicily, and the Parthenon of Athens is built on an eminence formed of it.

In the valley of the Rhine, in Alsace, and in many isolated parts of Europe, a particular sort of *diluvium* forms thick beds; it consists of a greyish-yellow mud, composed of argillaceous matter mixed with carbonate of lime, quartzose and micaceous sand, and oxide of iron. This mud, termed by geologists *löss*, attains in some places considerable thickness. It is recognizable in the neighbourhood of Paris. It rises a little both on the right and left, above the base of the mountains of the Black Forest and of the Vosges, and forms thick beds on the banks of the Rhine.

The fossils contained in the diluvial deposits consist generally of terrestrial shells, lacustrine or fluviatile, for the most part species still living. In parts of the valley of the Rhine, between Bingen and Basle, the fluviatile loam or löss, now under consideration, is seen forming hills several hundred feet thick, and containing here and there throughout that thickness land and fresh-water shells; from which it seems necessary to suppose, according to Lyell, first, a time when the löss was slowly accumulated, then a later period, when large portions of it were removed—movements of oscillation, consisting, first, of a general depression of the land, and then of a gradual re-elevation of the same.

We have already noticed the caverns in which such extraordinary accumulations of animal remains were discovered: it will not be out of place to give here a résumé of the state of our knowledge concerning the *bone caverns* and *bone breccias*.

The *bone caverns* are not simple cavities hollowed in the rock; they generally consist of numerous chambers or grottos communicating with each other by narrow openings, which can only be entered by creeping, these entrances often extending to considerable distances. One in Mexico extends to several leagues. Perhaps the most remarkable in Europe is that of Gaelenreuth in Franconia. The Harz mountains contain many fine caverns; among others *Bauman's cavern*, from which many bones have been taken. The *Kirkdale cave*, so well known from the description given of it by Dr. Buckland, lying about twenty-five miles north-north-east of York, was the burial-place, as we have seen, of about three hundred hyænas belonging to individuals of different ages: besides some other animal remains, among them, the bones of wolves, hares, water-rats and birds, mixed with those of some herbivorous animals. Buckland states that the bones of other than those of hyænas were gnawed. He also recognized numbers of coprolites of hyænas, as well as traces of the frequent passage of these animals from the entrance of the cavern. A modern naturalist visiting the Cavern of Adelsberg, in Carniola, traversed a series of chambers extending over three leagues in the same direction, and was only checked in his subterranean discoveries by coming to a lake which occupied its whole breadth.

The interior walls of the bone caverns are in general rounded off, and furrowed, presenting many traces of the erosive action of water, characteristics which frequently escape observation because the walls are covered with the calcareous concretion called *stalactite* or *stalagmite*, that is, masses of carbonate of lime proceeding from the deposits left by the infiltrating of water through the overlying limestone into the interior of the cavern. The formation of stalactite of Gaelenreuth. is thus described by Liebig. The limestone over the cavern is covered with a rich soil, in which the vegetable matter is continually decaying. This mould, containing humus acted on by moisture and air, evolves carbonic acid, which is dissolved by rain. The rain-water thus impregnated, permeating the porous limestone, dissolves a portion of it, and afterwards, when the excess of carbonic acid evaporates in the caverns, parts with the calcareous matter and forms stalactite; the stalactites being the pendant crystals of carbonate of lime, which hang suspended like icicles from the roof of the cave, through which water percolates : those formed on the flat surface of the floor being stalagmites. While caverns are still liable to be flooded, calcareous incrustations occur; but it is when no longer exposed to inundations that the solid floor, or *stalagmite*, is formed. These calcareous products ornament the walls of these gloomy caverns, giving them at once a brilliant and picturesque effect when exposed to light.

Under its covering of stalagmite the floor of the cavern frequently presents deposits of mud and ironstone. It was in excavating this soil that the bones of antediluvian animals, mixed with shells, with fragments of rocks and rolled stones, were discovered. The distribution of these bones in the middle of the gravelly argillaceous mud was as irregular as possible. The skeletons were rarely entire; the bones did not even occur in their natural position. The bones of small rodents were found accumulated in the cranii of great carnassiers. The teeth of bears, hyænas and rhinoceros were cemented with the *cubites*, or jaw-bones, of ruminants. The bones were very often polished and rounded, as if they had been suddenly transported from great distances; others are fissured; others, nevertheless, are scarcely altered. Their state of preservation varied with the position of the cavern.

The bones that are found most frequently in the caverns are those of the carnivora of the quaternary epoch : the bear, hyæna, the lion and tiger. The animals of the plain, and notably the great pachyderms—the Mammoths and rhinoceros—are only very rarely met with, and always in small numbers. From the cavern of Gaelenreuth more than 1000 skeletons have been taken, of which 800 belonged to the large *Ursus spelæus*, and sixty of the smaller species, with 200 hyænas, wolves, lions and gluttons. In the Kirkdale cave the remains, as we have seen, included about three hundred hyænas of all ages. Dr. Buckland concludes, from these circumstances, that the hyænas alone made this their den, and that the bones of other animals accumulated there had been carried thither by them as their prey : it is however now admitted that this part of the English geologist's conclusions do not apply generally. In the greater number of caverns the bones of the mammifera are broken and rubbed as with a long transport, *rolled*, according to the geological expression, and finally cemented by the same mud, and surrounded by the rocks of the neighbourhood. Beside bones of hyænas are found not only the bones of inoffensive herbivora, but the remains of lions and of bears. All these circumstances unite in establishing that the bones which fill the caverns have been floated at random into these anfractuosities by the rapid current of the diluvian waves. The ossiferous grottos are generally found near the entrance of the valley, in the plain or at a height which never exceeds the limits of the diluvial phenomena. We may, then, suppose that, in the greatest number of cases, the animals, surprised and killed by the sudden and impetuous torrents, have been drawn into the caverns by the currents, where they have been engulfed; and cavern, bones and earth itself buried in the diluvian mud.

We ought to note, in order that this explanation should be complete, that some geologists consider that these caverns served as a refuge for sick and wounded animals. It is certain that we see in our own days some animals, when attacked by sickness, seek refuge in the fissures of the rocks or in the hollows of trunks of trees, where they die: to this natural impulse it may probably be ascribed that the skeletons of animals are so rarely found in forests or plains. We may conclude then, that, besides the more general mode in which these caverns were filled with bones, which must, we think, be attributed to the current of the diluvial waters, the two other causes which we have enumerated may have been in operation; that is to say, they were the habitual sojourn of carnivorous and destructive animals, and they have become the retreat of sick animals on some particular occasions.

What was the origin of these caverns? How have these immense excavations been produced? With most modern naturalists, we consider them as clefts or fractures produced by the great geological phenomena, namely, the effect of the cooling of the earth. These clefts and fissures are usually filled by the injection of igneous matter from the interior, which generally fills up the voids; but from some peculiar circumstances the matter ejected from the centre has failed to reach them, and they remain empty. In support of this hypothesis, we may remark that nearly all these caverns occur in countries which have been the theatre of dislocations, and that they occur in the limestone rocks, particularly in the Jurassic and Neocomian formation, which presents many vast subterranean caverns. At the same time some fine caverns of the Silurian formation exist, such as the *Grotto des Demoiselles* of Herault, near Ganges. It should be added, in order to complete the explanation of the cavern formations, that the greater part of these vast internal excavations have been largely increased by subterranean water-courses, which have eroded, probably washed away, the walls, and in this manner greatly enlarged their primitive dimensions.

But there are other modes of accounting for these caverns than the above, which accounts in a more satisfactory manner for their existence. According to Sir Charles Lyell, there was a time when limestone rocks were dissolved, and when the carbonate of lime was carried out gradually by springs from the interior of the earth: that

Fig. 192.—Grotto des Demoiselles, Herault.

another era occurred, when engulfed rivers or occasional floods swept organic and inorganic débris into the hollows previously formed: finally, there were changes, in which engulfed rivers were turned into new channels, and springs dried up, after which the cave mud, breccia, gravel, and fossil bones were left in the position in which

they are now discovered. " We know," says that eminent geologist,
" that in every limestone district the rain-water is *soft* when it falls
upon the soil, whereas it is *hard*, or charged with carbonate of lime,
when it issues again to the surface in springs. The rain derives
some of its carbonic acid from the air, but much more from the decay
of vegetable matter in the soil which it percolates, and by the excess
of this acid, limestone is dissolved, and the water becomes charged with
carbonate of lime. The mass of solid matter silently and unceasingly
abstracted in this way from the rocks in every century is considerable,
and must in the course of thousands of years be so vast, that the space
it once occupied may well be expressed by a long suite of caverns."

The most celebrated of these bone caverns are those of Gaelenreuth,
in Franconia, of Nabeustein, and of Brumberg, in the same country;
the caves on the banks of the Meuse—of which the late Dr. Schmerling
examined forty; those of Yorkshire, Devonshire, and Derbyshire, in
England ; and several in Sicily, at Palermo, and Syracuse ; in France
at Herault, in the Cevennes and Franche Comte, and in the New
World in Kentucky and Virginia.

The *osseous breccia* differs from the bone-caverns only in their form.
They are masses of conglomerates, composed of the débris of various
rocks mingled with bones, cemented in a calcareous mud, which fill up
long trenches or fissures of the surface of the earth. The greater
part of the breccia which exists in Europe is disposed as a sort of
girdle round the Mediterranean, which indicates pretty clearly that
they are all connected with the same cleft in the terrestrial crust.
The most remarkable of them are seen at Cette, at Antibes, at Nice,
on the shores of Italy, and in the isles of Corsica and Sardinia.

Nearly the same bones are found in the *breccia* which we find in
the caverns, the chief difference being that fossils of the ruminants
are there in greater abundance. The proportions of bones to the
fragments of stone and cement vary considerably in different localities.
In the *breccia* of Cagliari, where the débris of ruminants are less
abundant than at Gibraltar and Nice, the bones, which are those of
the small rodentia, are, so to speak, more abundant than the mud in

which they are embedded. We find there, also, three or four species of birds which belong to the thrushes and larks. In the *breccia* at Nice the remains of some great carnivora are found, among which are detected two species, congeners of the lion and panther. At San-Siro, in Sicily, the *breccia* contains the bones of some species of dog.

But the *breccia* is not confined to Europe. We meet with it in all parts of the globe; and recent discoveries in Australia indicate a formation corresponding exactly to the *osseous breccia* of the Mediterranean, in which an ochreous-reddish cement binds together the fragments of rocks and bones, among which we find four species of kangaroos.

Fig. 193.—Baloptera Sepoideæ.

THE GLACIAL PERIOD.

THE two cataclysms of which we have spoken surprised Europe at the moment of its expansion by powerful convulsions. The whole scope of animated nature, the evolution of its creations, were suddenly arrested in that part of our hemisphere over which these gigantic convulsions spread, followed by the sudden submersion of whole continents. Organic life had scarcely recovered from the violent shock, when a second, and perhaps severer, blow assailed it. The northern and central parts of Europe, the vast countries which extend from Scandinavia to the Mediterranean and the Danube, were visited by a sudden and severe cold: the temperature of the glacial regions seized them. The plains of Europe, but now ornamented by the luxurious vegetation developed by the ardour of a burning climate, the interminable meadows on which herds of great elephants, active horses, the robust hippopotamus, and great carnassiers roamed and grazed, was comparatively on a sudden covered with a mantle of ice and snow.

To what cause are we to attribute a phenomenon so unforeseen, and exercising itself with such intensity ? In the present state of our knowledge no reliable explanation of the event can be given. Did the central planet, the sun, which was long supposed to distribute light and heat to the earth, lose during this period its calorific powers? This explanation would be insufficient, since at this period the solar heat is not supposed to have greatly influenced the earth's temperature. Were the marine currents, such as the *Gulf Stream*, which carries the Atlantic Ocean towards the west of Europe, warming and raising its temperature, suddenly turned in the contrary direction? No such hypothesis is sufficient to explain either the cataclysms or the glacial phenomena; and we need not hesitate to confess our ignorance of this strange, this mysterious episode in the history of the globe.

There have been attempts, and very ingenious ones too, to explain these phenomena, of which we shall give a brief summary, without committing ourselves to any further opinion, using for that purpose the information contained in M. Mangin's excellent work. "The most violent convulsions of the solid and liquid elements," says this able writer, "appear to have been themselves only the effects due to a cause much more powerful than the mere expansion of the pyrosphere; and it is necessary to recur, in order to explain them, to some new and bolder hypothesis than has yet been hazarded. Some philosophers have belief in an astronomical revolution which may have overtaken our globe in the first age of its formation, which may have modified its position in relation to the sun. They admit the poles have not always been as they are now, and that some terrible shock has displaced them, changing at the same time the inclination of the axis of the rotation of the earth." This·hypothesis, which is nearly the same as that propounded by the Danish geologist, Klee, has been ably developed by M. de Boucheporn. According to this writer, many multiplied shocks, caused by the violent contact of the earth with comets, produced the elevation of mountains, the displacement of seas, and perturbations of climate; phenomena which he ascribes to the sudden disturbance of the parallelism of the axis of rotation. The antediluvian equator, according to him, makes a right angle with the existing equator.

"Quite recently," adds M. Mangin, "a learned French mathematician, M. J. Adhemar, has taken up the same idea, but dismissing the more problematical elements of the concussion with comets as untenable, he seeks to explain the deluges by the laws of gravitation and celestial mechanics, and his theory has been supported by very competent writers. It is this:—We know that our planet is animated by two essential movements; one of rotation on its axis, which it accomplishes in twenty-four hours; the other of translation, which it accomplishes in 365 days. But besides these great and perceptible movements, the earth has a third, and even a fourth, movement, with one of which we need not occupy ourselves; it is that designated *nutation* by astronomers. It changes periodically, but within very restricted limits, the inclination of the terrestrial axis to the plane of the ecliptic by a slight oscillation, the duration of which is only eighteen hours, and its influence upon the relative length of day and night almost inappreciable. The other movement is that on which M. Adhemar's theory is founded.

"We know that the curve described by the earth in its annual revolution round the sun is not a circle, but an ellipse; that is, a circle slightly elongated, sometimes called a circle of two centres, one of which is occupied by the sun. This curve is called the ecliptic. We know also that in its movement of translation the earth preserves such a position that its axis of rotation is intersected at its centre by the plane of the ecliptic. But in place of being perpendicular, or at right angles with this plane, it crosses it obliquely in such a manner as to form on one side an angle of one-fourth, and on the other an angle of three-fourths of a right angle. This inclination is only altered in an insignificant degree by the movement of *nutation*. I need scarcely add that the earth, in its annual revolution, occupies periodically four principal positions on the ecliptic, which mark the limits of the four seasons. When its centre is at the extremity most remote from the sun, or *aphelia*, it is the summer solstice for the northern hemisphere. When its centre is at the other extremity, or *perihelia*, the same hemisphere is at the winter solstice. The two intermediate points mark the equinoxes of spring and autumn. The great circle of separation of light and shade passes, then, precisely through the poles, the day and night are equal, and the line of intersection of the plane of the equator and that of the ecliptic make part of the vector ray from the centre of the sun to the centre of the earth—what we call the *equinoctial line*.

"Thus placed, it is evident that if the terrestrial axis remained always parallel to itself, the equinoctial line would always pass through the point on the surface of the globe. But it is not absolutely thus. The parallelism of the axis of the earth is destroyed slowly, very slowly, by a movement which Arago ingeniously compares to the inclined turning of a top. This movement has the effect of making the equinoctial points on the surface of the earth retrograde towards the east from year to year, in such a manner that at the end of 25,800 according to some astronomers, but 21,000 years according to Adhemar, the equinoctial point has literally made a tour of the globe, and has returned to the same position which it occupied at the beginning of this immense period, which has been called the great year. It is this retrograde evolution in which the terrestrial axis describes round its own centre that revolution round a double conic surface which is known as the *precession of the equinoxes*. It was observed 2000 years ago by Hipparchus; its cause was discovered by Newton, and its complete evolution explained by D'Alembert and Laplace.

"Now, we know that the consequence of the inclination of the terrestrial axis upon the plane of the ecliptic is—

"1. That the seasons are inverse to the two hemispheres; that is to say, the northern hemisphere enjoys its spring and summer, while the southern hemisphere passes through autumn and winter.

"2. It is when the earth approaches nearest to the sun that our hemisphere has its autumn and winter, and that the regions near the pole, receiving none of the solar rays, are plunged into darkness approaching that of night during six months.

"3. It is when the earth is most distant from the sun, when much the greater half of the ecliptic intervenes between it and the focus of light and heat, that the pole, being now turned towards this focus, constantly receives its rays, and when the rest of the northern hemisphere enjoys its long days of spring and summer.

"Bearing in mind that, in going from the equinox of spring to that

of autumn of our hemisphere, the earth traverses a much longer
curve than it does on its return; bearing in mind, also, the accelerated
movement it experiences in its approach to the sun from the attraction, which increases in inverse proportion to the square of its
distance, we arrive at the conclusion that our summer should be
longer and our winter shorter than the summer and winter of our anti-
podes; and this is *actually* the case by about eight days.

"I say *actually*, because, if we now look at the effects of the precession of the equinoxes, we shall see that in a time equal to half of
the grand year, whether it be 12,900 according to the astronomers, or
10,500 years according to M. Adhemar, the conditions will be
reversed; the terrestrial axis, and consequently the poles, will have
accomplished the half of their bi-conical revolution of the centre of the
earth. It will then be the northern hemisphere which will have the
summers shorter and the winters longer, and the southern hemisphere
exactly the reverse. In the year 1248 before the Christian era, according to M. Adhemar, the north pole attained its maximum summer
duration. Since then, that is to say for the last 3112 years, it has
begun to decrease, and this will continue till the year 7388 of our
era before it attains its maximum winter duration.

"But the reader may ask, fatigued, perhaps, by these abstract
considerations, What is there here in common with the deluges?

"The *grand year* is here divided for each hemisphere into two great
seasons, which De Jouvencel calls the great summer and winter, which
will each, according to M. Adhemar, be 10,500 years.

"During the whole of this period one of the poles has constantly had
shorter winters and longer summers than the other. It follows that the
pole which submits to the long winter undergoes a gradual and continuous cooling, in consequence of which the quantities of ice and snow,
which melt during the summer, are more than compensated by that
which is again produced in the winter. The ice and snow go on
accumulating from year to year, and finish at the end of the period
by forming, at the coldest pole, a sort of crust or cap, voluminous,
thick, and heavy enough to modify the spheroidal form of the

earth. This modification, as a necessary consequence, produces a notable displacement of the centre of gravity, or—for it amounts to the same thing—of the centre of attraction, round which all the watery masses tend to restore it. The south pole, as we have seen, finished its great winter in 1248 B.C. The accumulated ice then added itself to the snow, and the snow to the ice, at the south pole, towards which the watery masses all tended until they covered nearly the whole of the southern hemisphere. But since that date of 1248 B.C., our great winter has been in progress. Our pole, in its turn, goes on getting cooler continually; ice is being heaped upon snow, and snow upon ice, and in 7388 years the centre of gravity of the earth will return to its normal position, which is the geometrical centre of the spheroid. Following the immutable laws of central attraction, the southern waters accruing from the melted ice and snow of the south pole will return to invade and overwhelm once more the continents of the northern hemisphere, giving birth to new continents, in all probability, in the southern hemisphere."

Such is a brief statement of the hypothesis which Adhemar has very ingeniously worked out. How far it explains the mysterious phenomena which we have under consideration we shall not attempt to say, our concern being with the facts. Does the evidence of upward and downward movements of the surface in tertiary times explain the great change? For if the cooling which preceded and succeeded the two European deluges still remains an unsolved problem, its effects are perfectly appreciable. The sudden cold which visited the northern and central parts of Europe resulted in the annihilation of organic life in those countries. All the watercourses, the rivers and rivulets, the seas and lakes, were suddenly frozen. As Agassiz says in his first work on 'Glaciers,' "A vast mantle of ice and snow covered the plains, the plateaus, and the seas. All the sources were dried up; the rivers ceased to flow. To the motions of a numerous and animated creation, the silence of death must have succeeded." Great numbers of animals perished from cold. The elephant and rhinoceros perished by thousands in the bosom of their grazing grounds, which were suddenly transformed into fields of ice and snow; and under this visitation these two species disappear, and seem to have been effaced from the creation. Other animals were overwhelmed, but their race did not perish entirely. The sun, which not long since lighted up the verdant plains, rising now on the frozen steppes, was only saluted by the whistling winds of the north, and the crashing of crevasses, which opened up on all sides under the radiating heat acting upon the immense glaciers which entombed so many animated beings.

How can we accept the idea that the plains, but yesterday smiling and fertile, have been already covered, and continue to be during a very long period, with an immense shroud of ice and snow? To satisfy the reader that this can be established on sufficient evidence, it is necessary to direct his attention to certain parts of Europe. It is necessary to visit, at least in idea, a country where the phenomena of *glaciers* still exist, and to show that this phenomena, now confined to countries of great elevation, extended itself, during geological times, over spaces infinitely more vast. We shall choose for this purpose the glaciers of the Alps. We shall show that the glaciers of Switzerland and Savoy have not always been confined to their present limits; that they are, so to speak, only miniature resemblances of the gigantic glaciers of other times; and that at the period under review they extended over all the great plains which approach the foot of the Alps.

To establish these proofs, we must enter upon some consideration of existing glaciers, upon their mode of formation, and the phenomena to which they belong.

The snow which during the whole year falls upon the mountains does not melt, but maintains its solid state, where their elevations exceed the height of 9000 feet or thereabouts. Where these snows accumulate to great thickness, in the valleys or in the deep mazy fractures in the soil, they harden under the influence of pressure resulting from their incumbent weight. But it always happens that a certain quantity of water, the result of momentary fusion of the superficial beds, traverses its substance, and this forms a crystalline

mass of ice, granulated in structure, which the Swiss naturalists designate *névé*. From the successive melting and freezing provoked by
the heat by day and the cold by night, the infiltration of air and
water in its interstices, the *névé* is slowly transformed into a homogeneous and sky-coloured block of ice, filled with an infinity of
little air-bubbles: this is what is called *glace bulleuse*, bubbled ice.
Finally, these masses are completely frozen; the water replaces the
air-bubbles: then the transformation is complete; the ice is homogeneous, and presents those fine azure tints so much admired by the
tourist who traverses the magnificent glaciers of Switzerland and
Savoy.

Such is the origin, such the mode in which the glaciers of the Alps
are formed. An important property of glaciers remains to be pointed
out. They have a general movement of translation in the direction of
their slope, under the influence of which they make a certain yearly
progress downward, according to the angle of the slope. The glacier
of the Aar, for example, advances at the rate of about 250 feet each
year.

Under the joint influence of the slope, the weight of the congealed
mass, and the fusion of the parts which touch the earth, the glacier
thus always tends downwards; but from the effects of a more ambient
temperature, the lower extremity melts most rapidly—it tends backwards. This is the difference between the two modes of action which
constitute the real progression of the glacier.

The friction which the glacier exercises upon the bottom and sides
of the valley ought necessarily to leave its traces on the rocks with
which it is in contact. In all places where a glacier has' passed, in
short, we remark that the rocks are polished, levelled, and rounded.
These rocks present, besides, striated marks running in the direction
of the slope of the glacier, which are produced by hard angular stones
or other substances embodied in the ice, which have left their marks
on the hardest rocks under the irresistible pressure of the heavy
descending mass. In a work of great merit which we have before
quoted, Mr. Charles Martins explains the physical mechanism by

which granitic rocks involved in the glacial movement have scratched, striped, and rounded the softer rocks which the glacier encountered in its descent. " The friction," says Mr. Martins, " which the glacier exercises upon the bottom and upon the walls in its passage is too considerable not to leave its traces upon the rocks with which it has been in contact; but its action varies according to the mineralogical nature of the rocks, and the configuration of the beds they occupy. If we penetrate between the soil and the bottom of the glacier, taking advantage of the icy caverns which sometimes open at its edge or extremities, we creep over a bed of pebbles and fine sand impregnated with water. If we raise this bed, we soon recognize that the subjacent rock is levelled, polished, ground down by friction, and covered with rectilinear stripes, resembling sometimes a small furrow, more frequently deep scratches, perfectly straight, as if they had been engraved by the aid of a burin, or even a very fine needle. The mechanism by which these stripes have been produced is that which industry employs in polishing stones and metal. By the aid of a fine powder, called *emery*, we rub the metallic surface until we give it a brilliant look, which proceeds from the reflection of the light from an infinity of minute stripes. The bed of pebbles and mud interposed between the glacier and the subjacent rock is here the emery. The rock is the metallic surface, and the mass of the glacier which presses on and displaces the mud in its descent towards the plain represents the hand of the polisher. These stripes generally follow the direction of the glacier; but as it is sometimes subject to small lateral deviations, the stripes sometimes cross themselves, forming a very small angle. If we examine the rocks on the side of a glacier, we find the same stripes engraved on them where they have been in contact with the congealed mass. I have often broken the ice where it thus pressed the rock, and have found under it surfaces polished and covered with stripes. The pebbles and grains of sand which had engraved them were still incased in the ice, fixed like the diamond of the glazier at the end of the instrument with which he marks his glass.

" The neatness and depth of the stripes depend on many circumstances: if the rock operated on is calcareous, and the emery is composed of pebbles and sand proceeding from hard rocks, such as gneiss, granite, or protogine, the stripes are very marked, as we can verify at the foot of the glaciers of Rosenlani, and of the Grindenwald, in the canton of Berne. On the contrary, if the rock is gneissic, granitic, or serpentinous, that is to say, very hard, the stripes will be less deep and less marked, as may be seen in the glaciers of the Aar, of Zermatt, and of Chamounix. The polish will be the same in both cases, and it is often as perfect as in marble polished for architectural purposes.

" The engraved stripes upon the rocks which enclose these glaciers are generally horizontal or parallel to the surface. Sometimes, owing to the contractions of the valley, these stripes are nearly vertical. This, however, need not surprise us. Driven onwards by the superincumbent weight the glacier forces itself through the narrow part, its bulk expanding upwards, in which case the flanks of the mountain which barred its passage are marked vertically. This is admirably seen near the Chalets of La Stieregg, a narrow defile which the lower glacier of the Grindenwald has to clear before it discharges itself into the valley. Upon the right bank of the glacier the stripes are inclined at 45° to the horizon. Upon the left bank they rise sometimes quite up to the neighbouring forest, carrying with them great clods of earth charged with tufts of rhododendrons and bouquets of alder, birches, and firs. The more tender rocks were broken up by the tremendous force of the glacier; the harder rocks resisted, but the surface is planed down, polished, and streaked, testifying to the prodigious force to which they had submitted. In the same manner the glacier of the Aar, at the foot of the promontory on which M. Agassiz' pavilion was erected, is polished to a great height, and on the base, towards the upper part of the valley, I observed stripes inclined to 64°. The ice, erect against this escarpment, seemed like a ladder, but the granite rock held its own, and the glacier was compelled to pass round it slowly.

" The very considerable pressure of a glacier, joined to its movement of progression, acts at once upon the bottom and flanks of the valley which it traverses: it polishes all the rocks too hard to be demolished

by it, and impresses upon them sometimes a form peculiar and characteristic. In destroying all the asperities of these rocks, it levels the
surface and rounds them up the stream, whilst down the stream they
sometimes preserve their abrupt, unequal, and rugged surface. We must
comprehend, in short, that the force of the glacier is borne principally
by the side turned towards the circle whence it descended, like the
piles of a bridge which have been more damaged up the river than
down by the icebergs which have been formed during the winter.
Seen from a distance, a group of rocks thus rounded and polished
remind us of the appearance of a flock of sheep; hence the name
roches moutonée given them by the Swiss naturalists."

Another phenomenon which plays an important part in the actual
glaciers, and those also which formerly covered Switzerland, is found
in the fragments of rock, often enormous, which have been transported
and deposited in their movement of progression.

The ridges of the Alps are exposed to continual degradations.
Formed of granitic rocks—rocks eminently alterable under the action
of air and water, they are disintegrated and often fall in fragments
more or less voluminous. "The masses of snow," continues Martins,
"which hang upon the Alps during winter, the rain which infiltrates
between their beds during summer, the sudden action of watery torrents,
and more slowly, but yet more powerfully, the chemical affinities,
degrading, disaggregating, and decomposing the hardest rocks ; these
débris fall from the summit, and from considerable heights, but within
the circle occupied by the glaciers, accompanied with frightful noises,
and amidst great clouds of dust. In the same manner, in the heart of
summer, I have seen avalanches of stone precipitated from the highest
ridges of the Schreckhorn, forming upon the immaculate snow a long
black train, consisting of enormous blocks and an immense number of
smaller fragments. In the spring a rapid melting of the winter snow
often engenders accidental torrents of extreme violence. If the fusion
is slow, water insinuates itself into the smallest fissures of the rocks,
congeals there and splits the most refractory. The blocks detached
from the mountains are sometimes of gigantic dimensions : we have

found them sixty feet in length, and those measuring thirty feet in every direction are by no means rare in the Alps."

Thus the action of aqueous infiltration followed by frost, the chemical decomposition which takes place when granite is subjected to the influence of humid air, degrades and disaggregates the rocks which constitute the mountains enclosing the glacier. Blocks of very considerable dimensions often fall on the surface of the glacier at the foot of the mountain. Were it immoveable, these débris would accumulate at its base, and would form there a mass of ruins heaped up without order. But the slow progression, the continuous displacement of the glacier, leads in the distribution of these blocks to a certain order of arrangement: the blocks falling upon its surface participate in its movement, and advance with it. But other downfalls happen daily, and the new débris combining with the first, the whole form a line along the outer edge of the glacier: these regulated trains of rocks bear the name of moraines. When these rocks fall from two mountains, and on each edge of the glacier, and two parallel lines of débris are formed, they are called *lateral moraines*. There are also *median moraines*, which are formed when two glaciers are confluent in such a manner that the *lateral moraine*, on the right of the one, leans on that of the left hand of the other. Finally, those moraines are *frontal*, or *terminal*, which repose, not upon the glacier, but have their point of termination in the valleys, and are due to the accumulation of fallen blocks from the terminal escarpments of glaciers, and arrested by some obstacle. In PL. XXXI. we have represented an actual Swiss glacier, in which are united the physical and geological peculiarities belonging to these enormous masses of congealed water: the moraines here are *lateral*, that is to say, formed of a double line of débris.

Transported slowly on the surface of the glacier, all the blocks from the mountain preserve their original form unaltered ; the cutting edge of the point is never altered by their gentle transport and almost imperceptible motion. The atmospheric agents alone can affect or destroy these rocks when formed of hard, resisting material. They then remain nearly of the same form and volume they had when they

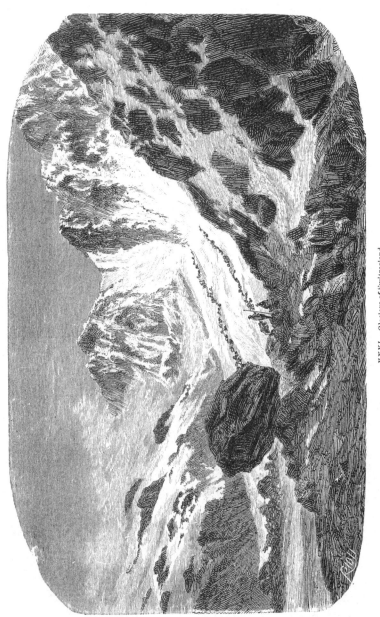

XXXI.—Glaciers of Switzerland.

fell on the surface of the glacier; but it is otherwise with blocks and débris enclosed between the rock and the glacier, whether it be on the bottom or between the glacier and the lateral walls. Some of these, under the powerful action of this gigantic rolling movement, will be reduced to an impalpable mud, others are shaped into facets, others ground into round balls, presenting a multitude of streaks crossing each other in all directions. These striped pebbles are of great importance in studying the extent of ancient glaciers: they testify on the spot to the existence of an anterior glacier which had fashioned, ground, and striped the pebbles, which water does not; on the contrary, it polishes and rounds them; it even effaces the natural stripes.

Thus the voluminous blocks transported to great distances from their true geological beds, that is, the *erratic blocks*, to use the consecrated term, with polished and striated surfaces and "sheep"-shaped summits, the *moraines*; finally, the pebbles, ground, polished, round or in facets, are all physical traces of glaciers in motion, and their presence alone is sufficient proof to the naturalist that a glacier has in other times existed in the locality where he finds them. The reader will now comprehend how it is possible to recognise in our days the existence of ancient glaciers in different parts of the world. Above all, where we find at once *erratic blocks* and *moraines*, and observe at the same time traces of rocks polished and striped in the same direction, we may pronounce with certainty as to the existence of a glacier in geological times. Let us take some instances.

In the Alps, at Pravolta, going towards Mont *Santo Primo*, upon a calcareous rock, we find the granitic block represented in Fig. 194. This erratic block exists, with thousands of others, on the slopes of the mountain. It is about fifty feet long, nearly forty feet broad, and five-and-twenty in height, and all its points and angles uninjured. Some parallel stripes occur along the surrounding rocks. All this demonstrates, and with evidence, that a glacier has extended itself in former times into this part of the Alps, where none appear at the present time, and that it has here deposited this enormous burden in its progressive movement.

In the Jura Mountains, on the hill of Fourvière, a limestone eminence at Lyon, blocks of granite are found, evidently detached from

Fig. 194.—Erratic blocks in the Alps.

the Alps, and carried here by the Swiss glaciers. The particular
mode of transport is represented theoretically in Fig. 195. A repre-

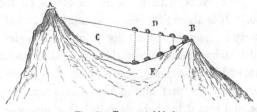

Fig. 195.—Transported blocks.

sents, for example, the summit of the Alps, B the Jura Mountains, or the hill of Fourvière, at Lyon. In geological times, the glacier A B C extended from the Alps to this mountain B. The granitic débris which was detached from the summit of the Alps fell on the surface of the glacier. The movement in progression transported these blocks as far as the summit B. At a later period the temperature of the globe is raised, and then the ice begins to melt, and the blocks D E have been quietly deposited in the spots where they have been found, without shock or injury sustained in this singular mode of conveyance.

Every day traces more or less recognizable are found on the Alps of ancient glaciers far distant from the limits of existing ones. Great masses of débris, comprehending blocks with sharp pointed angles, are found in the Swiss plains and valleys. *Perched* blocks, as in PL. XXXI., are often seen upon points of the Alps situated far below the existing glaciers, or dispersed over the plain which separates the Alps from the Jura, or even reposing in a manner quite incredible, when seen in their great mass at considerable heights on the eastern flank of this chain of mountains. It is with the assistance of these indices that the geologist has been able to trace to distances extremely remote, indications of the ancient glaciers of the Alps, follow them in their course, and fix their point of origin and where they stopped. Thus the humble Mount Zion, a softly swelling hill situated to the north of Geneva, was the point at which three great antediluvian glaciers had their confluence: the glacier of the Rhone, which filled all the basin of the Leman Lake, or Lake of Geneva; that of the Isere, which issued from the Annecy and Bourget Lake; and that of the Arve, which had its birth in the valley of Chamounix, all converged at this point; according to M. G. de Mortillet, who has carefully studied the extent and situation of these ancient glaciers, this had upon its northern flank the *glacier of the Rhine,* which occupied all the basin from Lake Constance to the borders of Germany; that of the *Linth,* which was arrested at the extremity of the Lake of Zurich—this city is built upon its terminal moraine; that of the *Reus,* which covered the lake of the four cantons with blocks torn from the ridges of Saint-Gothard;

2 D

that of the *Aar*, whose last moraine crowns the hills in the environs of Berne; that of the *Arve* and the *Isere*, which, as we have said, debouched from Lake Annecy and Lake Bourget; that of the *Rhone*, the most important of all. It was this glacier which deposited upon the flank of the Jura, at the lleight of 3400 feet above the level of the sea, the great *erratic block* already described. This mighty glacier occupied all the lateral valleys formed·by the two parallel chains of the Valais. It filled all the Valais, and covered the plain lying between the Alps and the Jura from Fort de L'Ecluse, near the spot where the Rhone disappears under the rocks, up to the neighbourhood of Aarau.

The débris of rocks transported by the icy sea which occupied all 'the Swiss plains takes, towards the north, the direction of the valley of the Rhine. On the other hand, the glacier of the Rhone, after reaching the plains of Switzerland, turned off obliquely towards the south, received the glacier of the Arve, then that of the Isere, passed between the Jura and the mountains of the Grand Chartreuse, covered La Bresse, nearly all Dauphiny, and terminated in the neighbourhood of Lyon.

Upon the southern flank of the Alps the ancient glaciers, according to M. de Mortillet's map, occupied all the great valleys from that of the Doire, on the west, to that of the Tagliamento, on the east. "The glacier of the *Doire*," says de Mortillet, whose text we abridge greatly, " debouches into the valley of the Po under the walls of Turin. That of the *Doira-Baltée* enters the plain of Ivrée, where it has left a magnificent semicircle of hills, which were its terminal moraine. That of the *Toce*, which would hurl itself into Lake Maggiore, against the glacier of the Tessina, and then throw itself into the valley of Lake Orta, at the southern extremity of which its terminal moraine is discovered. That of the Tessina filled the basin of Lake Maggiore, and established itself between Lugano and Varèse. That of the *Adda* filled the basin of Lake Como, and established itself between Mendrizio and Lecco, thus describing a vast semicircle. That of the *Oglio* terminated a little beyond Lake Iseo. That of the *Adige*, finding no passage through the narrow valley of Roveredo, took another course, and filled the immense valley of the Lake di Garda. At Novi it has left a magnificent moraine, of which Dante speaks in his 'Inferno.' That of the *Brenta* extended itself in a plateau of this commune. The *Drave* and the *Tagliamento* had also their glaciers. Finally, glaciers occupied all the valleys of the Austrian and Bavarian Alps."

Similar traces of the existence of ancient glaciers occur over all European countries. In the Pyrenees, in Corsica, in the Vosges, extensive ranges of country were covered, in geological times, by these vast plains of ice. The glacier of the Moselle was the most considerable of the Vosges, receiving numerous affluents; its lowest frontal moraine, which is situated below Remiremont, could not be less than a mile and a quarter in length.

But the phenomenon of glacier extension which we have examined in the Alps was not confined to Central Europe. The same traces of their ancient existence are observed in all the north of Europe, in Russia, in Iceland, in Prussia, in the British Islands, in part of Germany, in the north, and even in some parts of the south of Spain. In England, *erratic* blocks of granite have been found which proceed from the mountains of Norway. It is evident that these blocks were borne by a glacier which extended from the neighbourhood of the pole to England, over which they cleared the Baltic and the North Sea. In Prussia similar traces are observable.

Thus, during the quaternary epoch, glaciers which are now limited to the Polar regions, or to mountainous countries of considerable altitude, extended very far beyond anything now known; and taken in connexion with the deluge of the north, with which they are closely associated, and the vast amount of organic life of which they were the winding-sheet, they form, perhaps, the most striking and mysterious of all the geological phenomena.

M. Edouard Collomb, to whom we owe much of our knowledge of ancient glaciers, furnishes the following note explanatory of a map of Ancient Glaciers which he has prepared :—

" The space occupied by the ancient quaternary glaciers may be divided into two orographical regions :—1. The region of the north, from the 52° up to the North Pole.   2. The region of Central Europe and part of the south.

" The region of the north which has been covered by the ancient glaciers comprehends all the Scandinavian peninsula, Sweden, Norway, and a part of Western Russia, extending from the Nieman on the north, describing a curve which passed near the sources of the Dneiper and the Volga, thence directing itself towards the glacial sea. This region comprehends Iceland, Scotland, Ireland, the isles dependent on them, and, finally, part of England.

" This region is bounded on all its sides by a large band of from two to five degrees of breadth, on which is recognized the existence of erratic blocks of the north: it comprehended the middle region of European Russia—as Poland, a part of Prussia, of Denmark; losing itself in Holland on the Zuider Zee, it cut into the southern part of England, and we find a shred of it in France, upon the borders of the Contentin.

" The ancient glaciers of Central Europe consisted, first, of the grand masses of the Alps, stretching to the west and to the north, they extended themselves to the valley of the Rhone as far as Lyon, then passing the summit level of the Jura, they passed near Basle, covering Lake Constance, and stretching beyond Bavaria and into Austria.   Upon the southern slopes of the Alps they turned round the summit of the Adriatic, passed near to Udinet, covered Peschiera, Solferino, Coma, Varese, Ivree, stretching near to Turin, terminating in the valley of the Stura, near the Col de Tende.

" In the Pyrenees the ancient glaciers have occupied all the principal valleys of this chain, whether of the French or Spanish sides, and especially the valleys of the centre, which comprehend that of Suchon, of Aure, of Bareges, of Cauteret, and of Ossau.   In the Cantabrique, prolongments of the Pyrenees, the existence of the ancient glaciers has been recognized.

" In the Vosges and the Black Forest they covered all the parts south of these mountains. In the Vosges the principal traces are found in the valley of Saint-Amarin, of Giromaguy, of Munster, and of the Moselle.

"In the Carpathians and the Caucasus the existence of ancient glaciers of great extent has been observed.

"In the Sierra Nevada, in the south of Spain, mountains upwards of 11,000 feet, the valleys which descend from the Picacho de Veleta and Mulhacen have been encumbered with ancient glaciers in the quaternary epoch."

There is no reason to doubt that at this epoch all the British islands, at least all north of London, were covered by glaciers in their higher parts. "Those," says Professor Ramsay, "who know the Highlands of Scotland will remember that, though the weather has had a powerful influence upon them, rendering them in places rugged, jagged and cliffy, yet notwithstanding their general outlines are often remarkably rounded and flowing; and when the valleys are examined in detail, you find in their bottoms and on the sides of the hills that the mammellated structure prevails. This rounded form is known by those who study glaciers by the name of *roches moutonnées*, given to them by the Swiss writers. These mammellated forms are exceedingly common in many British valleys, and not only so, but the very same kind of grooving and striation so characteristic of the rocks in the Swiss valleys also marks those of the Highlands of Scotland, of Cumberland and Wales. Considering all these things, geologists, led by Agassiz some five or six-and-twenty years ago, have by degrees come to the conclusion that a very large part of our island was during the glacial period covered, or nearly covered, with a thick coating of ice in the same way that the north of Greenland is at present; and that from the long-continued grinding power of a great glacier, or set of glaciers nearly universal, the whole surface became moulded by ice."

Whoever traverses England, observing its features with attention, will remark in certain places traces of the ravages committed in this era. Some of the mountains present on one side a naked rock, and on the other a gentle slope, smiling and verdant, giving a character more or less abrupt, bold and striking to the landscape. Considerable portions of dry land were formerly covered by a bluish clay, which contained many fragments of rock or "boulders" torn from the old Cumbrian mountains, from the Pennine chain, from the moraines of the north of England, and from the chalk hills, hence called "boulder" clay, presents itself here and there, broken, wasted and ground up by the action of water. These erratic blocks have visibly been detached from the parent rock by violence and transported to considerable distances. They have been carried not only across plains, but over the tops of mountains; some of them being found 130 miles from the original matrix. We even find, as already hinted, some rocks of which no prototypes have been found nearer than Norway. There is, then, little room for doubting the fact of an extensive glacier system having covered the land, although the proofs have only been gathered laboriously and painfully in a long series of years. In 1840 Agassiz visited Scotland, and his eye, accustomed to glaciers in his native mountains, speedily detected the signs. Dr. Buckland became a zealous advocate of the same views. North Wales was soon recognized as an independent centre of a system which radiated from lofty Snowdon, through seven valleys, carrying with them large stones and grooving the rocks in their passage. In the pass of Llanberis, the principal of these, all the common proofs exist of the facts we state. "When the country was under water," says Professor Ramsay, "the drift was deposited which more or less filled up many of the Welsh valleys. When the land had risen a second time to a considerable height, the glaciers again increased in size : although they never reached the immense magnitude they attained in the earlier portion of the glacial period, still they became so large that such a valley as Llanberis, was a second time occupied by ice, which ploughed out the drift that more or less covered the valley. By degrees, however, as we approach nearer our own days, the climate began to be ameliorated and the glaciers to decline, till, growing less and less, they crept up and up, and here and there, as they died away, they left their terminal and lateral moraines still as well defined in some cases as moraines in lands where glaciers now exist. Frequently, too, masses of stone that floated on the surface of the ice were left perched upon the rounded *roches moutonnée* in a manner somewhat puzzling to those who are not glacial geologists."

"In short, they were let down upon the surface of these rocks so quietly and softly that there they will lie until an earthquake shakes them, or until the wasting of the rock on which they rest precipitates them to a lower level."

Among other proofs of glacier action and submersion in Wales may be mentioned the case of Moel Trafaen, a hill 1,400 feet high lying to the westward of Carnarvon Bay, and six or seven miles from Carnarvon. Mr. Trimmer had observed stratified drift near the summit of this mountain, from which he obtained some marine shells; but doubts were entertained as to their age until 1863, when a deep and extensive cutting was made in search of slate. In this cutting a stratified mass of loose sand and gravel was laid open near the summit, 35 feet thick, containing shells, some entire, but mostly in fragments. Sir Charles Lyell examined the cutting, and obtained twenty shells, and in the lower beds of the drift, "large heavy· boulders of far-transported rocks, glacially polished and scratched on more than one side:" underneath the whole, the edges of vertical slates, exhibiting "unequivocal marks of prolonged glaciation." The shells belonged to species still living in British or more northern seas. "A most important fact," Sir Charles adds, "when we consider we have scarcely a well-authenticated case as yet on record, beyond the limits of Wales, of marine shells having been found in glacial drifts at half the height."

It was the opinion of Agassiz, after visiting Scotland, that the Grampians had been covered by a vast thickness of ice, whence erratic blocks had been dispersed in all directions as from a centre; other geologists after a time adopted the opinion; Mr. Robert Chambers going so far as to maintain, in 1848, that Scotland had been at one time moulded in ice. Mr. T. F. Jamieson followed in the same track, adducing many new facts to prove that the Grampians once sent down glaciers in all directions towards the sea. "The glacial

grooves," he says, " radiate outward from the central heights towards
all points of the compass, although they do not strictly conform to the
actual shape and contour of the minor valleys and ridges." But the
most interesting part of Mr. Jamieson's investigations are undoubtedly
the ingenious manner in which he has worked out Agassiz' assertion
that Glenroy, whose remarkable parallel roads had puzzled so many
investigators, had been a frozen lake.

Glenroy is one of the many romantic glens of Lochaber, at the
head of the Spey, near to the great glen, now the Caledonian Canal,
which stretches obliquely across the country in a north-westerly direction from Loch Linnhe to Loch Ness, leaving Loch Arkeg, Loch
Aich, Glen Garry and many a highland loch besides on the left, and
Glen Spean, in which Loch Treig, running due north and south, has
its mouth on the south, while Glenroy opens into it from the north,
while Glen Gluoy opens into the great glen opposite Loch Arkeg.
Mr. Jamieson commenced his investigations at the mouth of Loch
Arkeg, which is about a mile from the lake itself. Here he found the
gneiss ground down as if by ice coming from the east. On the hill
north of the lake the gneiss was much worn, having markings which,
though much weathered, still exhibited well-marked striæ, directed up
and down the valley. Other markings showed that the Glen Arkeg
glacier not only blocked up Glen Gluoy, but the mouth of Glen Spean,
which lies two miles or so north of it on the opposite side.

At Bracklatter, on the south side of Glen Spean, near the junction
with Glen Lochy, glacial scores pointing nearly due west, but slightly
inclining to north were observed, as if caused by the pressure of ice
from Glen Leri. The south side of Glen Spean, from the mouth of
Loch Treig, is bounded by lofty hills, the spurs of Ben Nevis, the
highest of these peaks, being 3,000 feet. Numerous gullies intersect
their flanks, and the largest of these, Aonach More, present a series of
rocky amphitheatres, or rather large caldrons, with walls ground down
by long-continued glacial action : quartz veins had been shorn down to
the gneiss, which is also streaked with fine scratches, pointing down
the hollows and far up the rocks on either side. During all these

operations the great valley was probably filled up with ice, which would close Glen Gluoy and Glen Spean, and might also close the lowest of the lines in Glenroy.   But how about the middle and upper lines ?

A glacier crossing from Loch Treig, and protruding across Glen Spean, would cut off Glens Glaibu and Makoul, when the water in Glenroy could only escape over the Col into Strathspey, when the first level would be marked.

Now let the Glen Treig glacier shrink a little, so as to let out water to the level of the second line by the outline at Makoul, and the theory is complete.   When the first and greatest glacier gave way, Glenroy would be nearly in its actual state.

The glacier on issuing from the gorge at the end of Loch Treig would dilate immensely, the right flank spreading over a rough expanse of syenite, the neighbouring hills being mica-schists, with veins of porphyry.   Now the syenite breaks into large cubical blocks of immense size.   These have been swept before the advancing glacier along with other débris, and deposited into a semicircle of mounds having a sweep of several miles—circular bands which mark the edges of the glacier as it shrunk from time to time under an ameliorating climate.

This moraine, which was all that was wanting to complete the theory laid down by Agassiz, is found on the pony-road leading from the mouth of Loch Treig towards Badenoch.   A mile or so brings the traveller to the summit level of the road, and beyond the hill a low moor stretches away to the bottom of the plain.   Here, slanting across the slope of the hill towards Loch Treig, two lines of moraine stretch across the road.   At first it consists of mica-schists and bits of porphyry, but blocks of syenite soon mingle.   Outside these are older hillocks, rising in some places sixty and seventy feet high, forming narrow steep-sided mounds, with blocks fourteen feet sticking out of the surface, mixed with débris of mica-schist and gneiss.   The inner moraine consists almost wholly of large blocks of syenite, five, ten, fifteen, and five-and-twenty feet long.

The present aspect of Glenroy is that of an upper and lower glen opening up from the larger Glen Spean. The head waters of Lochaber gather in a wild mountain tract, near the source of the Spey. The upper glen is an oval valley, four miles long, by

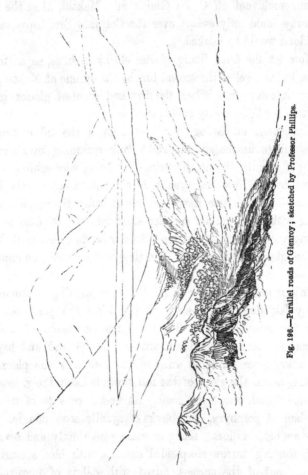

Fig. 196.—Parallel roads of Glenroy; sketched by Professor Phillips.

about one broad, bounded on each side by high mountains, which throw off two streams which divide the mica-schist from the gneiss systems; the former predominating on the west side, and the latter on the east. The united streams flow to the south-west for two miles, when the valley contracts to a rocky gorge which separates the upper from the lower glen. Passing from the upper to the lower glen, a line is observed to pass from near the junction of the two streams, on a level with a flat rock at the gorge, and also with the uppermost of the three lines of terraces in the lower glen. This line girdles the sides of the hills right and left, with a seemingly higher sweep, and is followed by two other perfectly parallel and continuous lines till Glenroy expands into Glen Spean, which crosses its mouth and enters the great glen a little south of Loch Lochy. At the point, however, where Glenroy enters Glen Spean, the two upper terraces cease, while the lower of the three appear on the north and south side of Glen Spean, as far as the pass of Glen Muckal, and southward a little way up the Gubban river and round the head of Loch Treig.

The phenomenon which so powerfully affected our hemisphere presents itself in a much grander manner in the New World. The glacier system appears to have taken in America the same gigantic proportions which other objects assume there. Nor is it necessary, in order to explain the permanent existence of this icy mantle in countries now flourishing, to assume any very extraordinary degree of cold. On this subject Mr. Ch. Martins thus expresses himself:— "The mean temperature of Geneva is 9° 5' Cent. Upon the surrounding mountains the limit of perpetual snow is found at 8,800 feet above the level of the sea. The great glaciers of the valley of Chamounix descend 5,000 feet below this line. Thus situated, let us suppose that the mean temperature of Geneva were lowered only 4°, and the average became 5° 5'; the decrease of temperature with the height being 1° for every 600 feet, the limit of perpetual snow would be lowered by 2,437 feet, and would be 6,363 feet above the level of the sea. We can readily admit that the glaciers of Chamounix would descend below this new limit, with a quantity at least equal to that which exists between their actual limit and their lower extremity. Now in reality the foot of the glacier is 5,000 feet above the ocean;

with a climate 4° colder, it would be 2,437 feet lower; that is to say, at the level of the Swiss plain. Thus, the lowering of the line of perpetual snow to this extent would suffice to bring the glacier of the Arve to the environs of Geneva. . . . Of the climate which has most favoured the prodigious development of glaciers we have a pretty correct idea; it is that of Upsal, of Stockholm, of Christiana, and of part of North America, in the State of New York. . . . To diminish by four degrees the mean temperature of a country in order to explain one of the grand revolutions is to venture on an hypothesis not bolder than geology has sometimes permitted to itself."

In proving that the glaciers covered part of Europe during a certain period, that they extended themselves from the North Pole to Italy and the Danube, we have sufficiently established the reality of this *glacier period*, which we must consider as a curious episode, however certain, in the history of the earth. Such masses of ice could only have covered the earth when the temperature of the air was lowered at least some degrees below zero; but organic life is incompatible with such a temperature; and to this cause must we attribute the disappearance of certain species of animals and plants—in particular, the rhinoceros and the elephants—which, before this sudden and extraordinary cooling of the globe, would appear to have confined themselves in immense bands to Northern Europe, and chiefly to Siberia, where their remains have been found in such prodigious quantities. Cuvier says, speaking of the bodies of the quadrupeds which the ice had seized, and in which they have been preserved in their hair, flesh, and skin, up to our own age, "If they had not been frozen as soon as killed, putrefaction would have decomposed them; and, on the other hand, this eternal congelation could not have occupied previously the place where they were seized; for they could not have lived in such a temperature. It was, therefore, in the same instant which saw them perish that the country they inhabited was rendered glacial. The event must have been sudden, instantaneous, and without any preparation."

How can we explain the *glacial period?*—to what cause attribute this sudden and excessive cold which covered great part of Europe, followed by a prompt return to its normal temperature ? We have explained M. Adhemar's hypothesis, to which it may be objected that the cold of the glacial period was so general throughout the Polar and temperate regions on both sides the equator, that mere local changes in the external configuration of our planet and disturbance of the centre of gravity scarcely affords an adequate cause for so great a revolution in temperature. Sir Charles Lyell, speculating upon the suggestion of Ritter and the discovery of marine shells spread far and wide over the Sahara desert by Messrs. Escher, Van der Linth, Desor, and Martins,—which seem to prove that the African desert had been under water at a very recent period,—infers that the Sahara desert constituted formerly a wide marine area, stretching several hundred miles north and south, and east and west. " From this area," he adds, " the south wind must formerly have absorbed moisture, and must have been still further cooled and saturated with aqueous vapour as it passed over the. Mediterranean. When at length it reached the Alps, and, striking them, was driven into the higher and more rarefied regions of the atmosphere, it would part with its watery burthen in the form of snow ; so that the aërial current which, under the name of the föhn, or sirocco, now plays a leading part with its hot and dry breath, sometimes, even in the depth of winter, in melting snow and checking the growth of glaciers, must, at the period alluded to, have been the principal feeder of Alpine snow and ice." Nevertheless, we repeat, no explanation presents itself which can be considered con- clusive; and in science its professors should never be afraid to say, *1 do not know.*

Fig. 197.—Fissurella nembosa.

CREATION OF MAN AND THE ASIATIC DELUGE.

IT was only after the glacial period, when the earth had resumed its normal temperature, that man was created. Whence came he?

He came whence the first blade of grass which grew upon the burning rocks of the Silurian seas came; from whence came the different races of animals which have from time to time replaced each other upon the globe, gradually rising in the scale of perfection. He emanated from the will of the Author of the worlds which constitute the universe.

The earth has passed through many phases since the instant when, according to the expression of the Sacred Writings, "the earth was without form and void; and darkness was upon the face of the deep. And the spirit of God moved upon the face of the waters." We have considered all these phases; we have seen the globe floating in space in the state of gaseous nebulosity, condensing into liquidity, and beginning to solidify at the surface. We have pictured its internal agitations, their distractions, the partial dislocations to which the earth has been subjected, almost without interruption, while it could not yet resist the force of the waves of the fiery sea imprisoned within its fragile crust. We have seen the envelope acquiring solidity, and the geological cataclysms losing their importance and frequency in proportion as this solid crust was increased in thickness. We have looked on, so to speak, while the work of organic creation was proceeding. We have seen life make its appearance upon the globe; the first plants and animals springing into existence. We have seen them increase and multiply, but constantly perfecting themselves as we advance in the progressive phases of the history of the earth—not in the sense which implies that the earliest of created beings is less adapted to its purpose than the most recent, but that later creations are organized with different objects and with a more complicated structure. We now arrive at the greatest epoch of this history, at the crowning of the edifice, *si parva licet componere magnis.*

At the close of the tertiary epoch, the continents and the seas have assumed the respective limits which they now present. The disturbances of the soil, the fractures of the crust and volcanic eruptions and earthquakes of which they are the consequences, only occurred now at rare intervals, occasioning only local and restricted disasters. The rivers and their affluents flowed between tranquil banks; the animated creation was that of our own days. An abundant vegetation, diversified by the existence of a climate which has now been acquired, embellishes the earth. A multitude of animals inhabit the waters, the dry land and the air. Nevertheless, creation has not yet achieved its greatest work; a being capable of comprehending these marvels and of admiring the sublime work—a soul is wanting to adore and give thanks to the Creator.

God created man.

What is man?

We may say that man is an intelligent and moral being; but this would give a very imperfect idea of his nature. Franklin says that man is he that can make tools. This is to reproduce a portion of the first proposition while depreciating it. Aristotle calls man the "wise being," ζῶον πολιτιχόν. Linnæus, in his 'System of Nature,' after designating him as wise, *homo sapiens,* follows it with the words *Nasco te ipsum.* The French naturalist and philosopher, Isidore Geoffroy Saint-Hilaire, says, " the plant *lives,* the animal *lives and feels,* man

lives, feels and thinks." A sentiment which Voltaire has amplified :
" the Eternal Maker," says the philosopher of Ferney, "has given to
man organization, sentiment and intelligence ; to the animals sentiment,
and what we call instinct ; to vegetables organization alone. His
power then acts continually upon these three kingdoms." It is probably
the animal which is here depreciated. The animal on many occasions
undoubtedly thinks, reasons, deliberates with itself, and acts in virtue
of a decision maturely weighed : it is not then reduced to a simple sensation.

To define exactly the human being, we believe that it is necessary
to characterize the nature and extent of his intelligence. In certain
cases the intelligence of the animal approaches nearly to that of man,
but his intelligence is armed with certain faculties which belong to him
exclusively ; in creating him God has added an entirely new step in
the ascending scale of animated beings. This faculty, special to the
human race, is *abstraction*. We shall say that man is an *intelligent*
being and gifted with the faculty of comprehending the *abstract*.

It is by this faculty that man is raised in an unheard-of degree to
material and moral power. By it he has subdued the earth to his
empire, and fitted his mind for the most sublime contemplations.
Thanks to this faculty, man has conceived the ideal, and realized
poesy. He has conceived the infinite, and created mathematics.
Such is the immense step which separates the human race from
animals—which make him, a creation apart and absolutely new upon
the globe. To comprehend the ideal and the infinite, to create poetry
and algebra, such is man ! To find and comprehend this formula—

$$(a + b)^2 = a^2 + 2ab + b^2,$$

or the algebraic idea of negative quantities, this belongs to man. It
is the great privilege of the human being to express and comprehend
thoughts like the following :—

> "J'étais seul près des flots, par une nuit d'étoiles :
> Pas un nuage aux cieux, sur les mers pas de voiles ;
> Mes yeux plongeaient plus loin que le monde réel,
> Et les vents et les mers, et toute la nature
> Semblaient interroger dans un confus murmure,
> Les flots des mers, les feux du ciel.

2 E

Et les étoiles d'or, légions infinies,
A voix haute, à voix basse, avec mille harmonies,
Disaient en inclinant leur couronne de feu ;
Et les flots bleus, que rien ne gouverne et n'arrête,
Disaient, en recourbant l'écume de leur crête :
C'est le Seigneur, le Seigneur Dieu !" *

The 'Mécanique Céleste' of La Place, the 'Principia' of Newton, Milton's 'Paradise Lost,' the 'Orientales' by Victor Hugo—such are the fruits of the *faculty of abstraction*.

In the year 1800, a being, half a savage, who had lived in the woods, clambered on the trees, slept all his life upon dried leaves, and fled on the approach of men, was brought to a physician named Pinel. Some sportsmen had found him ; he had no voice and no intelligence ; he was known at the time as the little savage of Aveyron. The Parisian *savants* for a long time disputed over this strange individual. Was it an ape ?—was it a wild man ?

The learned Dr. Isard has published an interesting history of the savage of Aveyron. "He would descend," he writes, "into the garden of the deaf and dumb, sometimes, and seat himself upon the edge of the fountain ; then, rocking himself until by degrees his body became quite still, when he would assume the character of decided melancholy reverie : he would remain thus for whole hours—appearing all the while to be attentively looking on the surface of the water— upon which he would from time to time throw blades of grass and dried leaves. In fine nights, when the clear moonlight penetrated into the chamber he occupied; he rarely failed to rise and place himself

---

* "Alone with the waves, on a starry night,
My thoughts far away in the infinite ;
On the sea not a sail, not a cloud in the sky,
And the wind and the waves with sweet lullaby
Seem to ask in murmurs of mystery
    Of the fires of heaven, of the waves of the sea.
And the golden stars in the heavens rose higher,
Harmoniously blending their crowns of fire,
And the waves which no ruling hand may know,
'Midst a thousand murmurs, now high, now low,
Sing, while curving their foaming crest to the sea,
    It is the Lord God ! It is He."

before the window, where he would remain the whole night, standing upright, immovable, the neck stretched out, the eyes fixed upon the landscape lit up by the moon, as if occupied in a sort of ecstasy of contemplation." This being was, undoubtedly, a man. No ape ever exhibited such signs of intelligence—such dreamy manifestations— vague conceptions of the ideal;—in other words, that faculty for the *abstract* which belongs to humanity alone. In order to usher in worthily the new inhabitant who comes to fill the earth with his presence—who brings with him intelligence to comprehend, to admire, to subdue, and rule the creation (PL. XXXII.), we require nothing less than the ancient and venerated language of Moses, whom Bossuet calls " the most ancient of historians, the most sublime of philosophers, the wisest of legislators." The words of the inspired legislator are as follows:—" And God said, Let us make man in our image, after our likeness: and let him have dominion over the fish of the sea, and over the fowl of the air, and over the cattle, and over all the earth, and over every creeping thing that creepeth upon the earth. So God created man in his own image, in the image of God created he him; male and female created he them."

" And God saw everything that he had made, that *it was* very good."

Volumes have been written upon the question of the unity of the human race; that is, whether there were many centres of the crea- tion of man, or if the parent of our race was the Adam of Scripture. We think, with many naturalists, that the stock of humanity is unique, and that the several races of negroes, black and yellow, are only the result of climate upon organism. We consider that the human race appeared for the first time with a divine mystery which is eternally impenetrable to us as to the mode of creation. In the rich plains of Asia, on the smiling banks of the Euphrates, as the traditions of the most ancient races teach us—in the midst of this rich and vigorous soil, under the brilliant climate and the radiant sky of Asia, in the shade of its luxuriant masses of verdure and its mild and perfumed atmosphere, man loves to represent to himself the father of his race issuing from the hand of his Creator.

We are, thus, far from sharing the opinion of those naturalists, if there be any such, who represent man at the beginning of the existence of his species as a sort of ape, of hideous face, degraded mien, and covered with hair, inhabiting caves like the bears and lions, and participating in instincts at once brutal and ferocious.* There is no doubt the primitive man had to pass through a period in which he had to contend for his existence with ferocious beasts, to live in the woods like the apes, and in the savannahs, where Providence had thrown him. But this period of probation came to an end, and man, an eminently social being, promptly combined in groups, animated by the same interests and the same desires, and alike destined to triumph over the elements, and to subdue to their rule the other inhabitants of the earth. "The first men," says Buffon, "witnessed the convulsive movements of the earth, still recent and frequent, having only the mountains for refuge against its inundations : and driven from this asylum by volcanoes and earthquakes, which trembled under their feet; uneducated, naked, and exposed to the elements, victims to the fury of ferocious animals, who could scarcely be avoided—impressed, also, with a common sentiment of gloomy terror, and urged by necessity—would they not unite, at first, for defence ?—then to assist in forming dwellings for each other, afterwards with arms ? They began by shaping into the form of hatchets those hard flints, hip-stones, and thunder-bolts, which are supposed to have been formed by thunder and fallen from the

---

* We say, if there be any such, for it will be observed that the boldest advocate of " progression " or development, only ventures to place the frame and crania of the most degraded human form in juxtaposition with the ape, and says, " behold," probably half in joke, for the learned geologists are not above one. It is told of a former distinguished and witty member of the Geological Society that, having obtained possession of the rooms on a certain day, when there was to be a general meeting, he decorated its walls with a series of cartoons in which the parts of the members were strangely reversed. In one cartoon Icthyosaurii and Plesiosaurii were occupied with the skeleton of Homo sapiens; in another, a party of Crustaceans were occupied with a cranium suspiciously like the same species; while in a third, a party of Pterichthys were about to dine on a biped suspiciously resembling a well-conditioned F.G.S.

skies, but which are, nevertheless, only the first examples of man's art in a pure state of nature. He will soon draw fire from these same flints, by striking them against each other ; he will seize the flames of the burning volcano, or profit by the fire of the burning lava to light his fire of brush-wood in the forest; for by the help of this powerful element he cleanses, purifies, and renders himself and his dwelling healthy. With his hatchet of stone he chops wood, cuts trees, shapes timber, and puts it together, fashions his arms, and makes instruments of first necessity ; and after having furnished himself with clubs and other weighty and defensive arms, did not these first men find means to reach the light-footed and distant stag? A tendon of an animal, a thread of the aloes-wood, or the tough bark of some ligneous plant, would serve as a cord to bring together the two extremities of an elastic branch of yew, forming a bow ; other and smaller flints, shaped to a point, arm the arrow. They will soon have snares, rafts and canoes ; they will form themselves into little nations composed of families of relations, as is still the custom with savage nations, who have their game, fish, and fruits, in common. But in all those countries where space is limited by water, or surrounded by high mountains, these small nations, becoming too numerous, have been in time forced to parcel out the land between them ; and from that moment the earth has become the domain of man : he has taken possession of it by his labour, he has cultivated it, and attachment to the soil follows the very first act of possession : the private interest makes part of the national interest : order, civilization, and laws succeed, and policy acquires form and consistency." We love to quote the pages of a great writer — but how much more eloquent would the words of the great naturalist have been if he had added to his own great qualities the knowledge which science has placed within the reach of writers in our day—if he could have painted man in the early days of his creation, accompanied by the immense animal population which then occupied the earth, and fighting with the ferocious beasts which filled the forests of the ancient world ! Man, weak comparatively in organization, destitute of natural weapons of

attack or defence, incapable of rising into the air like the birds, or
living under water like the fish and some reptiles, he seemed devoted
to speedy destruction. But he was marked on the forehead by the
divine seal. Thanks to his superior and exceptional intelligence, this
being, in appearance so helpless, has by degrees swept the most
ferocious of its occupants from the earth, leaving those only who cater
to his wants or desires, by whose aid he changes the primitive aspect
of whole continents.

The antiquity of man is a question which has largely engaged the
attention of geologists, and many ingenious arguments have been
hazarded, tending to prove that the human race and the great extinct
mammifera were contemporaneous. The circumstances bearing on the
question are usually ranged under three series of facts:—1. The
cavern deposits; 2. Peat and shell mounds; 3. Lacustrine habitations.
We have already briefly touched upon the cavern deposits. In the
Kirkdale Cave no remains or other traces of man's presence seem to
have been discovered. But in Kent's Hole, an unequal deposit of
loam and clay, along with broken bones much gnawed, and the teeth of
both extinct and living mammifera with implements evidently fashioned
by the human hand were found in the following order; in the upper
part of the clay, artificially-shaped flints; on the clay rested a layer of
stalagmite, in which streaks of burnt charcoal occurred, and charred
bones of existing species of animals. Above the stalagmite a stone axe,
of more finished appearance, with bone pins, metal plates, and other
remains, Celtic, British, and Roman, of very early date; the lower
deposits are those with which we are concerned. Mr. McEnery, the
gentleman who examined and described them, ascertained that
the flint instruments were covered by a deposit of stalagmite,
but he considered them to be of more recent date than the bones
found beneath them, and his opinion was that they had been made
by men who entered the cave and disturbed the original deposit
of sand and clay. Dr. Buckland refused his belief to the statement
that the flint implements were found under the stalagmite, and

always contended that they were the work of men who had broken up the sparry floor.

In 1858, Dr. Falconer heard of the newly-discovered cave at Brixham, on the opposite side of the bay to Torquay, and he took steps to prevent any doubts being entertained respecting its contents. This cave was composed of several branches, with four openings, formerly blocked up with breccia and earthy matter : the main opening being seventy-eight feet above the valley, and ninety-five feet above the sea, the cave itself being eight feet wide. The floor of the cave was a layer of stalagmite, from one to fifteen inches thick, sometimes containing bones, the horns of reindeer, a humerus of the cavern bear ; loam, reddish bone earth, with angular stones and some pebbles, from two to thirteen feet thick, containing the remains of elephants, rhinoceros, bears, hyænas, of felis, reindeer, horses, oxen, and several rodents ; beneath this, a layer of gravel, and rounded pebbles without fossils.

In these beds no human bones were found, but in almost every part of the bone bed were flint knives, one of the most perfect being found thirteen feet down in the bone bed. The most remarkable fact in connection with this cave was the discovery of an entire left hind leg of the cave bear lying in close proximity to this knife ; "not washed in a fossil state out of an older alluvium, and swept afterwards into this cave, so as to be mingled with the flint implements, but having been introduced when clothed in its flesh." In short, the implement and the bear's leg were evidently deposited about the same time, and it only required some approximative estimate of the date of this deposit to settle the question of the antiquity of man, at least in an affirmative sense.

Encouraged by the Brixham discoveries, a congress of French and English geologists met at Amiens in order to consider certain evidence on which it was sought to establish the fact that man and the mammoth were contemporaries.

The valley of the Somme, between Abbeville and Amiens, is occupied by beds of peat, some twenty or thirty feet deep, resting on a thin bed

of clay which covers other beds of sand and gravel, which again rest on
a white chalk with flints. Bordering the valley, some hills rise with a
gentle slope to the height of two or three hundred feet, and here and
there on their summit are patches of tertiary sand and clay with fossils,
and again more extensive layers of loam. The inference sought from this
geological structure is that the river, originally flowing through the
tertiary formation at the summit, has gradually cut its way through
the various strata to its present bed. From the depth of the peat, its
lower part lies below the sea level, and it is supposed that depression
of the region has occurred at some period : again, in land lying quite low
on the Abbeville side of the valley, but above the tide level, marine shells
occur, which indicates some elevation of the region. Again, about a
hundred feet above the valley on the right bank of the river, and on a
sloping surface, is the Moulin-Quignon, where shallow pits exhibit a
floor of chalk covered by gravel and sand, accompanied by gravel and
marly chalk and flints, more or less worn, well-rounded tertiary flints
and pebbles and tertiary sandstone fragments. Such is the general
description of a locality which has acquired considerable celebrity in
connection with this question.

The quaternary rocks of Moulin-Quignon and the peat beds of the
Somme formerly furnished Cuvier with some of the fossils he described,
and in later times chipped flint implements from the quarries and bogs
came into the hands of M. Boucher de Perthes : the statements were
received at first, not without suspicion, especially on the part of English
geologists who were familiar with similar attempts on their own credulity, that some, at least, of these were manufactured by the workmen. At
length, the discovery of a human jaw and tooth in the gravel beds of St.
Acheul, at Amiens, produced a rigorous investigation into the facts, and
it seems to have been established to the satisfaction of Mr. Prestwich,
that flint implements and the bones of the extinct mammalians exist in
the same beds, and in situations indicating very great antiquity. In the
sloping and irregular deposits overlooking the Somme, the bones of
elephants, of the rhinoceros, with land and fresh-water shells of existing
species are found mingling with flint implements. Shells like those now

found in neighbouring streams and hedgerows, with the bones of existing quadrupeds, have been obtained from the peat with flint tools of more than usual finish, and with them a few fragments of human bones. Of these reliquiæ, the Celtic memorials lie below the Gallo-Roman; above them, oaks, alders, and walnut trees occur, sometimes rooted, but no succession of a new growth of trees appear.

The problem of the St. Acheul beds stands thus: they would be deposited under fluvial action, and are probably the oldest in which human remains occur, older than the peat beds of the Somme, but what is their real age? Before laying before the reader the very imperfect answer this question admits of, a glance at the previous discoveries, which gave confirmation to the facts just narrated, may be useful.

Implements of stone and flint have been continually turning up during the last century and a half in all parts of the world. In the neighbourhood of Gray's-inn-lane, in 1715, a flint spear-head was picked up, and near it some elephants' bones. In the alluvium of the Wey, near Guildford, a wedge-shaped flint tool was found in the gravel and sand, in which elephants' tusks were also found. Under the cliffs at Whitstable, an oval-shaped flint tool was found in what had probably been a fresh-water deposit, and in which bones of the bear and elephant were discovered. Between Herne Bay and Reculver five other flint tools were found, and again three near the top of the cliff, all from fresh-water gravel. In the valley of the Ouse, at Beddenham, Bedfordshire, flint implements, like those of St. Acheul, mixed with the bones of elephants, rhinoceros, and hippopotamus, were found, and near them an oval and a spear-shaped implement. In the peat of Ireland great numbers of such implements have been found. But nowhere have they been so systematically sought for and classified as in the Scandinavian countries.

The peat deposit in these countries, and in Denmark especially, is formed in hollows and depressions, in the northern drift and boulder formation, from ten to thirty feet deep. The lower stratum of two or three feet consists of sphagum, over which lies another growth of peat formed of aquatic and swampy plants. On the edge of the bogs trunks of Scottish firs of large size are found—a tree which has not grown in the Danish islands within historic times and does not now succeed when planted, although it was evidently indigenous within the human period, since Steenstrup took with his own hands a flint implement from beneath the trunk of one. The sessile variety of the oak would appear to have succeeded the fir, and is found at a higher level in the peat. Still higher up, the common oak, *Quercus robur*, is found along with the birch, hazel, and alder. The oak has in its turn been succeeded by the beech.

Another source from which numerous relics of early humanity have been taken is the midden heaps found along the Scandinavian coast. The heaps consist of castaway shells mixed with bones of quadrupeds, birds, and fishes, which reveal in some respects the habits of the early races which inhabited the coast. Scattered through these mounds are flint knives, pieces of pottery, and ashes, but neither bronze nor iron. The knives and hatchets are said to be a degree less rude than those of older date found in the peat. Mounds corresponding to these, Sir Charles Lyell tells us, occur along the American coast, from Massachusetts and Georgia. The bones of the quadrupeds found in these mounds correspond with those of existing species, or species which have existed in the historic times.

By collecting, arranging, and comparing the flint and stone implements, the Scandinavian naturalists have succeeded in establishing a chronological succession of periods, which they designate, 1, the age of stone; 2, the age of bronze; 3, the age of iron. The first, or stone period, in Denmark, corresponded with the age of the Scotch fir, and, in part, of the sessile oak. A considerable portion of the oak period corresponded, however, with the age of *bronze*; swords made of that metal having been found in the peat on the same level with the oak. The *iron* age coincides with the beech. Analogous facts confirmatory of these statements occur in Yorkshire, and in the fens of Lincolnshire.

The traces left indicate that the aborigines went to sea in canoes scooped out of a single tree, bringing back deep-sea fishes. Skulls obtained from the peat and from tumuli believed to be contemporaneous with the mounds are small and round, with prominent supra-orbital ridges, somewhat resembling the skulls of Laplanders.

The third series of facts, or *lacustrine habitations*, consisted of the buildings on piles in lakes once common in Asia and Europe. They are first mentioned by Herodotus as being used among the Thracians of Pæona, in the mountain lake Prasius, where the natives lived in dwellings built on piles, connected with the shore by a narrow causeway, by which means they escaped the assaults of Xerxes. Buildings of the same description occupied the Swiss lakes, in the mud of which hundreds of implements like those found in Denmark have been dredged up. In Zurich, Moosedorf, near Berne, and Lake Constance, axes, celts, pottery, canoes made out of single trees, have been found; but of the human frame scarcely a trace. One skull dredgèd at Meilæn, in Zurich Lake, was intermediate between the Lapp-like skull of the Danish tumuli and the more recent European type.

The age of the different formations in which these records of the human race are found will probably ever remain a mystery. The evidence which would make the implements formed by man contemporaneous with the mammoth and other great mammalians would go a long way to prove that man was also pre-glacial. Let us see how that argument stands.

At the period when the upper Norwich crag was deposited, the general level of the British Isles is supposed to have been about six hundred feet above its present level, and so connected with the Eurorean continent as to have received the elements of its fauna and flora from thence.

By some great change, a period of depression occurs in which all the country north of the mouth of the Thames and the Bristol Channel are placed much below their present level, Moel Trafaen, now fourteen hundred feet high, being, in fact, submerged, during which period it received the erratic blocks and other marks indicative of floating icebergs which have been described in a former chapter. The country is raised again to something like its first level, and again occupied by plants, shells, fishes and reptiles, birds, and mammifera. Again subsidence occurs, and after several oscillations the level remains as we find it. The estimated time for these various changes is something enormous, but it is generally believed that the data for the calculations are somewhat fallacious—that, in short, the same guess might have extended it to double the number of years, and that equally good reasons might be found for reducing it to the four last figures, or 2000 years. The unit of the calculation is the upward rate of movement observed on the Scandinavian coast; applied to the oscillation of the ancient coast of Snowdonia, the figures presented are 224,000 years for the several oscillations of the glacial period. Adding the pre-glacial period, the computation gives an additional 48,000 years. But, let us repeat, the figures and data are purely imaginary.

With regard to the St. Acheul beds, said to be the most ancient formation in which the productions of human hands have been found, they are admittedly older than the peat beds, and the time required for the production of other peat beds of equal thickness has been estimated at 7000 years. The antiquity of the gravel beds of St. Acheul may be estimated on two grounds : 1. General elevation above the level of the valley. 2. By estimating the animal remains found in the gravel beds, and not in the peat. The first question implies the denudation of the valley below the level of the gravel, or the elevation of the whole plateau. Each of these operations would involve an incalculable time, for want of data. In the second case, judging from the slow rate at which quadrupeds have disappeared in historic times, the extinct Mammoth and other great quadrupeds must have consumed many centuries in dying out, for the notion that they died out suddenly, from sharp and sudden refrigeration, is not generally admitted.

With regard to the three ages of stone, bronze, and iron, M. Morlat has based some calculations upon the condition of the delta of Tinière, near Villeneuve, which lead him to assign to the oldest, or stone period, an age of 5000 to 7000 years, and to the bronze period from 3000 to 4000. We may, then, take leave of this subject with the avowal that, while admitting the probability that an immense lapse of time would be required for the operations described, we are altogether without reliable data for estimating its extent.

The opinion which places the creation of man on the banks of the Euphrates in Central Asia is confirmed by an event of the highest importance in the history of humanity, and by a crowd of concordant traditions, preserved by different races of men, all tending to confirm it. We speak of the Asiatic deluge.

Fig. 198.—Mount Ararat.

The Asiatic deluge, of which sacred history has transmitted to us the few particulars we know, was the result of the upheaval of a part of the long chain of mountains which diverge from the Caucasus. The earth opening by one of these fissures left in its crust in course of cooling, an eruption of volcanic matter escaped through the crater. Masses of watery vapour or steam accompanied the lava discharged from the interior of the globe, which, being dissipated in clouds and condensing, descended in torrents of rain, and the plains were drowned under the volcanic mud. The inundation of the plains in an extensive radius was the momentous result of this upheaval, and the formation of the volcanic cone of Mount Ararat, with the vast plateau on which it rests, altogether 17,323 feet above the sea, the permanent result. The event is graphically detailed in the seventh chapter of Genesis.

11. "In the six hundredth year of Noah's life, in the second month, the seventeenth day of the month, the same day were all the fountains of the great deep broken up, and the windows of heaven were opened.

12. "And the rain was upon the earth for forty days and forty nights."

17. "And the flood was forty days upon the earth; and the waters increased, and bare up the ark, and it was lifted up above the earth.

18. "And the waters prevailed and were increased greatly upon the earth; and the ark went upon the face of the waters.

19. "And the waters prevailed exceedingly upon the earth; and all the high hills, that were under the whole heaven, were covered.

20. "Fifteen cubits upward did the waters prevail; and the mountains were covered.

21. "And all flesh died that moved upon the earth, both of fowl, and of cattle, and of beast, and of every creeping thing that creepeth upon the earth, and every man:

22. "All in whose nostrils was the breath of life, of all that was in the dry land, died.

23. "And Noah only remained alive, and they that were with him in the ark.

24. "And the waters prevailed upon the earth an hundred and fifty days."

All the particulars here recited are only to be explained by the volcanic and muddy eruption which preceded the formation of Mount Ararat. The waters which produced the inundation of these countries proceeded from a volcanic eruption accompanied by enormous masses of vapour, which in due course were condensed and descended on the earth, inundating the extensive plains which stretch away from the foot of Ararat. The word "all the earth," which might be implied in the above extract, is explained by Mariel de Serres and other philologists as being an inaccurate translation. He shows that the Hebrew word *haarets* does not always mean "all the earth," but is often used in the sense of *region* or *country*, and that in this instance Moses uses it to express the part of the globe already peopled. In the same manner, *all the mountains* only extends to all the mountains known to him. In the same manner, M. Glaire, in the *Christomathie* which follows his Hebrew Grammar, quotes the passage in this sense: "The waters increased prodigiously, and the highest hills of the vast horizon were covered;" thus limiting the mountains covered by the inundation, to those bounded by the horizon.

Nothing occurs, therefore, in the description given by Moses to hinder us from seeing in the Asiatic deluge a means adopted by God to punish the human race still in its infancy. It seems to establish the countries lying at the foot of the Caucasus as the cradle of the human race, and it seems to establish also the upheaval of a chain of mountains, preceded by an eruption of volcanic mud, which drowned vast territories in these regions, consisting of plains of great extent. Of this deluge many races besides the Jews have preserved the tradition. Moses dates it from 1500 to 1800 years before the epoch in which he wrote. Bérose, the Chaldean historian, who wrote at Babylon in the time of Alexander, speaks of a universal deluge which he places immediately before the reign of Belus, the father of Ninus.

The *Vedas*, or sacred books of the Hindus, supposed to have been composed about the same time as Genesis, that is, about the year of the world 3300, make out that the deluge occurred 1500 years before their epoch. The *Ghebres* speak of the same event as having occurred about the same date.

Confucius, the Chinese philosopher and law-giver, born towards the year 551 before Christ, begins his history of China by speaking of the

Emperor named Jas, whom he represents as making the waters flow
back, which then *raised themselves to the heavens*, while they bathed the
foot of the highest mountains, covering the smaller hills and inundating the plains. Thus the Biblical deluge (PL. XXXIII.) is confirmed
in many respects, but it was local, like all phenomena of the kind, and
was the consequence of the upheaval of the mountains of western Asia.

A deluge quite of modern date conveys a tolerably exact idea of the
phenomena, and we recall the circumstances as assisting us to comprehend the true nature of the ravages the deluge inflicted upon Asia
in the quaternary period. At six days' journey from the city of
Mexico, there existed, in 1759, a fertile and well-cultivated district,
where abundance of rice, maize and bananas grew. In the month of
June frightful earthquakes disturbed the soil, and were continued
unceasingly during two whole months. On the night of the 28th
September the earth was violently convulsed, and a region of .many
leagues in extent was slowly raised until it attained a height of about
500 feet, forming a plateau of many leagues square. The earth
undulated like the waves of the sea in a tempest ; thousands of small
hills rose and disappeared in turn, and, finally, an immense gulf
opened, from which smoke, fire, red-hot stones and ashes were
violently discharged, and darted to prodigious heights. Six mountains
surged up from the gaping gulf ; among which the volcanic mountain
of Jorullo, which rises 2150 feet above the ancient plain, is the most
prominent.

At the moment when the earthquake commenced the two rivers of
Cuitimbo and San Pedro flowed backwards, inundating all the plain
now occupied by Jorullo ; but in the upheaving region, while it continued to' rise, a gulf opened and swallowed the rivers. They reap-
peared to the west, but at a point very distant from their ancient bed.
This inundation reminds us on a small scale of the phenomena which
attended the deluge of Noah.

Besides the deposits resulting from the partial deluges which we
have described as occurring in Europe and Asia in the quaternary

XXXIII.—The Asiatic Deluge.

period, there were produced in the same period many new formations resulting from the deposit of *alluvium* thrown down by seas and
rivers. These deposits were numerous and widely disseminated.
Their stratification is as regular as any which have preceded them;
they are distinguished from those of the tertiary epoch, with which
they are most likely to be confounded, by their situation, which is
very frequently upon the littoral of the sea, and by the predominance
of shells of a species identical with those now living in our seas.

A marine formation of this kind, constituting the coast of Sicily,
extends from Palermo to Catania, Gergenti and Syracuse, stretching
away to the centre of the island, where it rises to the height of 3,000
feet, being the most remarkable of the quaternary productions. It
is formed of two great beds; the lower is a bluish argillaceous marl,
the other a coarse, but very compact limestone, both containing shells
analogous to those which now occupy the Mediterranean coast. The
same formation is found in the neighbouring isles, especially Sardinia
and Malta; and the great sandy deserts of Africa, as well as the
argillaceous sands of the steppes of Eastern Russia, and the fertile
black earth of its southern plains, have the same geological origin;
so have the travertines of Tuscany, of Naples, and of Rome, and the
tufas which are the essential constituents of the Neapolitan soil. We
are now approaching so near to our own age, that we can almost
trace the hand of Nature in her works. Professor Ramsay shows in
the geological survey that beds nearly a mile in thickness have been
removed by denudation from the summit of the Mendip Hills, and
broad areas in South Wales and the neighbouring counties have been
denuded of their summits, and the materials transported elsewhere to
form the newer strata. Now no combination of causes have been
imagined which have not involved submersion during long periods, and
their subsequent elevation in periods of equal duration.

We can hardly walk any great distance along the coast, either of
England or Scotland, without remarking some flat terrace of an unequal breadth, and backed by an escarpment more or less steep: upon
such a terrace many of the towns along the coast are built. No

2 F

geologist now doubts that this fine platform, at the base of which is a field of loam or sandy gravel sown with marine shells, had been at some period the line of coast against which the waves of the ocean once broke at high water. At this period the sea rose twenty, and thirty, and some places a hundred feet higher than it does now: these ancient sea beaches in some places forming terraces of sand and gravel, with littoral shells, some broken, others entire, and corresponding with species in the seas below; in others they form bold projecting promontories or deep bays. Relatively to history, this coast line should be very ancient; though of yesterday, in a geological sense, its origin ascending far beyond all written tradition. The wall of Antoninus, built by the Romans to protect themselves from the attacks of the Caledonians, was built, in the opinion of the best authorities, not in connection with the old, but new coast line. We may, then, conclude that in A.D. 140, when the greater part of this wall was constructed, the zone of the ancient coast line had attained its present elevation above the actual level of the sea.

The same proofs of a general and gradual elevation of the country are observable almost everywhere; on the estuary of the Clyde canoes and other works of art have been exhumed, assignable to a recent period. Near St. Austell and at Carnon, in Cornwall, human skulls and other relics have been met with beneath marine strata, in which the bones of whales and land quadrupeds of still existing species were embedded. But in the countries where hard limestone rocks prevail: in the ancient Peloponnesus, along the coast of Argolis and Arcadia, three and even four ranges of ancient sea-cliffs are well preserved, which Messrs. Bablaye and Verlet describe as rising one above the other, at different distances from the actual coast, sometimes to the height of a thousand feet, as if the upheaving force had been suspended for a time, leaving the waves and currents to throw down and shape the successive ranges of lofty cliffs. On the other hand, another celebrated historical site may be adduced as evidence of subsidence of the coast-line of the Mediterranean. In the Bay of Baiæ, the celebrated temple of Serapis, which occupied the shore at the head of the bay in the third century, is now deeply embedded in the sandy littoral, portions of its pillars of marble being visible at a considerable depth in the clear waters. With respect to the littoral deposits of the quaternary period, they are of very limited extent, except in a few localities : they are found on the western coast of Norway; on the English coast ; in France, an extensive bed of quaternary formation covers the ancient littoral of Guienne, and on other parts of the coast, where it is sometimes concealed by trees and shrubs, or by blown sand, as at Dax, near Bourdeaux, where a steep bank may be traced about twelve miles inland, and parallel with the present coast, which falls suddenly about fifty feet from a higher platform of the land to a lower one extending to the sea. In making some excavations for the foundations of a building at Abesse, in 1830, it was discovered that this fall was drifted sand, filling up a steep perpendicular cliff about fifty feet high, consisting of a bed of tertiary clay extending to the sea, a bed of limestone with tertiary shells and corals, and, at the summit, the tertiary sand of the lands. The alluvium of the rivers, added to the marine deposits, have formed the littoral of the period, which occurs more especially near the mouths of rivers and water-courses.

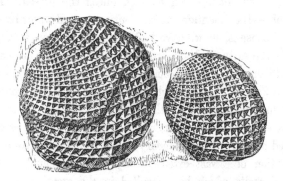

Fig. 199.—Ischadites Konigii.—Ludlow rocks.

EPILOGUE.

HAVING considered the past history of the globe, we may be permitted to give a glance of the future which awaits it.

Can the actual state of the earth be considered as definitive? The revolutions which have fashioned its surface, and produced the Alps in Europe, Mount Ararat in Asia, the Cordilleras in the New World—are they to be the last? In a word, will the terrestrial sphere ever preserve the form under which we know it—as it has been, so to speak, engrafted into our memories by the maps of the geographers?

It is difficult to reply with any confidence to this question; nevertheless, our readers will not object to accompany us a step further, while we express an opinion, founded on analogy and scientific induction.

What are the causes which have produced the present inequalities of the globe—the mountain ranges, continents, and water-basins? The primordial cause is, as we have had frequent occasion to repeat, the cooling of the earth, and the progressive solidification of the external nucleus still in a fluid or viscous state. These are the cause of the folds, furrows, and fractures which have led to the elevation of the great mountain ranges and the depression of the great valleys—which have caused some continents to surge from the bed of ocean and submerged others. The secondary causes which have contributed to the formation of a vast extent of dry land are found in the sedimentary deposits, which have resulted in the creation of new continents by filling up the basins of ancient seas.

Now these two causes, although in a much weakened form, continue in operation to the present day. The thickness of the terrestrial crust

EPILOGUE. 437

is only a fraction compared to the interior liquid mass. The principal cause, then, of the great dislocations of the soil is, so to speak, at our gates; it threatens us unceasingly. Of this the earthquakes and volcanic eruptions, which are still frequent in our day, give us disastrous and incontestable proofs. On the other hand, our seas are continually forming new land : the basin of the Baltic Sea, for instance, is gradually rising in consequence of the deposits which will obviously fill up its bed entirely in an interval of time which it would not be impossible to calculate.

It is, then, probable that the actual condition of the soil and the respective limits of seas and continents have nothing fixed or definite in them—that they are, on the contrary, open to great modifications in the future.

There is another problem much more difficult than the preceding, but for which neither induction nor analogy furnish us with any certain indications—it is the perpetuity of our species. Is man doomed to disappear from the earth one day, as all the races of animals which preceded him, and prepared the way for his coming, have done ? Will a new *glacial period,* analogous to that which, during the quaternary period, was felt so rigorously, come to put an end to his existence ? Like the trilobites of the Silurian period, the great reptiles of the lias, the mastodons of the tertiary, and the megatheriums of the quaternary epoch, is the human species to be one day annihilated—to disappear from the globe by a simple natural extinction ? Or must we believe that man, gifted with the attribute of reason, marked, so to speak, by the divine seal, is to be the last, the supreme end of creation ?

Science cannot pronounce upon these grave questions, which exceed the competence, and go beyond the circle of human reasoning. It is not impossible that man should be only a step in the ascending and progressive scale of animated beings. The Divine Power, which has thrown upon the earth, life, sentiment, and thought ; which has given to plants, organization; to animals, motion, sentiment, intelligence ; to man, in addition to these multiplied gifts, the faculty of reason, doubled in value by the ideal—reserves to Himself, perhaps, in His wisdom, the privilege of creating alongside of man, or after him, a being yet more perfect. This new being, religion and modern poesy would present in the ethereal and radiant type of the Christian angel, with moral qualities whose nature and essence would escape our perceptions—of which we could no more form a notion than one born blind could conceive of colours, or the deaf and dumb of sound. *Erunt æquales angelis Dei.* They will be as the angels of God, says Holy Scripture, speaking of man raised to life eternal.

During the primitive epoch the *mineral kingdom* existed alone ; the rocks, silent and solitary, were all that was yet formed of the burning earth. During the transition epoch, the vegetable kingdom, newly created, extended itself over the whole globe, which it soon covered from one pole to the other with an uninterrupted mass of verdure. During the secondary and tertiary epochs, the vegetable kingdom and the animal kingdom divided the earth between them. In the quaternary epoch the *human kingdom* appeared. Is it in the future destinies of our planet to receive yet another lord ? And after the four kingdoms which now occupy it, is there to be a *new kingdom* created, the attributes of which will ever be a mystery to us, but which will differ from man in as great a degree as man differs from animals, and plants from rocks ?

We must be contented with suggesting, without hoping to resolve this formidable problem. This great mystery, according to the fine expression of Pliny, " is hidden in the majesty of nature," *latet in majestate naturæ ;* or, to speak more in the spirit of Christian philosophy, in the thoughts of the Almighty Creator of the world, who also formed the universe.

THE END.

439

INDEX.

. *Lines in Italics are Woodcut illustrations.*

Coal measures, 116; time of formation, *ib.*; origin, 118; composition, *ib.*

Coal-mines of Treuil, 141.

Coccasteus, 108.

Columnar structure of basalt, 65.

Compact limestone, 152.

Common Anoplotherium, 286.

Comparative size of the earth, 31.

Condensation of vapour, 39.

Cone of ejection in volcanoes, 72.

Cone-shaped mountains, 151.

Confervæ of the chalk, 270.

Conglomerates, 374.

Coniferous trees, 166, 229.

Conifers of Jurassic times, 218.

Contortions of coal beds, 149.

Conybeare's account of Plesiosaurus, 202.

Copper slate of Thuringia, 153.

Coprolites, petrified excrements of antediluvian animals, 19.

Coprolites of Ichthyosaurus, enclosing bones, 196.

Coprolite of Icthyosaurus, showing the muscles of the intestines, 15, 197.

Coprolites of the hyæna, 380.

Coralline crag, 332; formations of the Jurassic period, 216.

Coral rag, 227: of the Jurassic period, 264.

Coral reefs of the Antilles, 264.

Coseguina, volcano of, 70.

Cornbrash, 219.

Corne mountains, 264.

Cornwall granite, 56.

Coralline stratum of the chalk hills, 227.

Coryphodon, 292.

Country round Lyme Regis, 190.

Crabs (*Pagurus*), 289, 310.

Crater of Vesuvius, 71.

Craven fault, 101.

Credneria, 258.

Cremopteris, 170.

Cretaceous period, 237; fauna, 248.

Cretaceous seas, 249; fishes of the, 257.

Cretaceous vegetation, 245.

Crinoides, 110.

Crioceratites Duvalli Tropæum, 236.

Crocodileimus, 310.

Crocodile of Maestricht, 266.

Cromarty, quarries of, 107.

Cromer forest, 370.

Cross Fell, 101.

Crust of the earth, 35.

Cryptogamæ, 178, 245.

Cryptomerias, 170.

Crystalline formation, 155.

Crystalline limestone, 152.

Crystalline strata, 47.

Crystallized rocks, 28.

Cucumites, 277.

Culm measures of Devonshire, 56.

Cupanioïdes, 277.

Cupressocrinus crassus, 110.

Cupressinaceæ, 170.

Cupriferous schists, 156.

Cuvier's account of Plesiosaurus, 203; account of Pterodactylus, 205; method of studying fossils, 14; on the destruction of species, 338; on the Mammoth, 307.

Cyathocrinus, 132.

Cycadaceæ, 172, 229.

Cycas Circinalis, 150.

Cypris spinigera and Waldensis, 260.

Cyrtoceras depressum, 154.

Dammaria, 170.

Danian beds, 271.

Dartmoor granite, 56.

Darwin on the coral formations, 226; on Megatheroid animals, 367.

Dasypus, 356.

De la Beche (Sir Henry), 18.

Deluge, confirmed by traditions of all ancient races, 431.

Diluvium, earth washed down by a river, and deposited in its bed, 373.

Descartes on incandescence, 23; tourbillon, 32.

Devonian fishes, 108.

Devonian period, 104; rocks, 111; their composition, 112.

Devonian plants, 106.

Diameter of the earth, 35.

Dicotyledons, 172.

Dicorn fossil, 343.

Diluvial fossils, 379.

Dinornis, 368.

Dinotherium restored, 300.

Dinotherium described, 311.

Diornis, 161.

Diplocanthus, 109.

Direct measurement of the earth, 33.

Dirt-bed fossils, 231.

Dodo, 161.

Donati's discovery, 13.

Downs (North and South), 240.

Downton sandstone, 95.

Dragons of mythology, 320; personified, 207.

Draco fimbriatus, 208.

Dryopithecus, 311.

Dueras limestone, 227.

Durdlestone Bay, 233.

Dykes or veins of metal injected through other strata, 64.

Early geologists, 13.

Earth in a gaseous state, 31.

Edentata, the, 356.

Echinoderms, 225, 265.

Ehrenberg's microscopic investigations, 239.

Electric fluid, 34.

LONDON: PRINTED BY WILLIAM CLOWES AND SONS, STAMFORD STREET, AND CHARING CROSS.

Printed in the United States
By Bookmasters